Protein Folding and Metal Ions

Mechanisms, Biology and Disease

Protein Folding and Metal Ions

Mechanisms, Biology and Disease

Edited by
Cláudio M. Gomes
Pernilla Wittung-Stafshede

CRC Press
Taylor & Francis Group
Boca Raton   London   New York

CRC Press is an imprint of the
Taylor & Francis Group, an **informa** business

CRC Press
Taylor & Francis Group
6000 Broken Sound Parkway NW, Suite 300
Boca Raton, FL 33487-2742

© 2011 by Taylor and Francis Group, LLC
CRC Press is an imprint of Taylor & Francis Group, an Informa business

No claim to original U.S. Government works

Printed in the United States of America on acid-free paper
10 9 8 7 6 5 4 3 2 1

International Standard Book Number: 978-1-4398-0964-8 (Hardback)

This book contains information obtained from authentic and highly regarded sources. Reasonable efforts have been made to publish reliable data and information, but the author and publisher cannot assume responsibility for the validity of all materials or the consequences of their use. The authors and publishers have attempted to trace the copyright holders of all material reproduced in this publication and apologize to copyright holders if permission to publish in this form has not been obtained. If any copyright material has not been acknowledged please write and let us know so we may rectify in any future reprint.

Except as permitted under U.S. Copyright Law, no part of this book may be reprinted, reproduced, transmitted, or utilized in any form by any electronic, mechanical, or other means, now known or hereafter invented, including photocopying, microfilming, and recording, or in any information storage or retrieval system, without written permission from the publishers.

For permission to photocopy or use material electronically from this work, please access www.copyright.com (http://www.copyright.com/) or contact the Copyright Clearance Center, Inc. (CCC), 222 Rosewood Drive, Danvers, MA 01923, 978-750-8400. CCC is a not-for-profit organization that provides licenses and registration for a variety of users. For organizations that have been granted a photocopy license by the CCC, a separate system of payment has been arranged.

Trademark Notice: Product or corporate names may be trademarks or registered trademarks, and are used only for identification and explanation without intent to infringe.

Library of Congress Cataloging-in-Publication Data

Protein folding and metal ions : mechanisms, biology and disease / editors, Cláudio M. Gomes, Pernilla Wittung-Stafshede.
   p. ; cm.
Includes bibliographical references and index.
Summary: "Interest in the uses and applications of metallo-chemistry in the life sciences continues to expand. This book provides a broad overview of metals and protein folding. It addresses fundamental aspects of the folding, stability, and design of metal binding proteins and peptides along with biological aspects related to the action of metallochaperones. The text features practical information related to methodologies, particular techniques, and bibliographic references. Readers will gain the molecular basis for protein-metal interactions and a good perspective on the applications of protein coordination chemistry in life sciences and biotechnology"--Provided by publisher.
   ISBN 978-1-4398-0964-8 (hardcover : alk. paper)
   1. Metalloproteins. 2. Protein folding. I. Gomes, Cláudio M., 1970- II. Wittung-Stafshede, Pernilla, 1968- III. Title.
   [DNLM: 1. Metalloproteins. 2. Metallochaperones. 3. Protein Conformation. 4. Protein Folding. 5. Protein Stability. QU 55.8]

QP552.M46P765 2011
572'.633--dc22

Contents

Preface vii
Acknowledgments ix
About the Editors xi
Contributors xiii

SECTION I Folding and Stability of Metalloproteins

Chapter 1 Metal Ions, Protein Folding, and Conformational States: An Introduction 3

Cláudio M. Gomes and Pernilla Wittung-Stafshede

Chapter 2 The Folding Mechanism of c-Type Cytochromes 13

Carlo Travaglini-Allocatelli, Stefano Gianni, and Maurizio Brunori

Chapter 3 The Mechanism of Cytochrome c Folding: Early Events and Kinetic Intermediates 37

Eefei Chen, Robert A. Goldbeck, and David S. Kliger

Chapter 4 Stability and Folding of Copper-Binding Proteins 61

Irina Pozdnyakova and Pernilla Wittung-Stafshede

Chapter 5 Iron-Sulfur Clusters, Protein Folds, and Ferredoxin Stability 81

Sónia S. Leal and Cláudio M. Gomes

Chapter 6 Folding and Stability of Myoglobins and Hemoglobins 97

David S. Culbertson and John S. Olson

SECTION II Mechanisms of Metal Transporters and Assembly

Chapter 7 Frataxin: An Unusual Metal-Binding Protein in Search of a Function 125

Annalisa Pastore

Chapter 8 Mechanism of Human Copper Transporter
Wilson's Disease Protein 145

Amanda Barry, Zara Akram, and Svetlana Lutsenko

SECTION III Metal Ions, Protein Conformation, and Disease

Chapter 9 α-Synuclein and Metals 169

Aaron Santner and Vladimir N. Uversky

Chapter 10 Zinc and p53 Misfolding 193

James S. Butler and Stewart N. Loh

Chapter 11 The Octarepeat Domain of the Prion Protein and Its Role
in Metal Ion Coordination and Disease 209

Glenn L. Millhauser

SECTION IV Metalloprotein Design, Simulation, and Models

Chapter 12 Metallopeptides as Tools to Understand Metalloprotein
Folding and Stability 227

Brian R. Gibney

Chapter 13 The Folding Landscapes of Metalloproteins 247

Patrick Weinkam and Peter G. Wolynes

Appendix 275

Index 287

Preface

Metals are essential for life, as well as for around one-third of the proteins that require the interaction with metal ions or metal containing cofactors to carry out a particular biological function. This interplay has been extensively explored in the past, as metalloproteins are involved in key biological processes, from energy production to the synthesis of DNA building blocks. Here we took the challenge to organize a book focusing on an essential aspect of the metal-protein interaction, relating to the role of metal ions in protein folding and structure, which we felt needed to be addressed, aiming at merging topical contributions in fundamental aspects of the folding, stability, and design of metal-binding proteins or peptides, with more biological aspects related the action of metallochaperones or with the intervention of metals in determining protein conformations associated with disease. With this purpose we have commissioned 13 chapters that provide a comprehensive, state-of-the-art perspective of the topic under discussion, which we have organized in four sections.

Section I, Folding and Stability of Metalloproteins, comprises contributions that provide case study examples of protein folding and stability studies in particular systems or proteins that comprise different metal ions (zinc, copper) or cofactors (iron-sulfur centers, heme). It aims to provide a view on more fundamental aspects such as for example those related to the influence of metal sites and respective redox state in the energetics of the folding reaction, interactions with nonnative protein states, or influence in folding pathways. Section II, Mechanisms of Metal Transporters and Assembly reviews the proteins that shuttle metal ions in the cell (iron and copper chaperones in the selected examples) to a particular target metalloprotein. A structural understanding of these proteins and of metal-mediated conformational changes is fundamental to characterize the *in vivo* insertion of metals into apoproteins, and protein-protein interactions. Section III, Metal Ions, Protein Conformation, and Disease, illustrates how metal binding can be connected to pathological protein conformations in unrelated diseases, from cancer to protein deposition disorders such as Parkinson's disease. This is an emerging field in which the role of metal ions is becoming increasingly more relevant, in a large number of different proteins and pathologies. Section IV, Metalloprotein Design, Simulation, and Models, gathers contributions addressing protein redesign of metal-containing proteins by computational methods, folding simulation studies, and work on model peptides, dissecting the relative energetic contribution of metals sites to protein folding and stability. It provides a key view on how computational studies are relevant to understanding these biological processes.

Such a combination of contributions is expected to be of interest to a larger audience of protein chemists, biochemists, and biophysicists, from postgraduate level onward, with broad interests in aspects relating to structural biology, bioinorganic chemistry, protein folding and disease, and molecular and cellular aspects of metals in biological systems. We hope that this overview on this key biological problem will make this volume a resourceful source of information as well as an attraction to new scientists to join the field.

Cláudio M. Gomes
Pernilla Wittung-Stafshede

Acknowledgments

We would like to express our gratitude to a number of people who played an influential role in the organization of this book. We thank Lance Wobus, our editor at Taylor & Francis, who launched the challenge of compiling a book on this theme and who was extremely helpful, professional, and diligent throughout all stages of the process. We are grateful to all our authors for their availability, effort, and contributions to this book, and also for their timely fulfillment of deadlines.

Cláudio Gomes would like to thank the several colleagues who have advised us on the do's and don'ts of book editing, and especially Pernilla, for having promptly agreed to join me in the adventure of co-editing this volume.

A final word to express gratitude to our mentors and professors who have guided us in shaping our careers.

About the Editors

Cláudio M. Gomes is group leader at the Instituto Tecnologia Química e Biológica, a research institute affiliated to the Universidade Nova de Lisboa, Oeiras, Portugal. Born in 1970, Cláudio earned his PhD from the Instituto Tecnologia Química e Biológica in 1999, as a Gulbenkian PhD program in biology and medicine graduate. From 2000 onward, he gradually switched his research interests from bioenergetics and metalloprotein structure-function toward protein folding and structural biophysics of misfolding diseases. Until setting up his independent laboratory in 2003, he was assistant professor at the Faculdade de Ciências e Tecnologia, Universidade Nova de Lisboa, where he started teaching in 2000. Cláudio has published more than 70 peer-reviewed papers.

Pernilla Wittung-Stafshede has been professor in biological chemistry in the Chemistry Department at Umeå University, Umeå, Sweden, since 2008. Born in 1968, she earned a PhD in physical chemistry from Chalmers University, Gothenburg, Sweden, in 1996. During 1997–1998, she did a postdoctoral period at Caltech, Pasadena, California. In 1999 she began her independent research career, with a focus on the role of metals in protein folding, as a chemistry professor at Tulane University, New Orleans, Louisiana. After 5 years at Tulane, she moved to Rice University, Houston, Texas, in 2004 and became a professor in biochemistry. Pernilla has graduated 10 PhD students to date, has obtained several awards, and has published over 150 peer-reviewed papers.

Contributors

Zara Akram
Department of Physiology
Johns Hopkins University
Baltimore, Maryland

Amanda Barry
Department of Physiology
Johns Hopkins University
Baltimore, Maryland

Maurizio Brunori
Department of Biochemical Sciences
Sapienza
University of Rome
Rome, Italy

James S. Butler
State University of New York
Syracuse, New York

Eefei Chen
Department of Chemistry
University of California
Santa Cruz, California

David S. Culbertson
Biochemistry and Cell Biology
  Department
Rice University
Houston, Texas

Stefano Gianni
Institute for Molecular Biology and
  Pathology, CNR
Rome, Italy

Brian R. Gibney
Brooklyn College
City University of New York
Brooklyn, New York

Robert A. Goldbeck
Department of Chemistry
University of California
Santa Cruz, California

Cláudio M. Gomes
Instituto Tecnologia Química e
  Biológica
Universidade Nova de Lisboa
Oeiras, Portugal

David S. Kliger
Department of Chemistry
University of California
Santa Cruz, California

Sónia S. Leal
Instituto Tecnologia Química e
  Biológica
Universidade Nova de Lisboa
Oeiras, Portugal

Stewart N. Loh
State University of New York
Syracuse, New York

Svetlana Lutsenko
Department of Physiology
Johns Hopkins University
Baltimore, Maryland

Glenn L. Millhauser
Department of Chemistry
University of California
Santa Cruz, California

John S. Olson
Biochemistry and Cell Biology
  Department
Rice University
Houston, Texas

Annalisa Pastore
Molecular Structure Division
MRC National Institute for Medical
  Research
London, UK

Irina Pozdnyakova
Chemistry Department
Umeå University
Umeå, Sweden

Aaron Santner
Molecular Kinetics, Inc.
Indianapolis, Indiana

Carlo Tavaglini-Allocatelli
Department of Biochemical Sciences
Sapienza
University of Rome
Rome, Italy

Vladimir Uversky
Department of Biochemistry and
  Molecular Biology
Indiana University School of Medicine
Indianapolis, Indiana

Patrick Weinkam
Department of Chemistry and
  Biochemistry
University of California–San Diego
La Jolla, California

Pernilla Wittung-Stafshede
Chemistry Department
Umeå University
Umeå, Sweden

Peter Wolynes
Department of Chemistry and
  Biochemistry
University of California–San Diego
La Jolla, California

Section I

Folding and Stability of Metalloproteins

1 Metal Ions, Protein Folding, and Conformational States
An Introduction

Cláudio M. Gomes and Pernilla Wittung-Stafshede

CONTENTS

Metals in Proteins—Which and Why?...3
How Are Metals Incorporated into Appropriate Proteins?..................................5
Roles of Metals in Protein Folding and Misfolding?..6
Metals as Modulators of Protein Structure and Conformation 7
References...9

METALS IN PROTEINS—WHICH AND WHY?

Life requires an interplay between organic (e.g., proteins, sugars, lipids) and inorganic (e.g., cations, often metals, and anions, such as phosphate) matter. At present, we are aware of at least 13 metals that are essential for plants and animals (Bertini et al. 1994; Lippard and Berg 1994). Four of these, sodium, potassium, magnesium, and calcium, are present in large quantities and are known as bulk metals (Fenton 1995). The remaining nine, which are present in small quantities, are the d-block elements vanadium, chromium, molybdenum, manganese, iron, cobalt, nickel, copper, and zinc, and are known as the trace metals. The bulk metals form 1–2% of the human body weight whereas the trace elements represent less than 0.01%. Even of iron, the most widely used trace metal, we need only in the order of 4–5 grams in a human body (Fenton 1995). The concentrations of metals in the cells are strictly regulated at their respective optimum levels: Too much or too little is often harmful and may even be lethal to the organism. Most of the trace metals are found as natural constituents of proteins. In this way, nature has taken advantage of the special properties of the metal ions and tuned them by protein encapsulation to perform a wide variety of specific functions associated with life processes (Bertini et al. 1994; Lippard and Berg 1994).

More than 30% of all proteins in the cells exploit one or more metals to perform their specific functions (Gray 2003); over 40% of all enzymes contain metals (Bertini et al. 1994; Lippard and Berg 1994). The amino acids that regularly act as metal ligands in proteins are thiolates of cysteines, imidazoles of histidines, carboxylates of glutamic and aspartic acids, and phenolates of tyrosines. Also, modified amino acids can serve as metal ligands; for example, γ-carboxyglutamate side chains often bind calcium ions (Warder et al. 1998). Each metal is considered a Lewis acid and favors different sets of protein ligands; the preferences are frequently dictated by the hard-soft theory of acids and bases. Discrete modifications on the residues involved in the first and second coordination sphere from softer to harder and vice versa result in preferential insertion of one metal in respect to another. This effect is illustrated by the different composition of metal ions in the binuclear sites found in different proteins that share the β-lactamase-like domain. Whereas zinc β-lactamases contain either a mono- or a di-zinc center, the catalytically active form of glyoxalase II contains a mixed iron-zinc binuclear center, and the flavodiiron proteins contain a di-iron site. These variations of metal site found within a common fold correlate with the subtle variations in the nature of the metal-coordinating amino acid side chains, where shifts from soft to hard favor iron rather than zinc binding (Gomes et al. 2002).

The coordination number and geometry of each metal site is determined by the metal's oxidation state, albeit substantial distortions from the idealized structures can and do occur in metalloproteins (Bertini et al. 1994; Lippard and Berg 1994). Due to specific chemical properties of each metal, different metals are apt for different types of biological functions, although there is overlap in most cases.

Iron and copper are redox-active metals and often participate in electron transfer (Fenton 1995). In respiration and photosynthesis processes, small redox-active metalloproteins facilitate electron-transfer reactions by alternately binding to specific integral membrane proteins that often contain several metal sites. Well-known examples of soluble redox-active metalloproteins are iron-sulfur-cluster proteins, heme-binding cytochromes, and blue-copper proteins. Iron and copper are also involved in dioxygen (O_2) storage and carriage via metalloproteins (e.g., hemoglobin, myoglobin, and hemocyanin). In contrast to iron and copper, zinc serves as a superacid center in several metalloenzymes, promoting hydrolysis or cleavage of a variety of chemical bonds. Representative proteins that use catalytic zinc ions are carboxypeptidases, carbonic anhydrase, and alcohol dehydrogenase. In addition, zinc ions often play structural roles in proteins (e.g., in superoxide dismutase and zinc finger motifs), and recent work has suggested a role for zinc in transcription via the protein Glut4 (Yazdani, Huang, and Terman 2008). Most of the other trace metals have been identified as parts of metalloenzymes (Gray 2003; Lippard and Berg 1994). For example, manganese is found as a cofactor in mitochondrial superoxide dismutase, inorganic phosphatase, and most notably photosynthesis system II. Nickel functions in enzymes such as urease and several hydrogenases. Both mobydenum and vanadium are found in nitrogenases, where they are present in larger clusters also containing iron and sulfur ions (Fenton 1995). Needless to say, most of the trace metals play key roles as protein/enzyme cofactors in human biology.

HOW ARE METALS INCORPORATED INTO APPROPRIATE PROTEINS?

Despite this wealth of information on high-resolution structures of many metalloproteins, the folding-binding pathways for biosynthesis of metalloproteins are mostly unknown (Wittung-Stafshede 2002), but several different scenarios can be outlined (Figure 1.1). Although unique features of different proteins result in selectivity for a particular metal, this selection is imperfect since proteins are dynamic molecules, something that is especially true for newly synthesized (unfolded) polypeptides. Protein affinity for metals tends to follow a universal order of preference, which for six-coordinated divalent metals is the Irving-Williams series (Williams 2002) (although this series excludes most zinc- and copper-protein complexes). Thus it remains a question as to how cells can contain proteins with weak-binding metals while simultaneously containing proteins requiring tight-binding metals. Intuitively, all metalloproteins would bind the most competitive metal. One solution to this

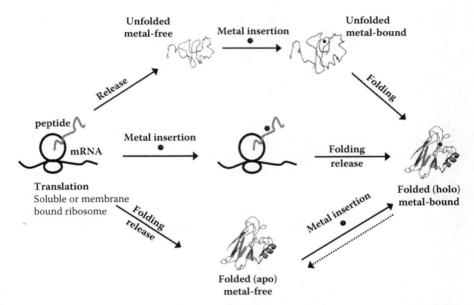

FIGURE 1.1 Illustration of protein biosynthesis and possible connections to metal binding. Proteins are translated by soluble or membrane-bound ribosomes (the latter can dictate what cellular compartment the protein is released into). Metals (or metal clusters) are inserted into the polypeptide either after transfer from a pool of free metal ions or via delivery by metallochaperones. For different metalloproteins, insertion of the metal is likely to occur at different steps of the protein biogenesis process: during translation; after polypeptide release, but before the polypeptide is completely folded; or, as a last step, after the polypeptide has adopted its folded conformation. Metal release from the native metal-bound form (holo) can result in either misfolding or unfolding as a result of conformational destabilization, or in a folded, metal-free conformation (apo) (dotted arrow).

paradox is the idea that the compartment (and its metal composition) in which the protein folds overrides the protein's metal-binding preference to control its metal content (Tottey et al. 2008). Another explanation is the fact that metal insertion into many proteins is strictly controlled by specific (e.g., copper chaperones) or unspecific (e.g., ferrochelatases) protein-based delivery system.

In the case of copper, such ions are almost nonexistent in their free form in the cytoplasm (O'Halloran and Culotta 2000) since copper's redox properties may result in oxidative damage of proteins, lipids, and nucleic acids. Instead, the cellular copper concentration is strictly controlled, and most copper ions are delivered to their destinations by copper chaperones (Lamb et al. 1999, 2001). In the case of iron, transferrins transport iron into cells and release it within the endosome; the protein hemopexin delivers apo-heme to the same compartment (Shipulina, Smith, and Morgan 2000). In the case of c-type cytochromes, it has been proposed that heme attachment is a required step before correct folding occurs *in vivo* (Bertini et al. 1994; Lippard and Berg 1994). The heme is covalently attached by a heme-lyase enzyme (Ramseier, Winteler, and Hennecke 1991) when the polypeptide is associated with the membrane. After heme insertion, the holoprotein is released from the membrane and is free to fold into its native configuration. While specific heme transporters and ligation enzymes appear necessary for c-type cytochrome assembly, a different, perhaps nonspecific or diffusional, mechanism for heme transport and insertion into perplasmic heme proteins has been proposed based on *Escherichia coli* studies (Goldman, Gabbert, and Kranz 1996). Also, the synthesis of iron-sulfur clusters *in vivo* requires a complex machinery encoded in prokaryotes by the *isc* (iron-sulfur-cluster) operon (Agar et al. 2000; Ollagnier-De Choudens et al. 2000, 2001). In eukaryotes, mitochondrial proteins, highly homologous to the bacterial ones, have been shown to be involved in Fe-S cluster assembly (Lill 2009). Although there are no known zinc chaperones, membrane transport and regulatory proteins specific for zinc have been identified.

ROLES OF METALS IN PROTEIN FOLDING AND MISFOLDING?

Since metal-binding proteins fold in a cellular environment where their cognate cofactors are present, either free in the cytoplasm or bound to delivery proteins, the question arises as to when metals bind to their corresponding proteins. Specifically, do they bind before, during, or after polypeptide folding? As demonstrated *in vitro*, many metalloproteins retain strong metalloligand binding after polypeptide unfolding (Robinson et al. 1997; Bertini et al. 1997; Wittung-Stafshede et al. 1998, 1999). This implies that *in vivo* metals may interact with their corresponding proteins before polypeptide folding takes place, which could impact the folding reaction. Local and nonlocal structure in the unfolded protein may form due to specific coordination of the cofactor (Pozdnyakova, Guidry, and Wittung-Stafshede 2000). Such structural restriction of the ensemble of conformations may dramatically decrease the entropy of the unfolded state, limiting the conformational search for the native state (Luisi, Wu, and Raleigh 1999). The metal may this way serve as a nucleation site that directs polypeptide folding along a specific pathway in the free-energy landscape. In this regard, considering the *in vivo* biosynthesis, one must take into account that proteins are made on ribosomes and that metals may bind before, during, or after release of the polypeptide from the ribosome. It has been proposed that polypeptides obtain structure as well as altered dynamics during the translational process (Ellis et al. 2008; Ellis, Culviner, and Cavagnero 2009). It is clear that metals in many cases stabilize final folded protein structures, but less is known about how they may modulate folding pathways, speed, and folding-transition state ensembles.

Metals are often yin-yang elements: They play essential roles as cofactors in protein, but are often toxic in large amount and/or when free in biological fluids. A number of diseases (for example, Menke's syndrome and Wilson's disease) as well as numerous neurodegenerative conditions and amyloid deposition syndromes have been linked to alterations in metal-protein interactions. In many described cases, metal binding to an oligomerization-prone protein modulates the aggregation and fibrillation pathways. For example, Cu^{2+} and Zn^{2+} ions have been shown to induce aggregation of the amylogenic peptide β-microglobulin, possibly via a mechanism involving a local increase in conformational flexibility upon metal binding, which then propagates to the core of the molecule, thus promoting global and slow fluctuations. This may contribute to the overall destabilization of the molecule, increasing the equilibrium population of the amyloidogenic and oligomeric intermediates (Villanueva et al. 2004; Blaho and Miranker 2009). This process is triggered only in the presence of stoichiometric amounts of Cu^{2+}, as the protein is not aggregation prone *in vivo* and under physiological conditions, thus illustrating the importance of the interplay between metal homeostasis and the generation of nonnative protein conformations.

The relevance of metal homeostasis is becoming more and more apparent in major neurodegenerative disorders, including Alzheimer's disease (AD) and Parkinson's disease (PD), which are characterized by elevated levels and miscompartmentalization of iron, copper, and zinc (Crouch et al. 2009; Zatta et al. 2009). Deposition of the amyloid β-protein (Aβ) is greatly influenced by metal ions, as metal binding is likely to interfere with the formation of the cross-β core structure of amyloid fibrils. For example, pulses of zinc in micromolar concentrations result in protein-metal interactions that specifically stabilize nonfibrillar pathogenic aggregate forms, which are more toxic than the insoluble fibrils that are formed upon Aβ self-assembly (Noy et al. 2008). This phenomenon has a physiological context, as transient high local concentrations of zinc and copper are indeed released in the glutamatergic synapse (Crouch et al. 2009). Ultimately, these examples underscore the fundamental importance of revealing the physical principles underlying metal interactions with folded, unfolded, and intermediate states of polypeptides.

METALS AS MODULATORS OF PROTEIN STRUCTURE AND CONFORMATION

Metal ions can also play a role as modulators of protein conformations. Binding of a metal ion to a folded polypeptide yields an effective cross-link between a specific set of residues within the protein, which results in a stabilizing effect and may be important to maintain a particular local structure. That is for example the case of Zn^{2+} coordination in zinc finger domains, which is the most abundant structural domain in the human genome and has the classical Cys_2His_2 motif (Petsko and Ringe

2004). In this case, the topology of this domain is kept in place by the zinc ion, which is strictly required as an essential structural determinant; in fact, removal of coordinating residues disrupts the fold (Lee and Subbiah 1991). However, direct action of Zn^{2+} ligands is also essential to drive the folding process of the zinc finger domain: Coordination of the ion to the cysteines is essential for a β-strand to α-helix transition that occurs with histidine binding and formation of the finger structure, which is not obtained if these ligand residues are switched within the sequence (Cox and McLendon 2000; Miura et al. 1987).

Also, more-subtle clipping effects may arise from metal binding. Studies on designed four-helix bundle proteins have illustrated how a metal ion can modulate protein dynamics: This protein comprising a Leu-rich hydrophobic core was found to fold into a conformation intermediate between that of the native and molten globule folds; although this conformation was globular and compact, and the α-helices stable, the apolar side chains exhibited increased dynamics. However, engineering of two Zn^{2+} binding sites within the core rigidified the structure and resulted in a protein with nativelike properties (Chamberlain, Handel, and Marqusee 1996). The same is in fact observed in natural proteins: Some dicluster ferredoxins contain an N-terminal extension that, depending on the isoform, may be stabilized either by a His/Asp Zn^{2+} site or by a local hydrophobic core (Rocha et al. 2006). These examples are illustrative that metal ion binding can be simultaneously important to fine-tune a particular native structure as well as to drive its correct folding.

One of the most dramatic examples of conformational changes induced by metal binding is given by Ca^{2+} binding proteins. Calcium is a major signaling ion, and evolution has designed particular protein folds that upon calcium binding undergo substantial conformational changes. That is the case of the EF-hand motif, or variations thereof, which consists of a helix-loop-helix topology, which is found in several proteins such as calmodulin or the S100 family members (Bayley, Martin, and Jones 1988; Fritz, Muller, and Mayhew 1973). The conformational change arising from Ca^{2+} binding to the EF-hand motif is rather substantial and results from the lateral motions of the two helices that flank the Ca^{2+} binding residues. This effect underlies the action of regulatory Ca^{2+} sensor proteins, as the effective binding of calcium to the sensor protein and the resulting conformational change can be propagated to another protein to which the calcium sensor is interacting. Noteworthy, this conformational change seems to be less dramatic among proteins containing a structural EF-hand motif, which are mainly involved in processes related to Ca^{2+} signal modulation such as the uptake, buffering, and transport of Ca^{2+}, like calbindin and parvalbumin. Among the calcium sensor proteins, which contain the afore-mentioned EF-hand motifs, are the proteins belonging to the S100 family (Fritz, Muller, and Mayhew 1973). This group of proteins consists of a large family of nearly 20 subtypes of S100s, with very diverse functions, ranging from regulation of cell cycle to cell growth and differentiation. The relevant feature for the purpose of this chapter is that, unlike other EF-hand proteins, many of the S100 proteins contain additional Zn^{2+} binding sites, whose metallation influences protein conformation, folding, and presumably function. That is the case of the S100A2 protein, a regulator of cell cycle that binds two Ca^{2+} and two Zn^{2+} ions per subunit, known to be associated with activation (Ca^{2+}) or inhibition (Zn^{2+}) of downstream signaling. Zinc binds at distinct sites, which have different metal-binding affinities, and it was found that this protein is destabilized by Zn^{2+} and stabilized by Ca^{2+}, suggesting a synergistic effect between metal binding, protein stability, and S100A2 biological activity (Botelho et al. 2009). According to this model, Ca^{2+} activates and stabilizes the protein, the opposite being observed upon Zn^{2+} binding, and this further illustrates how the binding of different metal ions results in conformational adjustments and modulation of protein folding and functions.

Finally, the effect of metal ions as modulators of protein conformations is particularly noteworthy among metal sensor proteins. These are proteins that upon metal binding interact with DNA and regulate gene expression. This implies conformational changes occurring that control open and closed conformation within the sensor proteins. These proteins are grouped within seven families, which comprise for example the ArsR familiy α5 and Fur sensors, *Bacillus subtilis* Mnt, and *E. coli* NikR (Giedroc and Arunkumar 2007). Altogether these sensors cover a broad range of metal ions, from transition metals to heavy metals. For many, the structural basis of metal induced allosteric switching is becoming clear with growing structural data that allows one to envisage how metal binding stabilizes particular conformations prone to DNA interactions (Giedroc and Arunkumar 2007). It is now becoming clear that metal binding results in a rigidification of the protein structure so as to maintain and stabilize both the dimerization domains and the alfa-turn-helix motif that makes specific interactions with DNA. Thus binding of metals to the metallosensor proteins switches on and off the DNA-binding conformations, via modulation of the structure of the native apo sensor proteins to which they bind.

REFERENCES

Agar, J. N., C. Krebs, J. Frazzon, B. H. Huynh, D. R. Dean, and M. K. Johnson. 2000. IscU as a scaffold for iron-sulfur cluster biosynthesis: sequential assembly of [2Fe-2S] and [4Fe-4S] clusters in IscU. *Biochemistry* 39 (27):7856–62.

Bayley, P., S. Martin, and G. Jones. 1988. The conformation of calmodulin: a substantial environmentally sensitive helical transition in Ca4-calmodulin with potential mechanistic function. *FEBS Lett* 238 (1):61–66.

Bertini, I., J. A. Cowan, C. Luchinat, K. Natarajan, and M. Piccioli. 1997. Characterization of a partially unfolded high potential iron protein. *Biochemistry* 36 (31):9332–39.

Bertini, I., H. B. Gary, S. J. Lippard, and J. S. Valentine. 1994. *Bioinorganic chemistry*. Mill Valley, CA: University Science Books.

Blaho, D. V., and A. D. Miranker. 2009. Delineating the conformational elements responsible for Cu(2+)-induced oligomerization of beta-2 microgbulin. *Biochemistry* 48 (28):6610–17.

Botelho, H. M., M. Koch, G. Fritz, and C. M. Gomes. 2009. Metal ions modulate the folding and stability of the tumor suppressor protein S100A2. *FEBS J* 276 (6):1776–86.

Chamberlain, A. K., T. M. Handel, and S. Marqusee. 1996. Detection of rare partially folded molecules in equilibrium with the native conformation of RNaseH. *Nat Struct Biol* 3 (9):782–87.

Cox, E. H., and G. L. McLendon. 2000. Zinc-dependent protein folding. *Curr Opin Chem Biol* 4 (2):162–65.

Crouch, P. J., L. W. Hung, P. A. Adlard, M. Cortes, V. Lal, G. Filiz, K. A. Perez, M. Nurjono, A. Caragounis, T. Du, K. Laughton, I. Volitakis, A. I. Bush, Q. X. Li, C. L. Masters, R. Cappai, R. A. Cherny, P. S. Donnelly, A. R. White, and K. J. Barnham. 2009. Increasing Cu bioavailability inhibits Abeta oligomers and tau phosphorylation. *Proc Natl Acad Sci U S A* 106 (2):381–86.

Ellis, J. P., C. K. Bakke, R. N. Kirchdoerfer, L. M. Jungbauer, and S. Cavagnero. 2008. Chain dynamics of nascent polypeptides emerging from the ribosome. *ACS Chem Biol* 3 (9):555–66.

Ellis, J. P., P. H. Culviner, and S. Cavagnero. 2009. Confined dynamics of a ribosome-bound nascent globin: cone angle analysis of fluorescence depolarization decays in the presence of two local motions. *Protein Sci* 18 (10):2003–15.

Fenton, D. E. 1995. *Biocoordination chemistry*. Oxford Chemistry Primers. Oxford: Oxford University Press.

Fritz, J., F. Muller, and S. G. Mayhew. 1973. Electron-nuclear double resonance study of flavodoxin from *Peptostreptococcus elsdenii*. *Helv Chim Acta* 56 (7):2250–54.

Giedroc, D. P., and A. I. Arunkumar. 2007. Metal sensor proteins: nature's metalloregulated allosteric switches. *Dalton Trans* (29):3107–20.

Goldman, B. S., K. K. Gabbert, and R. G. Kranz. 1996. Use of heme reporters for studies of cytochrome biosynthesis and heme transport. *J Bacteriol* 178 (21):6338–47.

Gomes, C. M., C. Frazao, A. V. Xavier, J. Legall, and M. Teixeira. 2002. Functional control of the binuclear metal site in the metallo-beta-lactamase-like fold by subtle amino acid replacements. *Protein Sci* 11 (3):707–12.

Gray, H. B. 2003. Biological inorganic chemistry at the beginning of the 21st century. *Proc Natl Acad Sci U S A* 100 (7):3563–68.

Lamb, A. L., A. S. Torres, T. V. O'Halloran, and A. C. Rosenzweig. 2001. Heterodimeric structure of superoxide dismutase in complex with its metallochaperone. *Nat Struct Biol* 8 (9):751–55.

Lamb, A. L., A. K. Wernimont, R. A. Pufahl, V. C. Culotta, T. V. O'Halloran, and A. C. Rosenzweig. 1999. Crystal structure of the copper chaperone for superoxide dismutase. *Nat Struct Biol* 6 (8):724–29.

Lee, C., and S. Subbiah. 1991. Prediction of protein side-chain conformation by packing optimization. *J Mol Biol* 217 (2):373–88.

Lill, R. 2009. Function and biogenesis of iron-sulphur proteins. *Nature* 460 (7257):831–38.

Lippard, S. J., and J. M. Berg. 1994. *Principles of bioinorganic chemistry*. Mill Valley, CA: University Science Books.

Luisi, D. L., W. J. Wu, and D. P. Raleigh. 1999. Conformational analysis of a set of peptides corresponding to the entire primary sequence of the N-terminal domain of the ribosomal protein L9: evidence for stable native-like secondary structure in the unfolded state. *J Mol Biol* 287 (2):395–407.

Miura, R., K. Yamaichi, K. Tagawa, and Y. Miyake. 1987. On the structure of old yellow enzyme studied by specific limited proteolysis. *J Biochem* (Tokyo) 102 (5):1311–20.

Noy, D., I. Solomonov, O. Sinkevich, T. Arad, K. Kjaer, and I. Sagi. 2008. Zinc-amyloid beta interactions on a millisecond time-scale stabilize non-fibrillar Alzheimer-related species. *J Am Chem Soc* 130 (4):1376–83.

O'Halloran, T. V., and V. C. Culotta. 2000. Metallochaperones, an intracellular shuttle service for metal ions. *J Biol Chem* 275 (33):25057–60.

Ollagnier-De-Choudens, S., T. Mattioli, Y. Takahashi, and M. Fontecave. 2001. Iron-sulfur cluster assembly: characterization of IscA and evidence for a specific and functional complex with ferredoxin. *J Biol Chem* 276 (25):22604–7.

Ollagnier-De Choudens, S., Y. Sanakis, K. S. Hewitson, P. Roach, J. E. Baldwin, E. Munck, and M. Fontecave. 2000. Iron-sulfur center of biotin synthase and lipoate synthase. *Biochemistry* 39 (14):4165–73.

Petsko, G. A., and D. Ringe. 2004. *Protein structure and function*. Primers in Biology. Mill Valley, CA: New Science Press.

Pozdnyakova, I., J. Guidry, and P. Wittung-Stafshede. 2000. Copper triggered b-hairpin formation. Initiation site for azurin folding? *J Am Chem Soc* 122:6337–38.

Ramseier, T. M., H. V. Winteler, and H. Hennecke. 1991. Discovery and sequence analysis of bacterial genes involved in the biogenesis of c-type cytochromes. *J Biol Chem* 266 (12):7793–7803.

Robinson, C. R., Y. Liu, J. A. Thomson, J. M. Sturtevant, and S. G. Sligar. 1997. Energetics of heme binding to native and denatured states of cytochrome b562. *Biochemistry* 36 (51):16141–46.

Rocha, R., S. S. Leal, V. H. Teixeira, M. Regalla, H. Huber, A. M. Baptista, C. M. Soares, and C. M. Gomes. 2006. Natural domain design: enhanced thermal stability of a zinc-lacking ferredoxin isoform shows that a hydrophobic core efficiently replaces the structural metal site. *Biochemistry* 45 (34):10376–84.

Shipulina, N., A. Smith, and W. T. Morgan. 2000. Heme binding by hemopexin: evidence for multiple modes of binding and functional implications. *J Protein Chem* 19 (3):239–48.

Tottey, S., K. J. Waldron, S. J. Firbank, B. Reale, C. Bessant, K. Sato, T. R. Cheek, J. Gray, M. J. Banfield, C. Dennison, and N. J. Robinson. 2008. Protein-folding location can regulate manganese-binding versus copper- or zinc-binding. *Nature* 455 (7216):1138–42.

Villanueva, J., M. Hoshino, H. Katou, J. Kardos, K. Hasegawa, H. Naiki, and Y. Goto. 2004. Increase in the conformational flexibility of beta 2-microglobulin upon copper binding: a possible role for copper in dialysis-related amyloidosis. *Protein Sci* 13 (3):797–809.

Warder, S. E., M. Prorok, Z. Chen, L. Li, Y. Zhu, L. G. Pedersen, F. Ni, and F. J. Castellino. 1998. The roles of individual gamma-carboxyglutamate residues in the solution structure and cation-dependent properties of conantokin-T. *J Biol Chem* 273 (13):7512–22.

Williams, R. J. 2002. The fundamental nature of life as a chemical system: the part played by inorganic elements. *J Inorg Biochem* 88 (3–4):241–50.

Wittung-Stafshede, P. 2002. Role of cofactors in protein folding. *Acc Chem Res* 35 (4):201–8.

Wittung-Stafshede, P., J. C. Lee, J. R. Winkler, and H. B. Gray. 1999. Cytochrome b562 folding triggered by electron transfer: approaching the speed limit for formation of a four-helix-bundle protein. *Proc Natl Acad Sci U S A* 96 (12):6587–90.

Wittung-Stafshede, P., B. G. Malmstrom, J. R. Winkler, and H. B. Gray. 1998. Electron-transfer triggered folding of deoxymyoglobin. *J Phys Chem* 102:5599–5601.

Yazdani, U., Z. Huang, and J. R. Terman. 2008. The glucose transporter (GLUT4) enhancer factor is required for normal wing positioning in Drosophila. *Genetics* 178 (2):919–29.

Zatta, P., D. Drago, S. Bolognin, and S. L. Sensi. 2009. Alzheimer's disease, metal ions and metal homeostatic therapy. *Trends Pharmacol Sci* 30 (7):346–55.

2 The Folding Mechanism of c-Type Cytochromes

*Carlo Travaglini-Allocatelli,
Stefano Gianni, and Maurizio Brunori*

CONTENTS

Introduction 13
Equilibrium Studies 14
Kinetic Studies on c-Type Cytochromes 17
Folding Mechanism of Cytochrome c_{551} from *Pseudomonas aeruginosa* 19
Collapse in the sub-ms Time Window 22
Cytochrome c_{552} from *Thermus thermophilus* 23
Cytochrome c_{552} from *Hydrogenobacter thermophilus* 25
A Consensus Folding Mechanism 26
Engineering Folding Pathways 30
Future Perspective 33
References 33

INTRODUCTION

Cytochromes c (cyts c) are small monomeric proteins of 80–120 residues involved in different and crucial aspects of cellular life, from electron-transport processes to apoptosis. These heme-proteins show a typical α-helical fold that is recognized as a structural superfamily in protein classification tools such as SCOP (Andreeva et al. 2008) or CATH (Orengo et al. 1997). The three major α-helices (generally referred to as N-terminal helix, 60s helix, and C-terminal helix, following numbering of amino acidic residues of horse heart cyt c) wrap around the heme group that is covalently linked to the protein via two thioether bonds between its vinyl groups and two cysteine residues in the conserved CysXaaXaaCysHis motif (Figure 2.1).

Attachment of the heme group to the apoprotein *in vivo* is a complex posttranslational process that involves different enzymatic activities (Ferguson et al. 2008). It is therefore not surprising that attempts to obtain properly folded recombinant holo cytochromes c by heterologous expression in *Escherichia coli* were unsuccessful for a long time; efficient production of recombinant prokaryotic holo cyt c in *E. coli* is now generally accomplished under control of the *E. coli* enzymatic apparatus for heme attachment (Thöny-Meyer et al. 1995).

The heme iron is always axially coordinated to the His residue of the CysXaaXaaCysHis motif on the proximal side. While the His ligand is maintained

FIGURE 2.1 Three-dimensional structures of cytochrome c (cyt c) proteins discussed in the text. From left to right: *Thermus thermophilus* cyt c_{552} (PDB 1C52), horse cyt c (PDB 1OCD), *Hydrogenobacter thermophilus* cyt c_{552} (PDB 1YNR), and *Pseudomonas aeruginosa* cyt c_{551} (PDB 351C). The heme and the two amino acid side chains coordinating the Fe^{3+} iron in the fifth (histidine) and sixth (methionine) positions are shown in stick representation. The structure of TT-cyt c_{552} has two additional α-helices and a β-sheet. The N- and C-terminal α-helices, whose role is crucial for the folding of cyt c proteins (see text), are similarly oriented in each case and are shown on top of the structures.

even under denaturing conditions, the distal ligand, generally Met, is inherently labile and readily displaced by other side chains, such as a deprotonated His or Lys, which can become trapped during folding (Babul and Stellwagen 1971). The presence of a covalently bound heme can be considered as the privilege and disgrace of these small single-domain proteins: On one hand, it is an ideal natural quencher of the intrinsic fluorescence of the protein, it enables the breakage or formation of some individual bonds to be monitored (Gianni et al. 2003), and it represents a perfect tool for the design of photo-induced experiments to follow ultrafast conformational transitions (Jones et al. 1993; Hagen et al. 1996; Yeh and Rousseau 1998; Chang et al. 2003). On the other hand, the presence of this large prosthetic group has often hindered a generalization of the rules emerging from *in vitro* folding studies on this system.

In this chapter we report some of the key findings on the folding pathway of c-type cytochromes. We first describe the equilibrium studies on this protein family and then extend our discussion to kinetic measurements. Finally, we present the hypothesis of a common folding mechanism for all c-type cytochromes, and discuss the possibility of tuning their folding pathways along different parallel routes by mutagenesis. Each section briefly outlines experimental methodologies and analytical approaches classically employed in protein-folding studies.

EQUILIBRIUM STUDIES

A valuable method to assess the stability of proteins in solution relies on analysis of equilibrium denaturation transitions induced by increasing concentrations of chaotropic denaturants (such as urea and guanidine hydrochloride [GdnHCl]). We briefly describe the general rules classically employed for the analysis of equilibrium denaturation, and comment on the peculiar features of c-type cytochromes.

Understanding the forces that control protein stability is a difficult task. There are many factors to be considered: the number of populated species at equilibrium, the forces stabilizing these states (in terms of both inter- and intramolecular forces), and the analytical methodologies employed. Equilibrium and kinetic studies of protein folding most often take advantage of spectroscopy such as absorbance, fluorescence, circular dicroism, small-angle x-ray scattering, and nuclear magnetic resonance (NMR). Key to the successful use of any of these spectroscopic techniques is a resolvable difference in signal between states populated in the transition between the folded and the unfolded protein.

Since the native state is by definition populated under physiological conditions, the study of protein folding requires the perturbation of the distribution of states. Variables affecting protein stability, such as temperature, pH, ionic strength, and solvent composition (Nicholson and Scholtz 1996), are generally varied to induce a perturbation. Very often, urea or GdnHCl are employed to change solvent composition and thereby protein stability.

It has been empirically determined that the stability of a globular protein can be expressed as a linear function of denaturant concentration (Tanford 1968), and therefore the linear extrapolation method is routinely used in isothermal protein-folding studies. Following this approach, determination of the fraction of unfolded state at different denaturant concentrations allows one to estimate, with some confidence, the protein stability in the absence of denaturant (Pace and Shaw 2000; Parker, Spencer, and Clarke 1995) by applying simplified assumptions on the effect of a denaturant (Myers, Pace, and Scholtz 1995).

Early solvent denaturation studies suggested that c-type cytochromes, like most other single-domain proteins in their size class (about 100 residues), while displaying variable thermodynamic stabilities, exhibit two-state equilibrium behavior (Knapp and Pace 1974), insofar as only the native and fully denatured states are populated throughout the transition isotherm. Generally, equilibrium denaturation curves obtained for a given cytochrome c allow for the calculation of similar thermodynamic folding parameters, e.g., free energy of denaturation and apparent cooperativity, when followed by different spectroscopic probes. More recently however, spectroscopic and calorimetric studies revealed deviations from a fully cooperative, ideal two-state transition for horse cyt c at denaturant concentrations below those leading to major unfolding, in both the oxidized and reduced states (Russell and Bren 2002; Latypov et al. 2008). These observations suggest that, in some cases, cyt c may (un)fold via at least one stable intermediate, which accumulates at equilibrium. Romesberg and coworkers recently attempted to characterize the structure of the equilibrium folding intermediate of horse cyt c (Sagle et al. 2006). This proved very difficult because although NMR and site-specific isotope labeling have the potential for high structural resolution, no signals from the intermediate were observable because the exchange rates were too fast on the NMR timescale. However, the time resolution of the infra-red (IR)-based experiments carried out by Romesberg and coworkers provided an unprecedented direct observation of the equilibrium folding intermediate, which appears to have a significantly altered 60s helix.

Over and above standard spectroscopic methods, hydrogen exchange (HX) is an attractive method for detecting small populations of partially unfolded states. Hence, some authors suggested that these states resemble folding intermediates, so that HX may represent a shortcut to resolve folding pathways using equilibrium measurements only.

Chemical exchange of labile protons with deuterons from the solvent can be followed by NMR. In a native protein, many amide protons are protected from exchange by hydrogen bonding or burial in a "close" configuration. Exchange occurs only when the protected proton transiently populates an "open" conformation with an apparent rate constant of:

$$k_{ex} = \frac{k_o k_{int}}{k_c + k_{int}} \quad (2.1)$$

where k_o and k_c are the microscopic rate constants for the opening and closure processes and k_{int} is the intrinsic rate constant of exchange of a given amino acid (it must be noted that k_{int} depends not only on the nature of the amino acid, but also on temperature and pH). Given these premises, hydrogen exchange measurements may be performed under two different limiting conditions. In fact, at relatively low pH (pH ≤ 7), $k_{int} \ll k_c$, so that Equation 2.1 simplifies to:

$$k_{ex} = \frac{k_o k_{int}}{k_c} = K_p k_{int} \quad (2.2)$$

Under these conditions, known as the EX2 limit, the observed rate of exchange of each protected amide proton will reflect its protection factor K_p, which is a measure of the equilibrium constant for the opening-closure reaction for a particular proton. K_p resembles the equilibrium constant for folding, measured at a residue level.

On the other hand, at pH ≥7, some proteins may display $k_{int} \gg k_c$. Under such condition, known as the EX1 limit, Equation 2.1 simplifies to:

$$k_{ex} = k_o \quad (2.3)$$

so that the exchange rate constant is a direct measure of the unfolding (opening) rate constant, measured under conditions favoring folding.

The analysis of the HX method and its applicability to protein folding pathways has been widely discussed elsewhere (Clarke, Itzhaki, and Fersht 1998; Englander 1998) and will not be addressed in any further detail.

C-type cytochromes have been subjected to HX experiments under both EX1 and EX2 limiting conditions (Hoang et al. 2002; Rumbley et al. 2001). We outline next some of the more significant experimental findings obtained under the EX2 limit.

Englander and coworkers (Maity et al. 2005) showed that the distribution of protection factors in horse cyt c highlights the presence of five cooperative folding-unfolding units (called foldons) that explore the closed and open (or folded and unfolded) configurations under native conditions. It was therefore argued that cyt c folding proceeds by the stepwise assembly of the foldon units rather than by formation of a sparse interacting network of residues. A sequential stabilization process determines the folding pathway: previously formed foldons guide and stabilize subsequent foldons to progressively build the native protein. According to the work of Englander

FIGURE 2.2 Unfolding by native-state HX. Cyt c with color-coded unfolding units and its suggested folding/unfolding pathway. Equilibrium NHX experiments suggested the unfolding sequence of events to proceed from the white to the black regions.

and coworkers, analysis of the stabilization free-energy pattern, as revealed by HX equilibrium experiments, suggests that the folding pathway of horse cyt c involves the formation of an early foldon, represented by the interaction between the N- and C-terminal helices, followed by the consolidation of the 60's helix and concluded by the locking in place of the heme pocket and the distal coordination of the heme iron by Met80 (Figure 2.2).

It is yet unclear how to test the relationship between the kinetic intermediates described next and the equilibrium species, identified by HX experiments.

KINETIC STUDIES ON c-TYPE CYTOCHROMES

While equilibrium studies provide useful information about folding in terms of stability and cooperativity of a protein, information about the reaction dynamics has to be acquired by kinetics. Understanding the mechanism of folding demands a kinetic analysis of the relaxations induced by a perturbation of the equilibrium, imposed by changing the experimental conditions. The challenge of the kinetic approach is to provide direct information about the actual folding pathway of a protein and to verify if an apparently two-state system at equilibrium may, or may not, be kinetically complex (Matouschek et al. 1990), with a folding reaction involving transiently populated marginally stable intermediate(s), undetectable at equilibrium.

As previously described, kinetic folding studies are based on rapid perturbations of the equilibrium conditions in order to change the energetics of the system.

In general, folding is followed *in vitro* by first denaturing a protein with urea or GdnHCl, and then diluting the denaturant to favor refolding. In a two-state model, when only the native (N) and the unfolded (U) states are populated, the time course is described by an exponential function with an apparent rate constant given by

$$U \underset{K_u}{\overset{K_f}{\rightleftarrows}} N \qquad k_{obs} = k_F + k_U \tag{2.4}$$

Since denaturants affect the stability of the different conformational states of a polypeptide by selective stabilization of the relevant solvent exposed surface area, for a protein obeying a two-state transition the activation energy for the folding and unfolding processes will also be linearly dependent on denaturant concentration (Jackson 1991). This accounts for the typical V-shape of the semilogarithmic plot of k_{obs} versus denaturant concentration (called chevron plot).

In some cases, proteins that exhibit a cooperative two-state equilibrium transition display complex kinetics. The population of a folding intermediate reflects the presence of more than one energy barrier, and folding would be described by the sum of two or more exponential processes (Roder and Colon 1997). In these cases, the presence of folding intermediates may result either in a complex multiexponential folding time course, or in a curvature of the chevron plots (i.e., a deviation from the ideal V-shaped behavior or "rollover effect"), or eventually be deduced by burst-phase analysis of the observed amplitudes (see later discussion). In the case of the c-type cytochromes, all these signatures of complex folding scenarios have been observed, as briefly discussed next.

Nowadays, concepts like "kinetic traps" or "rollover effects" in chevron plots are fairly common in folding studies, but it is interesting to recall that they were originally introduced/observed in kinetic folding studies on cyt c. Indeed, early studies on horse cyt c revealed that this protein folds via a complex mechanism involving multiexponential kinetics (Colon and Roder 1996; Colon et al. 1996). An essential point is to distinguish between kinetic traps competing with productive folding, and genuine reaction intermediates on the path to the native state. In fact, when a denatured cyt c is rapidly mixed with a solution favoring folding, several misfolded conformations compete with productive folding. Among others, since the sixth coordination of the heme iron to Met is lost upon unfolding, several side chains may compete with the native ligand coordination. Hence, a first type of misfolded conformation, which leads to multiexponential kinetics in the ms time range, involves the nonnative coordination to the heme iron of histidines, lysines, or even the N-terminal amino group, instead of the native distal ligand. These events have been demonstrated both by performing refolding experiments at low pH (which prevents His coordination) and by site-directed mutagenesis (Elove, Bhuyan, and Roder 1994; Colon et al. 1997; Pierce and Nall 2000). A second type of misfolding event (which is quite general) arises from *cis-trans* isomerization of prolyl peptide bonds, as documented in many other proteins (Schmid and Baldwin 1978).

As described in the following, over and above the formation of kinetic traps the folding pathway of several c-type cytochromes clearly consists of more than one step, occurring from the μs to the ms time range, and well above. It is thus extremely difficult to depict a general and comprehensive scenario for the folding of c-type cytochromes when considering all the complexities of these systems. Therefore we have taken an alternative approach based on comparative characterization the folding pathway of different c-type cytochromes from mesophilic and thermophilic organisms. As detailed in the sections that follow, the results recapitulate many of the key features of the folding of this protein family. Hence, we describe some of our findings on selected cytochromes, in the perspective of current and past literature from other laboratories.

FOLDING MECHANISM OF CYTOCHROME c_{551} FROM *PSEUDOMONAS AERUGINOSA*

The three-dimensional structure of the cytochrome c_{551} from *P. aeruginosa* (PA-cyt c_{551}; Figure 2.1; Detlefsen et al. 1991; Matsuura et al. 1982) shows conservation of the typical fold in spite of low (~30%) sequence homology with eukaryotic cyt c. A major structural peculiarity of PA-cyt c_{551} is its smaller size (82 versus 104 residues) due to the absence of a loop facing the heme propionate groups. In addition this protein has an acidic isoelectric point (pI = 4.7 versus pI = 10.5 of horse cyt c), which allows investigation of refolding at zero net charge. Moreover, the presence of only one histidine residue (His16, coordinating the heme iron on the proximal side) reduces the chances of miscoordination events observed at neutral pH in the case of horse cyt c (Elove, Bhuyan, and Roder 1994; Colon et al. 1997) and simplifies the analysis of the folding time course.

A logarithmic plot of the observed unfolding and refolding rate constant as a function of urea concentration (chevron plot) at pH 4.7 is shown in Figure 2.3. Under all conditions, the observed time course as monitored by stopped-flow and Trp fluorescence can be fitted with a single-exponential process (neglecting proline *cis/trans* isomerization, which accounts for <10% of the observed amplitude). Inspection of the chevron plot reveals the presence of a pronounced rollover effect in the refolding branch. In view of what was seen for other proteins (Silow and Oliveberg 1997), we showed that the rollover at pH 4.7 was related to the formation of transient aggregates, rather than to accumulation of a genuine intermediate state(s). In the case of PA-cyt c_{551}, this interpretation was substantiated by the results of refolding experiments carried out as a function of protein concentration, whereby it was possible to show that such rollover disappears at very low protein concentrations (Gianni et al. 2001; Travaglini-Allocatelli et al. 1999). Interestingly, however, the deviation from linearity observed in refolding experiments carried out at other pH values (i.e., pH 7.0 and 3.0) could not be explained solely on the basis of aggregation effects, but required a more complex folding model.

The presence of a rollover effect in the refolding branch of the chevron plot has been often interpreted in terms of accumulation of folding intermediates (N. Ferguson et al. 1999; Khorasanizadeh, Peters, and Roder 1996); however, an alternative explanation takes into account a change in the rate-limiting barrier along the reaction coordinate (Oliveberg et al. 1998). In particular, the amplitude behavior predicted by the latter model postulates that the refolding kinetics should be single-exponential. In the case of PA-cyt c_{551}, even in the presence of a burst phase below

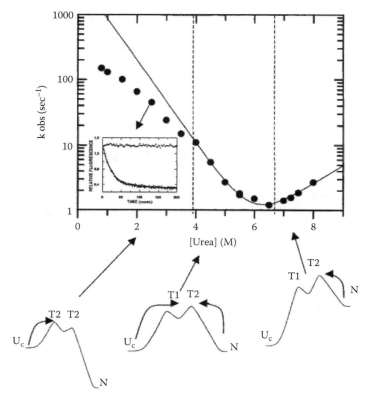

FIGURE 2.3 Chevron plot of wt PA-cyt c_{551} at pH 7.0. Below the plot the energy profiles at low, medium, and high denaturant concentration are depicted. We suggest that the change in slope in the refolding branch and the decrease in the value of m_F seen at [urea] <4 M is due to an additional transition state (*T1*). This transition state *T1* appears under stabilizing condition and, according to Hammond's postulate, is less compact than transition state *T2*, which becomes dominant at [urea] >4 M. The reaction coordinate and the free energy of transition state *T1* were estimated using kinetic parameters measured at low protein concentration (0.13 µM), in order to reduce the influence of aggregation effects (see the text). A representative refolding time course obtained at 2.5 M [urea] is shown along with its matching unfolded control obtained after normalization for the signal of the unfolded protein at the same urea concentration; the time course is described by a single exponential, and no "burst phase" amplitude is seen under these conditions. (Adapted from Gianni, S. et al. 2001. Snapshots of protein folding: a study on the multiple transition state pathway of cytochrome c(551) from *Pseudomonas aeruginosa*. *J Mol Biol* 309:1177–87. With permission.)

2 M urea (data not shown), we noticed that the rollover was still present at higher [urea] (2 M < [urea] < 4 M) where the refolding kinetics is single-exponential with signal recovery matching the unfolded control (see Figure 2.3, inset). Because of the complexity of the chevron plot in urea at pH 7.0, we assumed that the PA-cyt c_{551} folding pathway proceeds through two different transition states, called *T1* and *T2*. In Figure 2.3 we depict the situation where the rate-limiting barrier at the

lower denaturant concentrations is associated to an unfolded-like transition state (t_1), consistently with the Hammond postulate (Hammond 1955). The observation that the rollover effect disappears at very low protein concentrations only at pH 4.7, where the net charge of the protein is zero and the stability of the native state reaches a maximum (data not shown), suggests that the transition state t_1 originates from electrostatic repulsions that may limit the observed refolding rate. Therefore it is reasonable to assume that the net charge above and below the pI controls the degree of compactness of the transition state t_1. On the other hand, in an attempt to characterize the structural features of t_2, we studied the folding kinetics of several site-directed mutants and monitored the effect of mutation on the folding process. Data revealed that the stabilization of t_2 is mediated by the salt bridge interaction between Glu70 and Lys10 at the interface between the N- and C-terminal helices (Gianni et al. 2001; Travaglini-Allocatelli et al. 1999). These observations parallel the earlier suggestions by several groups indicating such interaction as a critical step in the folding of c-type cytochromes (Colon et al. 1996; Marmorino and Pielak 1995; Colon and Roder 1996).

Further kinetic analysis on PA-cyt c_{551} revealed complexities indicative of a folding mechanism involving parallel pathways (Gianni et al. 2003). In particular, using interrupted refolding experiments we explored whether all the molecules achieve the native state via the same path or whether some of the molecules may refold via a faster route, as initially shown to be the case for lysozyme (Kiefhaber 1995). We therefore carried out a double-jump interrupted refolding experiment in which (1) unfolded PA-cyt c_{551} is mixed against a refolding buffer (first mix), and (2) after a controlled delay time, refolding is interrupted by rapid addition of high concentrations of the same denaturant (second mix). This approach is quite powerful because it allows one to distinguish partially folded intermediates from native molecules since these states are characterized by different unfolding rates: in fact, since the native protein is separated from the unfolded one by the highest energy barrier, it should unfold more slowly than any partially folded intermediate. According to Shastry and Roder (1998), in the case of horse cyt c the calculated unfolding rate for both the early and late intermediates at high denaturant concentration (i.e., [GdnHCl] = 3 M) would be too fast (>10,000 s^{-1}) for stopped-flow detection. Thus the fractional population of native molecules formed during the delay time between the first and second mix is represented by the relative amplitude of the slowest unfolding event. Surprisingly, we found the formation of ~50% of the native protein already at the shortest delay time (10 ms), suggesting the presence of a fast refolding to explain the kinetic partitioning between fast- and slow-refolding channels (Kiefhaber 1995; Figure 2.4). A first model postulates the presence of a mixture of different unfolded states (Lee et al. 2002). In the case of heme proteins, another potential source of heterogeneity is the presence of alternative axial heme ligands. In fact, interconversion between different unfolded states that leads to parallel folding pathways has also been identified for reduced horse cyt c in which, after CO photodissociation, the unfolded ensemble is represented by a distribution of different coordination states (Goldbeck et al. 1999). An alternative scenario predicts parallel pathways even if folding starts from a homogenous unfolded state. According to this model, dilution of the denaturant may give rise to formation of kinetically trapped intermediates that fold more slowly, while the remainder of the unfolded molecules can fold rapidly resulting in the observed kinetic partitioning. Altogether, analysis of the observed kinetics and refolding amplitudes suggested that in the case of PA-cyt c_{551}, the model postulating a heterogeneous denatured state would fit best the experimental data (Gianni et al. 2003).

COLLAPSE IN THE SUB-MS TIME WINDOW

As recalled earlier, refolding kinetics is generally followed in a stopped-flow apparatus (dead time ≥2 ms) by rapid dilution of a protein that has been dissolved in high denaturant. In this type of experiment, the kinetically resolved amplitude is sometimes different from that expected from the total signal change between the denatured and the native state. This phenomenon, known as the "burst phase collapse," implies that there is an early, very rapid event, which is lost in the dead time of stopped-flow instruments. Various proteins, including PA-cyt c_{551} and other members of the cytochrome c family, show this behavior. Using a super-rapid continuous-flow apparatus, Shastry and Roder (Shastry, Luck, and Roder 1998; Shastry and Roder 1998) have provided direct evidence that these early events in the folding of horse cyt c follow exponential kinetics, suggesting the presence of an energy barrier. Consistent with this, laser T-jump experiments have shown that the cyt c collapse is a thermally activated process (Hagen and Eaton 2000). The development of a circular dichroism (CD) spectrometer with a dead time of 0.4 ms enabled Akyiama et al. (2000) to demonstrate that the helical content of the collapsed species formed in the burst phase is about 20% of that of native horse cyt c. More recently, using time-resolved (<0.2 ms), small-angle x-ray scattering, the same group (Akiyama et al. 2002) assessed the radius of gyration of this early species, which suggests that cyt c folding proceeds with a collapse around a specific region.

In an effort to observe the rapid folding phase of PA-cyt c_{551} directly, we carried out continuous-flow fluorescence experiments (dead time of about 60 μs) in collaboration with Roder and coworkers. PA-cyt c_{551} was fully unfolded in 1 M urea at pH 2.0, and refolding was triggered by a sixfold dilution of the denaturant with buffer at a final pH of 4.7. Representative continuous-flow refolding traces at urea concentrations from 0.17–1.5 M and pH 4.7 are combined with stopped-flow traces recorded under matching conditions (Figure 2.5). The traces are scaled with respect to the fluorescence of the unfolded protein. The combined kinetics is well described by a sum of three exponential phases. The results are consistent with the transient accumulation of an on-pathway folding intermediate on the 100 μs timescale (fast phase), followed by the rate-limiting conversions of the intermediate into the native state (intermediate phase in the ms timescale). The presence of an additional phase with minor amplitude (5–25%) and a rate constant slower than the main folding phase can be explained by a minor population of molecules folding along a parallel pathway with a 5- to 10-fold slower rate. The burst phase observed by stopped-flow fluorescence therefore underlies a more complex picture than previously assumed, which may not be explained solely by the sequential accumulation of partially structured intermediates.

As recalled previously, rapid photochemical methods have been employed to study the rate of collapse of reduced horse cyt c. Such techniques took advantage of the fact that in this protein, the Fe^{2+} of the covalently bound heme is coordinated to a His and a Met residues, on the proximal and distal side, respectively. Unfolding

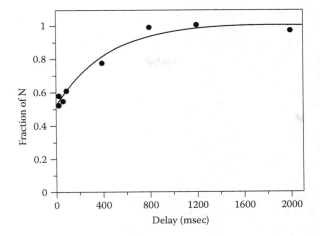

FIGURE 2.4 Double-mixing interrupted refolding on wt PA-cyt c_{551} measured in GdnHCl at pH 3.0. Time course of appearance of native cyt c_{551} during refolding in 0.4 M GdnHCl at pH 3.0 and 10°C. Continuous line corresponds to $k = 2.5$ s^{-1}. (Adapted from Gianni, S. et al. 2003. Parallel pathways in cytochrome c(551) folding. *J Mol Biol* 330:1145–52. With permission.)

by GdnHCl or urea weakens the Fe-Met bond, and the reduced unfolded cyt c easily binds CO and other heme ligands, which react very slowly or not at all with the native protein. Therefore in the presence of CO, reduced cyt c unfolds at lower denaturant concentrations than in the absence of this ligand, and rapid photochemical removal of CO from unfolded cyt c is expected to trigger at least a partial refolding (Jones et al. 1993). As we have shown, however, this approach is complicated by the breakage of the proximal His-Fe bond that may occur as a consequence of CO photodissociation in the unfolded cytochrome c because of the so-called base elimination mechanism. Rebinding of CO to the four-coordinate heme yields kinetic intermediates unrelated to folding. This hypothesis was further supported by parallel observations carried out with protoheme and microperoxidase (Arcovito et al. 2001).

CYTOCHROME c_{552} FROM *THERMUS THERMOPHILUS*

Cytochrome c_{552} from *Thermus thermophilus* (TT-cyt c_{552}) is a thermostable class I cyt c with an unusually long amino acid sequence (131 residues) that is distantly related to other members of this class. Inspection of its three-dimensional structure (Soulimane et al. 1997) shows that it contains some unique features that are absent in the canonical cyt *c* fold, such as an additional C-terminal portion consisting of two α-helices representing a characteristic extra clamp, and a β-sheet covering one edge of the heme (Figure 2.1). Because of its extraordinary thermodynamic stability ($\Delta G = -21.4 \pm 2.0$ kcal/mol^{-1} at pH = 7.0, T = 10°C), a detailed analysis of its (un) folding kinetics by GdnHCl could be carried out only at pH = 2.1.

In addition to a burst phase collapse of the unfolded chain in the sub-ms time window, we observed a multiphasic refolding time course. In particular, apart from proline *cis/trans* isomerization, the GdnHCl dependence of the folding and unfolding rates measured for TT-cyt c_{552} showed a faster phase and a distinctly resolved slower phase with a well-characterized rollover effect, whereby the logarithm of the observed refolding rate constant as a function of denaturant deviates from linearity and tends to level off at low denaturant concentrations (Travaglini-Allocatelli et al. 2003). As discussed previously, this rollover effect is generally attributed to the presence of an intermediate species in the folding mechanism. However, detection of a folding intermediate is not sufficient, per se, to distinguish between two fundamentally different mechanisms that may involve (1) a productive *on-pathway* obligatory intermediate or (2) a misfolded *off-pathway* state that must unfold to achieve the native protein. As has been shown for other proteins, such an issue may be solved only if all the microscopic rate constants are measured over a wide range of denaturant concentration (Bai 2000; Jemth 2004). In the case of TT-cyt c_{552}, interrupted refolding experiments allowed us to measure the unfolding limb of the intermediate, thereby defining its kinetic role. By this approach, the intermediate was populated to a considerable extent in the first mixing step and thereafter challenged with high and variable denaturant concentrations in the second mix to measure its unfolding rates. Quantitative analysis of the four microscopic rate constants clearly indicated that accumulation of the intermediate is an obligatory step in the folding process of TT-cyt c_{552} (Figure 2.5; Travaglini-Allocatelli et al. 2003); the method for discriminating between on- and off-pathway intermediates is extensively described in

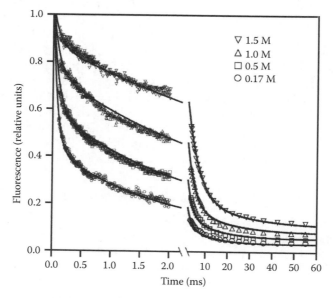

FIGURE 2.5 Tryptophan fluorescence changes during refolding of PA-cyt c_{551} (20 µM) at different urea concentrations (0.17 M circles, 0.5 M squares, 1.0 M upper triangles, 1.5 M lower triangles in 0.1 M phosphate, pH 4.7) measured in a matching continuous-flow and stopped-flow experiments at 15°C. The data were normalized relative to the signal of the unfolded state at 1 M urea, pH 2. The continuous lines show the time course of folding at each urea concentration.

Gianni et al. (Gianni et al. 2007). Furthermore, the denaturant dependence of each microscopic rate constant allowed us to calculate the position of all of the species (intermediate and transition states) along the reaction coordinate; this is expressed in terms of a parameter called the Tanford β-value (β_T), which reflects the relative position of each species expressed as a fraction of native-like structure (from 0 to 1). The β_T values calculated for TT-cyt c_{552} were β_{T1} = 0.40, β_I = 0.56, and β_{T2} = 0.70 (Travaglini-Allocatelli et al. 2003).

CYTOCHROME c_{552} FROM *HYDROGENOBACTER THERMOPHILUS*

Characterization of the folding mechanism of *Hydrogenobacter thermophilus* cytochrome c_{552} (HT-cyt c_{552}) is particularly interesting for the purpose of comparative folding, because contrary to *T. thermophilus* cytochrome c_{552}, HT-cyt c_{552} has a typical class I cytochrome c fold (Figure 2.1) and shares high sequence identity (57%) and structural homology with its mesophilic counterpart PA-cyt c_{552} from the bacterium *P. aeruginosa*. As discussed previously, the latter was found to refold through a broad energy barrier with two transition states (*T1* and *T2*) separated by a high-energy intermediate (Gianni et al. 2001). On the basis of the substantial difference in thermodynamic stability between these two structurally homologous cytochromes ($\Delta\Delta G$ = 3.0 kcal mol^{-1} at pH 4.7), we speculated that an intermediate species with properties similar to those proposed for the high-energy intermediate of PA-cyt c_{551} (and other evolutionary distant c-type cytochromes) would be populated in the folding of HT-cyt c_{552}.

Kinetic characterization was carried out at pH 4.7 to facilitate comparison of the results with those previously obtained for PA-cyt c_{551} (Gianni et al. 2001), monitoring both far-ultraviolet (far-UV) CD and Trp fluorescence signals. The kinetics of folding and unfolding followed by far-UV CD were single exponential at all denaturant concentrations, with no evidence for protein concentration dependence (from 0.5–5.0 µM) of the first-order rate constant (data not shown), suggesting the absence of transient aggregation events (Silow and Oliveberg 1997). Hence the chevron plot obtained from far-UV CD data corresponds to a simple V-shape, apparently consistent with a two-state folder (data not shown; see Travaglini-Allocatelli et al. 2005). The thermodynamic stability derived from quantitative analysis of the kinetic experiments (G_{DN} = −8.6 ± 0.1 kcal mol^{-1}) is in very good agreement with the value obtained from equilibrium experiments. However, when the refolding time course of HT-cyt c_{552} was monitored by Trp fluorescence (Figure 2.6), in addition to an extensive emission quenching (k = 75 s^{-1} at 0.7 M GdnHCl) an additional slower phase (~3 s^{-1}) with very small amplitude (~5% of the signal) was detected. This slower process may have escaped detection by far-UV CD because of the lower signal-to-noise ratio with respect to fluorescence quenching. The shallow denaturant dependence and the small amplitude of this slower process initially suggested that it might be assigned to proline *cis-trans* isomerization. However, it has been stressed (Daggett and Fersht 2003) that with proteins "the devil lies in the details": Additional experiments allowed us to conclude that this slow phase reflects, in fact, the formation of native molecules from an intermediate state, as detailed next.

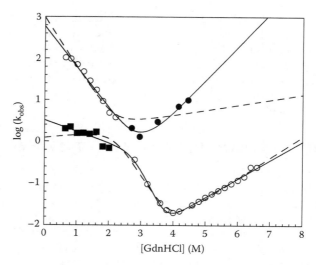

FIGURE 2.6 Folding kinetics of HT-cyt c_{552} at pH 4.7, 10°C. Chevron plot of the observed folding and unfolding kinetics followed by fluorescence. Single-mixing stopped-flow experiments (open circles and filled squares). The filled squares represent the slower rate constants observed only by fluorescence at low [GdnHCl]. Double-mixing experiments are presented with filled circles. Continuous and broken lines represent the best global fit for an on- and off-pathway model, respectively.

As shown by Kiefhaber (Kiefhaber 1995) for lysozyme and discussed earlier for PA-cyt c_{551}, interrupted refolding experiments allow the fraction of native protein formed during the refolding process to be monitored. In the case of HT-cyt c_{552}, unfolded molecules were allowed to refold by denaturant dilution in the first mix and then were unfolded again in a second mix after a variable delay time. The amplitude behavior of the single exponential unfolding reaction as a function of delay time between the first and the second mix reflects the time course of formation of native molecules. The results showed that the major fraction of native protein is formed in a first-order kinetic reaction (Figure 2.7) with $k \sim 5$ s^{-1}, similar to the slower process seen in fluorescence-monitored single-mixing experiments (see previous paragraph). Therefore at the same [GdnHCl], the faster process ($k \sim 80$ s^{-1} at 1 M GdnHCl) associated with the major amplitude, should reflect formation of a partially folded intermediate, in which the fluorescence and CD properties are very similar to those of the native state (implying only a small additional fluorescence quenching associated to the transition I → N). Again, interrupted refolding experiments allowed for the demonstration that this intermediate state is an obligatory species on the path to the native state.

A CONSENSUS FOLDING MECHANISM

To test whether the folding mechanism proposed earlier for PA-cyt c_{551}, TT-cyt c_{552}, and HT-cyt c_{552} might be relevant to other members of this family, we have

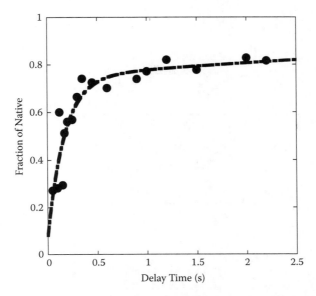

FIGURE 2.7 Time course of appearance of native protein of HT-cyt c_{552} during refolding at 0.67 M GdnHCl, as obtained by interrupted refolding experiments; exponential fit of the data yields a value of $k = 5.4 \pm 0.5$ s^{-1}, inconsistent with the faster rate constant ($k = 100 \pm 5.5$ s^{-1}) measured by fluorescence single-mixing experiments under the same conditions. A small fraction of native molecules (20%) forms through a kinetically trapped *cis*-prolyl-denatured state, as previously observed for a variety of proteins.

considered folding kinetic data that are available for some eukaryotic and prokaryotic cyt c proteins from mesophiles. The chevron plots calculated for horse cyt c (104 residues) (Colon et al. 1996), a stabilized site-directed mutant (N52I) of yeast iso-2 cyt c (113 residues) (McGee and Nall 1998), and *Rhodobacter capsulatus* cyt c_2 (116 residues) (Sauder, MacKenzie, and Roder 1996) are shown in Figure 2.8, where they are compared with that of TT-cyt c_{552} (131 residues). All of these chevron plots, which show a rollover effect at low denaturant concentrations, have been globally fitted to a three-state model sharing a common position for the intermediate state ($\beta_I = 0.66 \pm 0.05$). The relatively large value of β_I indicates that about two-thirds of the buried surface area is already formed in the intermediate, which is therefore a fairly compact species. Moreover, the position of the late transition state ($T2$) is rather conserved ($\beta_{T2} = 0.69 - 0.77$), in spite of the large differences in the sequence, size, charge, and thermodynamic stabilities of the proteins.

According to this analysis, the folding mechanism of all these different cyts c is compatible with a three-state model if the additional rollover effect that is detected during unfolding at very high denaturant concentrations is ignored. The latter effect, which might become evident only under very destabilizing conditions, has been assigned to breakage of the native Met-Fe^{3+} coordination bond and has been discussed extensively (Gianni et al. 2003; Colon et al. 1996; Sauder, MacKenzie, and Roder 1996). In principle, such a specific feature should be included in a complete

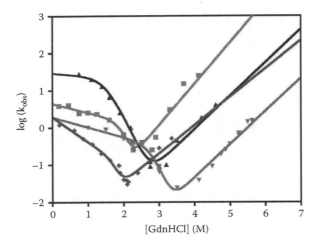

FIGURE 2.8 Chevron plots of four different cyt c proteins (TT-cyt c_{552}, lower triangles; horse cyt c, squares; yeast iso-2 N52I mutant, rhombi; *Rhodobacter capsulatus* cyt c_2, upper triangles). The continuous lines are the best fit for a kinetic three-state on-pathway model (U↔I↔N), in which all data sets are fitted globally to a shared Tanford β-value for the intermediate ($β_I$), assuming a fast preequilibrium between U and I. We were unable to obtain satisfactory fits by restraining more parameters, which is not surprising considering the expected Hammond effect induced by the different stabilities of the intermediate, transition, and native states.

reaction scheme, leading to a four-state mechanism; nevertheless, it has been already shown that the formation of the Met-Fe^{3+} coordination bond, although important for the overall thermodynamic stability of the native state, has no role whatsoever in the kinetics and mechanism of refolding.

The proposed folding mechanism can be compared with the kinetics of PA-cyt c_{551}. We previously showed that, even if the folding of this protein under some conditions involves parallel pathways to the native state (Gianni et al. 2003), refolding kinetics in the presence of a concentration of more than about 1.5 M urea can be well described by a single-exponential behavior even in the presence of a strongly pronounced refolding rollover effect. Over and above the possible alternative routes to the native state, the folding of PA-cyt c_{551} thus occurs through a rough energy barrier involving two transition states, called *T1* and *T2*. Such a model, with two consecutive transition states on a sequential pathway, implies the presence of a local minimum in the transition barrier between *T1* and *T2*. This high-energy intermediate is, however, intrinsically unstable and thus precluded from direct kinetic and structural characterization; nevertheless, the positions of *T1* and *T2* along the reaction coordinate can be assessed ($β_{T1}= 0.32 ± 0.04$ and $β_{T2}= 0.71 ± 0.05$) (Gianni et al. 2001), and indeed they are similar to those obtained for TT-cyt c_{552} and HT-cyt c_{552}. It therefore seems that a minimum three-state folding scheme, characterized by similar intermediate and transition states, can be of general validity for this protein family. Of all of the folding data published on the cyt c family, only the chevron plot for cyt c_{553} from *Desulfovibrio vulgaris* (79 residues) seems consistent with a true two-state

The Folding Mechanism of c-Type Cytochromes

mechanism (Guidry and Wittung-Stafshede 2000). It is possible, however, that even for this protein a rollover effect might be observed under more stabilizing conditions or if urea were used instead of GdnHCl, as has been observed for PA-cyt c_{551}.

The best experimental tool to assess the structure of transiently populated species, such as transition and intermediate states, is the Φ-value analysis pioneered by Fersht and coworkers. In this approach, the effect of specific mutations on the energetics of all of the species involved is determined in an attempt to define the role of individual residues in the folding process (Fersht, Matouschek, and Serrano 1992). In the case of recombinant c-type cytochromes, however, the difficulties in the expression protocols (due to the covalent attachment of the heme) have probably limited this fruitful approach. The late transition state (*T2*) in the folding of PA-cyt c_{551} represents one of the best structurally characterized transient states in the whole cyt c family (Gianni et al. 2001). The emerging picture indicates that structure formation is driven by some key tertiary contacts at the interface between the N- and C-terminal helices (Figure 2.9), which provide stabilization of *T2* ($\Phi_{T2} \geq 0.7$). Structure gradually tapers off through the remaining two helices and the heme pocket with very low Φ-values; in particular, the Φ-value for the Met-Fe^{3+} coordination bond formation is essentially zero, as shown by the refolding kinetics of a PA-cyt c_{551} variant with a chemically modified methionine residue (Gianni et al. 2003).

Notably, a crucial role for the two terminal helices in the folding of cyt c has been previously claimed in mutational studies carried out on horse and yeast cyt c (Colon et al. 1996; Marmorino, Lehti, and Pielak 1998). Moreover, on the basis of the structural comparison of a large set of cyt c proteins, Ptitsyn (Ptitsyn 1998) proposed that some conserved residues at the same interface between the N- and C- terminal α-helices might act as a nucleus for folding in this protein family. Finally, the structure of the transition state *T2* of PA-cyt c_{551} is in close agreement with the overall folding pathway of horse cyt c described in great detail by Englander and coworkers (Rumbley et al. 2001), who used isotope exchange kinetics. In fact, the order of structure formation proposed by these authors closely resembles the overall picture of the transition state obtained by Φ-value analysis on PA-cyt c_{551}.

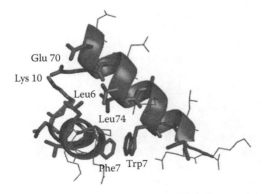

FIGURE 2.9 Detailed representation of the interface between the N- and C-terminal helices in PA-cyt c_{551}. The key residues of this interaction networks are Leu6, Phe7, Lys10, Glu70, Leu74, and Trp77, represented in sticks.

ENGINEERING FOLDING PATHWAYS

If the proposed consensus mechanism outlined previously is valid, it should be possible to stabilize the high-energy intermediate state of PA-cyt c_{551} by a few critical mutations in order to make it kinetically detectable, as observed for HT-cyt c_{552}. The underlying hypothesis is that the higher stability and the different distribution of helical propensity of HT-cyt c_{552}, compared with PA-cyt c_{551}, stabilizes the folding intermediate sufficiently to be populated and detected.

Among different multiple and single site directed mutants of PA-cyt c_{551} designed on the basis of HT-cyt c_{552} primary structure, the Phe7Ala mutant was selected as a promising combination of (1) an approximate fourfold increase in the helical propensity of the N-terminal helix (as calculated by AGADIR; Muñoz and Serrano 1997), (2) a minor structural perturbation, and (3) an increased thermodynamic stability (Borgia et al. 2006).

Contrary to the wild-type (wt) protein, the refolding time course of Phe7Ala mutant, measured at pH 4.7 and low GdnHCl concentration (≤1.75 M), is clearly biphasic and is satisfactorily fitted to a double-exponential decay, although a third and much slower phase (k is approximately 0.03 s^{-1}) with a small amplitude (<10%) was also observed. Interrupted unfolding experiments (data not shown) demonstrated that this slow phase reflects a minor fraction of molecules refolding along a pathway rate limited by prolyl-peptide bond isomerization processes (Schmid and Baldwin 1978). Interestingly, the chevron plot reported in Figure 2.10 shows that while the logarithm of the faster refolding rate constant linearly decreases with increasing [GdnHCl], the slower one displays a clear curvature (rollover effect) at low denaturant concentrations (between 0.5 and 2.0 M GdnHCl). Both features, which are generally interpreted in terms of multistate folding, were not observed in the refolding of the wt PA-cyt c_{551} from the GdnHCl denatured state (Travaglini-Allocatelli et al. 1999). To exclude the possibility that the refolding rollover is caused by formation of transient aggregates, we followed the protein concentration dependence (between 0.1 and 50 μM after mixing) of the refolding time course at 1 M GdnHCl, without any detectable effects (data not shown).

To demonstrate that for the Phe7Ala mutant an intermediate species is accumulated under refolding conditions, we carried out interrupted refolding experiments. Contrary to the wt protein, where this type of experiment did not reveal any additional process over and above the unfolding of N, in the case of the Phe7Ala mutant a fast-unfolding species was evident using short delay times. Fitting the unfolding time course observed at 4.15 M GdnHCl to a double-exponential decay yields a rate constant for the slow phase (0.20 s^{-1}), which is virtually identical to that measured under the same conditions in single-mix unfolding experiments (Figure 2.10) and represents the unfolding of the native protein formed during the delay time. The additional fast phase (k = 34 s^{-1}) observed only for the mutant indicates the existence of a population of partially structured molecules not observed in the case of the wt protein. At longer delay times the amplitude of the fast unfolding phase disappears progressively, and at delay times >5 s, a single-exponential process is observed, as expected if only the native state were present.

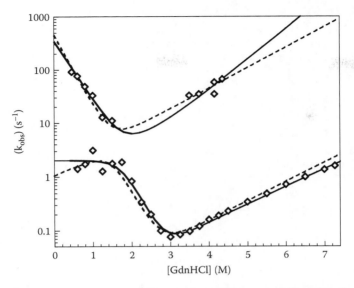

FIGURE 2.10 Folding kinetics of Phe7Ala mutant of PA-cyt c_{551} followed by fluorescence at pH 4.7, 10°C. All data points are from single-mixing experiments with the exception of the fast-unfolding limb (empty squares), followed by double-mixing experiments. Continuous and dashed lines represent the best global fit for the on- and off-pathway model, respectively.

The observation that a detectable amount of native state can be formed in the short time between the first and the second mixing (100 ms) suggested the presence of parallel refolding routes. To determine whether the native state is formed only in a sequential reaction, involving an obligatory on-pathway intermediate (U-I-N), or if other mechanisms are operative (Wildegger and Kiefhaber 1997), we plotted the amplitudes obtained in the same interrupted refolding experiment as a function of the delay time. For the Phe7Ala mutant the amplitude dependence of the slow unfolding reaction on the delay time at pH 4.7 is well described by a double-exponential process (Figure 2.11), with no evidence of native protein formation at the shortest delay time (10 ms). The two phases have rate constants and relative amplitudes close to those measured in classical single-mix dilution experiments, at the same pH and GdnHCl concentration (i.e., at 1 M GdnHCl: $k_1 = 27$ s^{-1} versus 36 s^{-1}, $a_1 = 85\%$ versus 87%; $k_2 = 1.40$ s^{-1} versus 1.45 s^{-1}, $a_2 = 15\%$ versus 11%; see also Figure 2.10). This result clearly indicates that the two phases detected in single-mix stopped-flow experiments both represent processes leading to the native state and imply a triangular folding mechanism characterized by parallel pathways. When the interrupted refolding experiments were carried out at pH 3.0, no fast-unfolding species was detected (data not shown); moreover, a plot of the amplitudes against the delay time can be fitted to a single exponential (Figure 2.11), yielding a rate consistent with that of the main phase observed in single-mix dilution experiments under the same conditions (approximately 3 s^{-1}). Both these observations suggest that at the lower pH the alternative refolding route is probably abolished and the intermediate state is no longer populated. It was proposed (Borgia et al. 2006) that at pH 3.0, where the

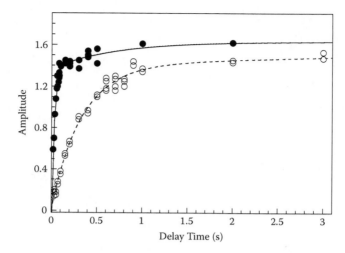

FIGURE 2.11 Time course of appearance of the native Phe7Ala mutant of PA-cyt c_{551} during refolding at 1 M GdnHCl and pH 4.7 (filled circles) or at 0.27 M GdnHCl at pH 3.0 (open circles). The denatured protein was allowed to refold for different delay times before being denatured again. The relative amplitude of the slowest unfolding reaction is a measure of the formation of native protein during the delay time. At pH 4.7 the continuous line is a double-exponential curve required to describe the data ($k_1 = 35$ s^{-1}; $k_2 = 1.4$ s^{-1}). The data at pH 3.0 were satisfactorily fitted to a single-exponential decay (dashed line), yielding a value of $k = 2.7$ s^{-1} consistent with the rate constant measured by single-mixing experiments at this pH (data not shown).

mutant Phe7Ala is considerably less stable than at pH 4.7, the intermediate is likely to be high energy, probably because of protonation of Glu70. This residue, indeed, forms a salt bridge with Lys10, an interaction shown to be important for the docking of the N- and C-terminal helices, as discussed earlier (Travaglini-Allocatelli et al. 1999; Gianni 2001).

The data shown in Figure 2.10, fitted with a three-state model, involve the presence of an intermediate; however, discrimination between the on- and off-pathway model was not possible because of the limited number of points describing the refolding rollover and unfolding of the intermediate.

Over and above the presence of a fast unfolding process, which suggests a sequential mechanism, the biphasic character of the native-state formation as seen in the interrupted refolding experiment suggests a triangular mechanism as a more plausible scenario: Following this view the native state in Phe7Ala is formed via two parallel reactions, i.e., a faster one directly from the denatured state and a slower one involving the transient population of an intermediate en route to the native state. Such a triangular model is consistent with evidence for parallel pathways in the wt protein (Gianni et al. 2003) and the results obtained for HT-cyt c_{552}, which refolds through an on-pathway intermediate (Travaglini-Allocatelli et al. 2005).

Furthermore, determination of the crystal structure of Phe7Ala mutant allowed us to propose a structural interpretation for the stabilization of the native state and of the consensus intermediate. The replacement of the Phe ring with the methyl group of Ala leads to an enhancement of the helical propensity of the N-terminal helix and allows the strengthening of a main-chain H-bond in the same helix. Moreover, the internal cavity resulting from the mutation is filled by an array of three water molecules representing the "core" of an H-bond network, which could concur to the stabilization of the native state (Borgia et al. 2006).

These results illustrate how protein engineering, structure determination, and folding kinetics concurred to unveil a folding intermediate and the structural determinants of its stability; more generally we showed how it is possible to change the shape of the energetic landscape of a protein-folding process by rational mutagenesis.

FUTURE PERSPECTIVE

In our opinion, a consensus folding mechanism could be operative in very different members of the cyt c family that are characterized by distinct sequences, molecular weights, and thermodynamic stabilities. The analysis outlined herein shows that the calculated Tanford β-values for the different species are conserved and indicates that the positions along the reaction coordinate of the intermediate and transition states are similar.

The observed variations in β_{12} calculated for different members of the cyt c family can be explained by the Hammond effect, whereby a destabilization of the native state relative to the intermediate leads to a shift in the transition state position toward the native state, as observed for several other proteins. Moreover, as we have described herein the hypothesis of a consensus folding mechanism allowed us to rationally tune the folding pathway of a selected c-type cyt by site-directed mutagenesis.

Finally, the recent developments of the expression protocols for c-type cytochromes in *E. coli* pave the way to a detailed phi-value analysis, the last piece of the puzzle in the folding of these fascinating proteins.

REFERENCES

Akiyama, S., S. Takahashi, K. Ishimori, and I. Morishima. 2000. Stepwise formation of alpha-helices during cytochrome c folding. *Nat Struct Biol* 7:514–20.

Akiyama, S., S. Takahashi, T. Kimura, K. Ishimori, I. Morishima, Y. Nishikawa, and T. Fujisawa. 2002. Conformational landscape of cytochrome c folding studied by microsecond-resolved small-angle x-ray scattering. *Proc Natl Acad Sci U S A* 99:1329–34.

Andreeva, A., D. Howorth, J. M. Chandonia, S. E. Brenner, T. J. Hubbard, C. Chothia, and A. G. Murzin. 2008. Data growth and its impact on the SCOP database: new developments. *Nucl Acids Res* 36:419–25.

Arcovito, A., S. Gianni, M. Brunori, C. Travaglini-Allocatelli, and A. Bellelli. 2001. Fast coordination changes in cytochrome c do not necessarily imply folding. *J Biol Chem* 276:41073–78.

Babul, J., and E. Stellwagen. 1971. The existence of heme-protein coordinate-covalent bonds in denaturing solvents. *Biopolymers* 11:23569–661.

Bai, Y. 2000. Kinetic evidence of an on-pathway intermediate in the folding of lysozyme. *Protein Sci* 9:194–96.

Borgia, A., D. Bonivento, C. Travaglini-Allocatelli, A. Di Matteo, and M. Brunori. 2006. Unveiling a hidden folding intermediate in c-type cytochromes by protein engineering. *J Biol Chem* 281:9331–36.

Chang, I. J., J. C. Lee, J. R. Winkler, and H. B. Gray. 2003. The protein-folding speed limit: intrachain diffusion times set by electron-transfer rates in denatured Ru(NH3)5(His-33)-Zn-cytochrome c. *Proc Natl Acad Sci U S A* 100:3838–40.

Clarke, J., L. S. Itzhaki, and A. R. Fersht. 1998. Hydrogen exchange at equilibrium: a short cut for analysing protein-folding pathways? *Trends Biochem Sci* 22:284–87.

Colon, W., G. A. Elove, L. P. Wakem, F. Sherman, and H. Roder. 1996. Side chain packing of the N- and C-terminal helices plays a critical role in the kinetics of cytochrome c folding. *Biochemistry* 35:5538–49.

Colon, W., and H. Roder. 1996. Kinetic intermediates in the formation of the cytochrome c molten globule. *Nat Struct Biol* 3:1019–25.

Colon, W., L. P. Wakem, F. Sherman, and H. Roder. 1997. Identification of the predominant non-native histidine ligand in unfolded cytochrome c. *Biochemistry* 36:12535–41.

Daggett, V., and A. Fersht. 2003. The present view of the mechanism of protein folding. *Nat Rev Mol Cell Biol* 4:497–502.

Detlefsen, D. J., V. Thanabal, V. L. Pecoraro, and G. Wagner. 1991. Solution structure of Fe(II) cytochrome c_{551} from *Pseudomonas aeruginosa* as determined by two-dimensional 1H NMR. *Biochemistry* 30:9040–46.

Elove, G. A., A. K. Bhuyan, and H. Roder. 1994. Kinetic mechanism of cytochrome c folding: involvement of the heme and its ligands. *Biochemistry* 33:6925–35.

Englander, S. W. 1998. Native-state HX. *Trends Biochem Sci* 23:379–81.

Ferguson, N., A. P. Capaldi, R. James, C. Kleanthous, and S. E. Radford. 1999. Rapid folding with and without populated intermediates in the homologous four-helix proteins Im7 and Im9. *J Mol Biol* 286:1597–1608.

Ferguson, S. J., J. M. Stevens, J. W. Allen, and I. B. Robertson. 2008. Cytochrome c assembly: a tale of ever increasing variation and mystery? *Biochem Biophys Acta* 1777:980–84.

Fersht, A. R., A. Matouschek, and L. Serrano. 1992. The folding of an enzyme. I. Theory of protein engineering analysis of stability and pathway of protein folding. *J Mol Biol* 224:771–82.

Gianni, S., Y. Ivarsson, P. Jemth, M. Brunori, and C. Travaglini-Allocatelli. 2007. Identification and characterization of protein folding intermediates. *Biophys Chem* 128:105–13.

Gianni, S., C. Travaglini-Allocatelli, F. Cutruzzolà, M. G. Bigotti, and M. Brunori. 2001. Snapshots of protein folding: a study on the multiple transition state pathway of cytochrome c(551) from *Pseudomonas aeruginosa*. *J Mol Biol* 309:1177–87.

Gianni, S., C. Travaglini-Allocatelli, F. Cutruzzolà, M. Brunori, M. C. Shastry, and H. Roder. 2003. Parallel pathways in cytochrome c(551) folding. *J Mol Biol* 330:1145–52.

Goldbeck, R. A., Y. G. Thomas, E. Chen, R. M. Esquerra, and D. S. Kliger. 1999. Multiple pathways on a protein folding energy landscape: kinetic evidence. *Proc Natl Acad Sci U S A* 96: 2782–87.

Guidry, J., and P. Wittung-Stafshede. 2000. Cytochrome c(553), a small heme protein that lacks misligation in its unfolded state, folds with rapid two-state kinetics. *J Mol Biol* 301:769–73.

Hagen, S. J., and W. A. Eaton. 2000. Two-state expansion and collapse of a polypeptide. *J Mol Biol* 301:1019–27.

Hagen, S. J., J. Hofrichter, A. Szabo, and W. A. Eaton. 1996. Diffusion-limited contact formation in unfolded cytochrome c: estimating the maximum rate of protein folding. *Proc Natl Acad Sci U S A* 93:11615–17.

Hammond, G. S. 1955. A correlation of reaction rates. *J Am Chem Soc* 77:334–39.

Hoang, L., S. Bedard, M. M. Krishna, Y. Lin, and S. W. Englander. 2002. Cytochrome c folding pathway: kinetic native-state hydrogen exchange. *Proc Natl Acad Sci U S A* 99:12173–78.

Jackson, S. E., and A. R. Fersht. 1991. Folding of chymotrypsin inhibitor 2. 1. Evidence for a two-state transition. *Biochemistry* 30:10428–35.

Jemth, P., S. Gianni, R. Day, B. Li, C. M. Johnson, V. Daggett, and A. R. Fersht. 2004. Demonstration of a low-energy on-pathway intermediate in a fast-folding protein by kinetics, protein engineering, and simulation. *Proc Natl Acad Sci U S A* 101:6450–55.

Jones, C. M., E. R. Henry, Y. Hu, C. K. Chan, S. D. Luck, A. Bhuyan, H. Roder, J. Hofrichter, and W. A. Eaton. 1993. Fast events in protein folding initiated by nanosecond laser photolysis. *Proc Natl Acad Sci U S A* 90:11860–64.

Khorasanizadeh, S., I. D. Peters, and H. Roder. 1996. Evidence for a three-state model of protein folding from kinetic analysis of ubiquitin variants with altered core residues. *Nat Struct Biol* 3:193–205.

Kiefhaber, T. 1995. Kinetic traps in lysozyme folding. *Proc Natl Acad Sci U S A* 92:9029–33.

Knapp, J. A., and C. N. Pace. 1974. Guanidine hydrochloride and acid denaturation of horse, cow, and *Candida krusei* cytochromes c. *Biochemistry* 13:1289–94.

Latypov, R. F., K. Maki, H. Cheng, S. D. Luck, and H. Roder. 2008. Folding mechanism of reduced cytochrome c: equilibrium and kinetic properties in the presence of carbon monoxide. *J Mol Biol* 383:437–53.

Lee, J. C., K. C. Engman, A. Tezcan, H. B. Gray, and J. R. Winkler. 2002. Structural features of cytochrome c folding intermediates revealed by fluorescence energy-transfer kinetics. *Proc Natl Acad Sci U S A* 99:14778–82.

Maity, H., M. Maity, M. M. Krishna, L. Mayne, and S. W. Englander. 2005. Protein folding: the stepwise assembly of foldon units. *Proc Natl Acad Sci U S A* 102:4741–46.

Marmorino, J. L., M. Lehti, and G. J. Pielak. 1998. Native tertiary structure in an A-state. *J Mol Biol* 275:379–88.

Marmorino, J. L., and G. J. Pielak. 1995. A native tertiary interaction stabilizes the A state of cytochrome c. *Biochemistry* 34:3140–43.

Matouschek, A., J. T. Kellis Jr., L. Serrano, M. Bycroft, and A. R. Fersht. 1990. Transient folding intermediates characterized by protein engineering. *Nature* 346:440–45.

Matsuura, Y., T. Takano, and R. E. Dickerson. 1982. Structure of cytochrome c_{551} from *Pseudomonas aeruginosa* refined at 1.6 A resolution and comparison of the two redox forms. *J Mol Biol* 156:389–409.

McGee, W. A., and B. T. Nall. 1998. Refolding rate of stability-enhanced cytochrome c is independent of thermodynamic driving force. *Protein Sci* 7:1071–82.

Muñoz, V., and L. Serrano. 1997. Development of the multiple sequence approximation within the AGADIR model of α-helix formation: comparison with Zimm-Bragg and Lifson-Roig formalisms. *Biopolymers* 41:495–509.

Myers, J. K., C. N. Pace, and J. M. Scholtz. 1995. Denaturant m values and heat capacity changes: relation to changes in accessible surface areas of protein unfolding. *Protein Sci* 4:2138–48.

Nicholson, E. M., and J. M. Scholtz. 1996. Conformational stability of the *Escherichia coli* HPr protein: test of the linear extrapolation method and a thermodynamic characterization of cold denaturation. *Biochemistry* 35:11369–78.

Oliveberg, M., Y. J. Tan, M. Silow, and A. R. Fersht. 1998. The changing nature of the protein folding transition state: implications for the shape of the free-energy profile for folding. *J Mol Biol* 277:933–43.

Orengo, C. A., A. D. Michie, D. T. Jones, M. B. Swindells, and J. M. Thornton. 1997. CATH: a hierarchic classification of protein domain structures. *Structure* 5:1093–1108.

Pace, C. N., and K. L. Shaw. 2000. Linear extrapolation method of analyzing solvent denaturation curves. *Proteins* Suppl 4:1–7.

Parker, M. J., J. Spencer, and A. R. Clarke. 1995. An integrated kinetic analysis of intermediates and transition states in protein folding reactions. *J Mol Biol* 253 (5):771–86.

Pierce, M. M., and B. T. Nall. 2000. Coupled kinetic traps in cytochrome c folding: his-heme misligation and proline isomerization. *J Mol Biol* 298:955–69.
Ptitsyn, O. B. 1998. Protein folding and protein evolution: common folding nucleus in different subfamilies of c-type cytochromes? *J Mol Biol* 278:655–66.
Roder, H., and W. Colon. 1997. Kinetic role of early intermediates in protein folding. *Curr Opin Struct Biol* 7:15–28.
Rumbley, J., L. Hoang, L. Mayne, and S. W. Englander. 2001. An amino acid code for protein folding. *Proc Natl Acad Sci U S A* 98:105–12.
Russell, B. S., and K. L. Bren. 2002. Denaturant dependence of equilibrium unfolding intermediates and denatured state structure of horse ferricytochrome c. *J Biol Inorg Chem* 7:909–16.
Sagle, L. B., J. Zimmermann, P. E. Dawson, and F. E. Romesberg. 2006. Direct and high resolution characterization of cytochrome c equilibrium folding. *J Am Chem Soc* 128:14232–33.
Sauder, J. M., N. E. MacKenzie, and H. Roder. 1996. Kinetic mechanism of folding and unfolding of *Rhodobacter capsulatus* cytochrome c2. *Biochemistry* 35:16852–62.
Schmid, F. X., and R. L. Baldwin. 1978. Acid catalysis of the formation of the slow-folding species of RNase A: evidence that the reaction is proline isomerization. *Proc Natl Acad Sci U S A* 75:4764–68.
Shastry, M. C., S. D. Luck, and H. Roder. 1998. A continuous-flow capillary mixing method to monitor reactions on the microsecond time scale. *Biophys J* 74:2714–21.
Shastry, M. C., and H. Roder. 1998. Evidence for barrier-limited protein folding kinetics on the microsecond time scale. *Nat Struct Biol* 5:385–92.
Silow, M., and M. Oliveberg. 1997. Transient aggregates in protein folding are easily mistaken for folding intermediates. *Proc Natl Acad Sci U S A* 94:6084–86.
Soulimane, T., M. von Walter, P. Hof, M. E. Than, R. Huber, and G. Buse. 1997. Cytochrome-c_{552} from *Thermus thermophilus:* a functional and crystallographic investigation. *Biochem Biophys Res Comm* 237:572–76.
Tanford, C. 1968. Protein denaturation. *Adv Protein Chem* 23:121–282.
Thöny-Meyer, L., F. Fischer, P. Künzler, D. Ritz, and H. Hennecke. 1995. *Escherichia coli* genes required for cytochrome c maturation. *J Bacteriol* 177:4321–26.
Travaglini-Allocatelli, C., F. Cutruzzolà, M. G. Bigotti, R. A. Staniforth, and M. Brunori. 1999. Folding mechanism of *Pseudomonas aeruginosa* cytochrome c_{551}: role of electrostatic interactions on the hydrophobic collapse and transition state properties. *J Mol Biol* 289:1459–67.
Travaglini-Allocatelli, C., S. Gianni, V. K. Dubey, A. Borgia, A. Di Matteo, D. Bonivento, F. Cutruzzola, K. L. Bren, and M. Brunori. 2005. An obligatory intermediate in the folding pathway of cytochrome c_{552} from *Hydrogenobacter thermophilus*. *J Biol Chem* 280:25729–34.
Travaglini-Allocatelli, C., S. Gianni, V. Morea, A. Tramontano, T. Soulimane, and M. Brunori. 2003. Exploring the cytochrome c folding mechanism: cytochrome c_{552} from *Thermus thermophilus* folds through an on-pathway intermediate. *J Biol Chem* 278:41136–40.
Wildegger, G., and T. Kiefhaber. 1997. Three-state model for lysozyme folding: triangular folding mechanism with an energetically trapped intermediate. *J Mol Biol* 270:294–304.
Yeh, S. R., and D. L. Rousseau. 1998. Folding intermediates in cytochrome c. *Nat Struct Biol* 5:222–28.

3 The Mechanism of Cytochrome c Folding
Early Events and Kinetic Intermediates

Eefei Chen, Robert A. Goldbeck, and David S. Kliger

CONTENTS

Introduction 37
Protein-Folding Triggers 38
Time-Resolved Polarization Spectroscopy 42
    Early Events in Reduced Cytochrome *c* Folding 45
Future Studies 56
References 56

INTRODUCTION

Fundamental questions of how a protein or peptide folds to its native state have been probed using multiple spectroscopic techniques and multiple protein/peptide systems. As a result there has been progress toward a better understanding of folding concepts such as polypeptide collapse, the dynamics of the unfolded and partly unfolded states, folding heterogeneity, classical versus energy landscape folding mechanisms, and on-pathway versus off-pathway intermediates. These advances were achievable in part because of technological improvements that allowed for fast detection and fast initiation of protein-folding/unfolding reactions. Thus it became possible to capture the earliest dynamics of protein folding that occurred during the burst phase of conventional stopped-flow instruments.

    The use of fast reaction initiation and detection methods was well established for studies of protein function before its application to protein folding. Since flash photolysis methods were introduced in the 1940s (Porter 1950), many biomolecular reactions have been initiated with triggers such as photo-induced electron transfer, ligand photolysis, and photo-induced chromophore isomerization and monitored *in vitro* using time-resolved methods. However, proteins studied under physiological conditions are not generally expected to undergo the large conformational changes that are necessary to probe mechanisms of protein folding/unfolding. More recently, however, researchers have been able to find denaturing conditions that capitalize on the aforementioned triggers, as well as on temperature-jump (T-jump), pH-jump, and rapid-mixing initiation methods, to substantially perturb the folding/unfolding equilibrium and bring the timescale of folding triggering on par with nanosecond time-resolved detection methods.

The different folding triggers have been coupled with optical probes such as time-resolved circular dichroism (TRCD), infrared (TRIR), resonance Raman (TR^3 or RR), and fluorescence (TRFL) spectroscopies, which are sensitive to different parts of a biomolecule's structure. Because of the complementary nature of such optical probes, the combined data are able to paint a clearer, more structurally detailed description of the different stages of protein folding/unfolding. Such structural details are important for understanding not only the folding/unfolding mechanisms of proteins, but also their function in physiologically relevant biochemical processes. For structural information it is difficult to surpass the atomic-level detail revealed by x-ray crystallography. However, such structural detail has typically not been available with the high time resolution afforded by optical spectroscopy. Recently, this gap between the two approaches has been closing because of the increasingly structure-specific information available from time-resolved optical methods and recent improvements in the time-resolution of traditional structural methods such as x-ray crystallography and nuclear magnetic and electron spin resonance spectroscopies (Srajer and Royer 2008; Qu et al. 1997). Together, these exciting advances promise to provide unprecedented details about the mechanisms of both folding and functional reactions in biochemical systems.

This chapter will discuss various methods used to initiate and monitor folding in cytochrome c with a focus on the nanosecond polarization detection methods used in this lab, placing the results obtained from those methods in the context of findings from other time-resolved studies and current mechanistic thinking about the folding of this paradigmatic protein.

PROTEIN-FOLDING TRIGGERS

Because biomolecules are dynamic structures, time resolution is of significant importance to mechanistic studies. Nanosecond time-resolved detection methods would be of limited use in protein-folding studies without a means to trigger protein-folding/unfolding reactions on an equally fast or faster timescale. The trigger methods mentioned previously allow folding of two general classes of proteins: those that have a photoactivatable group, either exogenous or endogenous, and those that do not have a photosensitive chromophore.

Laser-Induced Temperature Jump. In the latter category, perturbation of the folding/unfolding equilibrium via external factors such as pH and temperature presents fewer limitations on the proteins that can be studied. The laser-induced thermal denaturation method rapidly shifts the equilibrium between folded and unfolded states of a protein by introducing a pulse of laser light (1.4 to 2 μm) that is absorbed by water in an aqueous sample. Using this approach it is possible to generate a 10–30 K T-jump within the ~10 ns pulse width of the laser. The time resolution for this method of generating a T-jump is considerably faster than early T-jump methods that used a capacitive electrical discharge across the sample cell to achieve T-jumps within 50 ns to a microsecond (Hoffman 1971). Another method to achieve rapid increases in temperature used the absorption of high-powered laser pulses by a dye (Phillips, Mizutani, and Hochstrasser 1995). However, the use of a dye can potentially interfere with the folding/unfolding/biochemical reaction of interest, as can the high salt concentrations required for the capacitive discharge approach. Today, the method that uses the absorption of laser energy by an overtone of the water (or D_2O) OH-stretch is thus the most common way to generate a rapid temperature change and has been coupled with TRFL, TR^3, TRIR, and time-resolved absorption (TROD or TROA), and more recently, optical rotatory dispersion (TRORD) (Chen et al. 2005) detection methods. For more details about the laser-induced T-jump methods, readers are encouraged to go to Callender et al. (1998) and references therein.

Laser-Induced pH Jump. In addition to temperature changes, a rapid pH change is another way to study protein-folding/unfolding reactions. This can be achieved by rapid mixing (see next subsection) or by laser-induced photoexcitation of molecules that have different ground versus excited-state acid-base properties (Gutman and Nachliel 1997). In the latter method, the major limitation for protein-folding/unfolding studies is that upon photoexcitation of the photoactive molecule the pH change is short-lived, returning to its initial value on a timescale around that of the molecule's excited-state lifetime. Although the lifetime of the pH change can be tailored by the choice of the proton-transfer species, the generally short excited-state lifetimes do not allow for an appreciable time window during which protein folding/unfolding can be studied. An approach that allows rapid formation of persistent pH changes uses laser-induced release of a proton from a photoactivatable caged proton compound, such as ortho-nitrobenzaldehyde (*o*-NBA) and the more water-soluble *p*-hydroxyphenacyl esters (Abbruzzetti et al. 2001; Callender and Dyer 2006). The rate constants for release of protons from *o*-NBA and *p*-hydroxyphenacyl esters are 2×10^9 s^{-1} and 10^8 s^{-1}, with formation efficiencies of 50% and 10–20%, respectively. The utility of caged compounds is illustrated by studies using *o*-NBA, in which a pH drop can be achieved within ~10 ns and maintained for hundreds of milliseconds after the photoexcited release of protons.

Rapid Mixing. Typical stopped-flow mixers that are coupled with conventional CD and fluorescence instruments have mixing dead times on the order of milliseconds. As a result, the kinetic events that occur on the nanosecond and microsecond timescale are buried in the "burst" phase of these classical experiments. However, using a design by Regenfuss et al. (1985), it is possible to obtain mixing times of 45 µs (Shastry, Luck, and Roder 1998) and 100 µs (Takahashi et al. 1997). That mixer used concentric quartz capillaries with a platinum bead located at the tip of the inner capillary to introduce highly turbulent flow conditions to achieve these microsecond mixing times. A more recent microfluidic mixer reportedly achieved a mixing time of <10 µs, by optimization of a turbulence-free mixing method wherein four channels taper in width and intersect at the center of a mixing chip (Knight et al. 1998). Fast mixing is achieved because of the small length scales (as narrow as 50 nm), which allow the buffer molecules to diffuse rapidly across the inlet stream (denatured protein). Amidst the many protein/peptide studies that have reported results of folding dynamics during the burst phase, the interpretations of those results have triggered debates as to whether the data indicate the presence of a true intermediate species or merely reflect a non-folding-specific biopolymer compaction expected during denaturant dilution (Jennings 1998; Qi, Sosnick, and Englander 1998).

Photoexcitation with Exogenous Photosensitive Groups. A different approach to initiation of protein folding involves synthesis of a protein/peptide sequence with a photoactive molecule. When incorporated into a peptide sequence, the photosensitive group can be used to control the content of secondary structure. For example, the CD spectra for poly(L-glutamic acid) linked to spiropyran units show the formation of α-helix from random coil structure upon the closed to open ring interconversion of spiropyran induced by visible light. Another example is the use of azobenzene groups to modulate: the conformation of cyclic peptides, the sense of a peptide helix, peptide aggregation, and the helix, β-sheet, and coil content of peptide structures.

Although many of these photosensitive peptides show significant structural changes under equilibrium conditions, in relatively few of these systems have the nanosecond dynamics of peptide conformational changes been successfully observed. Peptide-folding reactions triggered by photolysis of the aryl disulfide linkers thiotyrosine and aminothiotyrosine are limited by the geminate reformation of the disulfide bond, which occur within 1 ns and 10 μs, respectively. TRIR studies of the aminothiotyrosine cross-linked peptide reported that propagation toward the helical, linear peptide had not taken place by 2 ns, which was the longest measurable time delay reported. The azobenzene cross-linker, however, has been coupled with engineered, 16-residue peptides having different stabilities, different helical contents, and different levels of light-induced conversion to disordered structure. By varying the spacing of the two cysteine residues that link the azobenzene group, the compatibility of the helix conformation with the *trans* and *cis* geometries can be regulated so that significant increases or decreases in helix secondary structure can be induced by photoirradiation. Two such 16-residue peptides, with the $i, i + 11$ and $i, i + 7$ cysteine spacings, have been successfully probed using TRORD and TRIR spectroscopy. All references for this section can be found in Chen et al. (2008).

The Endogenous Photosensitivity of the Heme Group in Cytochrome c. T-jump, pH jump, and rapid mixing are very general trigger techniques that can be used to initiate folding/unfolding in either iron oxidation state of cyt *c*. In practice, however, these methods have been applied mainly to the ferric state (oxcyt *c*), partly because of the experimental challenges imposed by the anaerobic conditions necessary to study ferrous cyt *c* (redcyt *c*). The presence of the heme prosthetic group in cyt *c* provides additional methods, photodissociation of an exogenous ligand bound to ferrous heme or photo-initiated reduction of oxcyt *c*, by which redcyt *c* folding can be triggered under anaerobic conditions. Both methods rely on the differences in folding stabilities of oxcyt *c* and redcyt *c*.

The use of ligand photolysis as a folding trigger was introduced in 1993, in a study where photodissociation of CO-bound redcyt *c* resulted in the formation of CO-deligated redcyt *c* within the time resolution of the TROA apparatus (Jones et al. 1993). Under physiological conditions, the CO ligand will not bind oxcyt *c* or redcyt *c* at the sixth axial position of the heme moiety. However, in the presence of guanidine hydrochloride (GuHCl), the bond formed between the heme Fe(II) and the natively occupied Met80 is destabilized, allowing the CO ligand to displace Met80 and form a photosensitive Fe(II)-CO ligation. Laser-induced photolysis leads to formation of CO-unbound redcyt c from the initially unfolded CO-bound redcyt c. Because of their different folding stabilities, CO-bound redcyt c is almost completely unfolded in 3–5 M GuHCl, whereas CO-unbound redcyt c has near-native secondary structure. A similar folding free-energy difference is the impetus behind the photoreduction trigger method (discussed later). Figure 3.1 shows that the differences in the folding free energies are even greater between oxcyt c and redcyt c than between the latter and the CO-bound redcyt c species. Between 2 and 4 M GuHCl, oxcyt c is largely unfolded and redcyt c is largely folded. Therefore either phototrigger can be used to generate an immediate redcyt c photoproduct with the largely unfolded conformation of the initial oxcyt c or CO-bound redcyt c species. However, because this unfolded photoproduct is formed under conditions that favor the folded conformation, redcyt c folding is facilitated and can be probed at the earliest (hundreds of nanoseconds) stages. An advantage of the photoreduction trigger is that the folding process can be followed up to formation of the natively folded state (>10 ms). In contrast, the folding pathway initiated by the photolysis trigger competes with the bimolecular CO recombination reaction and thus limits the monitoring time window to hundreds of microseconds at 1 atm CO. Nonetheless, the CO photolysis method coupled with TRCD and TRMCD spectroscopies has provided important information about early events in redcyt c folding (Chen et al. 1998; Goldbeck et al. 1999; Abel et al. 2007).

Of the three reducing agents first described by Gray and colleagues for this purpose, $Ru(2,2'-bypyridine)_3^{2+}$, $Co(C_2O_4)_3^{3-}$, and nicotinamide adenine dinucleotide

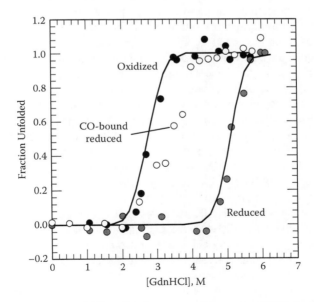

FIGURE 3.1 Denaturant unfolding titration curves for oxcyt c, CO-bound redcyt c, and CO-unbound redcyt c.

(NADH), NADH is perhaps the most versatile for initiating folding studies in heme proteins (Pascher et al. 1996; Mines et al. 1996; Telford et al. 1998). Reduction by photo-activated NADH provides the longest time window in which protein folding can be probed, as redcyt c is formed within hundreds of nanoseconds and reoxided to oxcyt c after hundreds of milliseconds (due to the presence of trace O_2). Upon irradiation of a denatured solution of oxcyt c in the presence of $Ru(2,2'-bypyridine)_3^{2+}$, it is possible to form redcyt c within a microsecond, but measurements are limited to times shorter than ~1 ms by the reoxidation of cyt c by $Ru(2,2'-bypyridine)_3^{3+}$. To probe the slower phase of folding (>1 ms), $Co(C_2O_4)_3^{3-}$ may be used as the reducing agent, forming redcyt c within 1 ms. The competition between folding and reoxidation of cyt c by $Co(C_2O_4)_3^{3-}$ and trace O_2 on this timescale depends on the denaturant concentration.

TIME-RESOLVED POLARIZATION SPECTROSCOPY

Structure-sensitive probes that have contributed to our understanding of folding/unfolding of proteins in general, and cyt c in specific, include IR, RR, fluorescence, and CD/ORD spectroscopies. Although absorption spectroscopy provides a convenient way to monitor kinetic behavior and has been frequently applied to folding reactions, specific and accurate protein conformational assignments typically require information from more structure-sensitive probes. Not only is RR spectroscopy sensitive to the ligation state of heme proteins, but with UV excitation it can also provide structural details on protein vibrations from the polypeptide backbone and the different amino acid side chains (e.g., amide, tryptophan, tyrosine groups) (Balakrishnan et al. 2008). In contrast to IR spectroscopy, a distinct advantage of RR spectroscopy is that biomolecules can more easily be probed in an aqueous environment because there is minimal spectral interference from the deformation modes of water. However, because of the spectroscopic selection rules for Raman and IR they provide complementary information. That is, vibrational transitions are Raman active if there is changing molecular polarizability, whereas IR-active modes correspond to vibrational modulation of a permanent molecular dipole moment. For complex biopolymers, the spectral intricacy of the large number of overlapping IR modes is mitigated because vibrational absorptions arise from transitions that are relatively localized on a molecule and because of the regularity of specific structural motifs such as those found in proteins, nucleic acids, and lipid membranes. For proteins, IR measurements focus on the C=O and N-H stretching and the N-H deformation vibrations of the protein backbone structure and provide markers for the helical, sheet, and random coil secondary structures. Fluorescence spectroscopy using intrinsic tryptophan residues or extrinsic fluorophores is also a common tool to probe structure changes (Callender et al. 1998). Depending on the fluorescence parameter, such as fluorescence λ_{max}, quantum yield, or integrated emission intensity, local or global information can be obtained about the protein structure and kinetics. In general, the changes in fluorescence intensity as a result of extrinsic or intrinsic quenching groups will report on global folding events. However, more localized structural information is available by studying protein variants in which distances between the donor and acceptor pair are varied.

CD/ORD is sensitive to global secondary structure and has long been used to report on the π → π* and n → π* transitions of the major secondary structures (Woody 1968). Different chromophores found in biomolecules will have different CD absorption bands. In the far-UV range, the large CD bands reflect the n → π* transition of the polypeptide backbone conformation, whereas CD bands in the near-UV region are characteristic of the aromatic amino acid residues, phenylalanine, tyrosine, and tryptophan. For proteins with endogenous prosthetic groups, such as heme proteins, the structural asymmetry of the heme moiety induced by its protein environment also results in CD absorption bands. Thus CD data measured over a broad spectral range will give global information about biomolecular structures. In this section, we will focus on the TRCD and TRORD methods used in protein-folding studies within our lab. For further details on these spectroscopies, for a description of the TRMCD system, and for a review of other polarization methods see Goldbeck et al. (1997).

Nanosecond Circular Dichroism Spectroscopy. Since it was initially coupled with stopped-flow and flash photolysis methods in 1974, the time resolution of CD measurements has improved from milliseconds to as fast as picoseconds. All references for this and the next section can be found in Chen et al. (2005). The evolution of the nanosecond TRCD instrument from single-wavelength to multiwavelength measurements and then the extension of these measurements into the far-UV spectral region and to TRMCD techniques have been particularly useful for the study of protein folding. The subsequent development of nanosecond TRORD and TRMORD methods provided a signal-to-noise advantage that addressed the limitation of light intensity as a result of absorption by denaturant.

The most sensitive way to measure CD is to use ellipsometric methods, where a sample is placed between crossed linear polarizers (LPs). The sensitivity advantage gained by using this null-detection method is, however, compromised by the simultaneous detection of circular birefringence (CB), or ORD. The CB artifact is avoided in conventional CD measurements by directly probing the differential absorption of left circularly polarized and right circularly polarized light. For nanosecond time resolution, CD spectra without the distortions associated with CB artifacts can be measured by monitoring changes in the beam polarization rather than in the sample absorbance. The TRCD apparatus (Figure 3.2A), having the pump, probe, and detection components that are fundamental to a typical TROD configuration, implements a quasi-null approach, where both the sign and magnitude of the CD signals are determined, by using a mechanically strained, fused silica plate positioned between two crossed LPs. This strain plate (SP) introduces a slight phase retardance (δ, 1–2°), along an axis oriented ±45° relative to the linear polarization axis, that converts the incident linearly polarized light into highly eccentric, elliptically polarized light. Left elliptically polarized (LEP, −45°) and right elliptically polarized (REP, +45°) light is achieved by rotating the SP by 180°. The major axis of the REP and LEP light lies along the original linear polarization light axis, and the minor axis has an intensity that is about 10^{-4} times lower than that of the major axis. With the second, analyzing polarizer, the minor-axis intensity of the ellipse is passed to the detector. The ellipsometric signal is defined by the differential intensity of LEP and REP light normalized to the sum of the detected intensities, $S = (I_R - I_L)/(I_R + I_L)$. Because the sample's CD adds

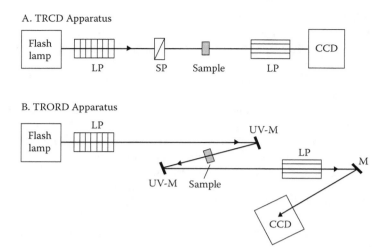

FIGURE 3.2 The TRCD (A) and TRORD (B) systems essentially differ by one element, the strain plate (SP). Because CD and ORD are Kramers-Kronig mates, they should report the same information. However, experimentally each method has a distinct advantage: CD measurements provide better spectral characterization, whereas ORD measurements offer better kinetic details. In (B) the ORD system has been modified to allow for experiments on small sample volumes.

to or subtracts from that minor-axis intensity, the difference between measurements with oppositely oriented δ's will be proportional to the CD in the sample. In contrast, CB in the sample simply rotates the axis of the ellipse, and therefore the effect of CB will cancel.

Nanosecond Optical Rotatory Dispersion Spectroscopy. In 1953, a quasi-null polarimetric method was introduced for high-sensitivity static ORD measurements. This method was later adapted for nanosecond TRORD spectroscopy by this lab, with the first application of the technique to studies of carbon monoxide–bound hemoglobin. Because ORD and CD are Kramers-Kronig transform mates, they should, in theory, report the same information. However, experimentally each method offers distinct advantages, which arise from the way that CD and ORD are measured. That is, there is a signal-to-noise advantage in light-limited kinetics experiments because ORD signals can be measured outside of absorption bands. However, for this reason interpretation of OR data may be complicated by multiple Cotton contributions to the signal. In contrast, CD measurements are localized to absorption bands and thus the CD spectra are generally easier to interpret in applications such as protein secondary-structure analysis. Thus whereas TRORD signals are easier to measure in studies that probe the kinetics of folding reactions, TRCD spectra are preferred when the goal of the experiment is to interpret the accompanying secondary-structure changes of that reaction.

The far-UV TRORD apparatus (Figure 3.2B) is essentially the same as the TRCD system, but without the SP. In addition, it was modified for measurements of small sample volumes. From the probe source, the initially unpolarized light is directed to the first MgF$_2$ LP and then a UV-enhanced aluminum mirror (UV-M), which focuses the light beam diameter down by a factor of ~20 from ~6 mm. A second UV-M, positioned off-axis, collects the light, which then propagates to a second MgF$_2$ LP. The probe beam is focused onto the slit of a spectrograph that is equipped with a 600-groove/mm (200 nm blaze) grating and a charge-coupled device (CCD) detector. Intensities are measured when the first LP is rotated by a small angle ($\beta = 2°$ for redcyt c studies), either $+\beta$ or $-\beta$, off the crossed position of the two LPs. The recorded signal represents the difference of the intensities measured at $\pm\beta$ normalized by their sum. In the same way that CB effects are minimized in CD measurements, the effect of CD on the CB signal is canceled using this method. That is, a chiral sample will have an optical rotation that either adds to or subtracts from β, the reference rotation. And, the measured signal is proportional to the CB and amplified by a factor of $1/\beta$. In the case where measurements are made as a function of wavelength (ORD), the signal is proportional to ORD/2β.

Magnetic Time-Resolved CD and ORD. With appropriate modifications of the aforementioned TRCD and TRORD systems, magnetic CD (MCD) and magnetic ORD (MORD) can also be measured with similar time resolutions as natural CD and ORD. In this way, TRMCD/MORD spectroscopy has been used to study the native and nonnative ligations of the heme group in redcyt c during folding (Kliger, Chen, and Goldbeck 2004).

EARLY EVENTS IN REDUCED CYTOCHROME c FOLDING

In 1998, we coupled nanosecond TRCD spectroscopy with photodissociation triggering to follow the secondary-structure kinetics of redcyt c (Chen et al. 1998). This work launched several subsequent papers that both correlated early folding events with specific protein structural features and pointed to their broader significance in the context of the energy landscape model (Chen, Wittung-Stafshede, and Kliger 1999; Goldbeck et al. 1999; Abel et al. 2007; Chen, Goldbeck, and Kliger 2003, 2004; Chen et al. 2007). The remainder of this chapter will focus on the results obtained from redcyt c folding studies within the last 10 years using nanosecond TRCD, TRORD, TRMCD, and TRMORD polarization spectroscopies. In addition, those data will be placed in the context of other cyt c folding studies, which have focused largely on the oxidized species.

Earliest Secondary-Structure Formation Events. The initial observations implying fast secondary-structure formation in oxcyt c came from stopped-flow far-UV CD studies, wherein 44% of the native CD signal (222 nm) was detected within the ~4 ms dead time (Elöve et al. 1992). Subsequently, the kinetics of the burst phase of redcyt c folding was probed directly with nanosecond TRCD measurements. These showed that 8% of the far-UV CD spectrum (relative to the native state signal) was formed with a time constant ≤0.5 μs after photodissociation of the CO ligand (Chen et al. 1998). Further changes in secondary structure were obscured by the competing CO ligand recombination reaction. However, with rapid reduction of oxcyt c, substantial (~20–35%) helical secondary-structure formation was observed for redcyt c on the submicrosecond to microsecond timescale (Chen, Wittung-Stafshede, and Kliger 1999; Chen, Goldbeck, and Kliger 2003). A noncanonical behavior was observed for these redcyt *c* submillisecond kinetics, where the time constant for fast helix formation decreased with increasing denaturant concentration. That is, in 2.7 M GuHCl (pH 7, 25°C) there was no apparent formation of secondary structure until after 5 μs (τ = 12 μs), but in 4 M GuHCl (pH 7, 25°C) about 20% secondary structure was observed within the instrument time resolution ($\tau \leq 0.4$ μs) (Figure 3.3). Given that the average time constant for photoreduction in this system is about 5 μs, the fastest time constant of ≤ 0.4 μs for secondary-structure formation presented a paradox. How can the reduction-triggered folding process proceed faster than the bulk of the reduction reaction used to trigger folding? A resolution to this paradox appears to lie in the heterogeneous nature of the folding dynamics on such short timescales, a consequence of the finite conformational diffusion dynamics of the unfolded protein chains, and the heterogeneous mixture of reductants that is produced by the photo-excitation of NADH (Orii 1993).

The results of these TRCD/TRORD redcyt *c* studies were consistent with those obtained from continuous-flow mixing experiments monitored with far-UV CD (Akiyama et al. 2000). Those experiments, which used a denaturant jump from 4.4 to 0.7 M GuHCl, observed the appearance of ~20% helical secondary structure within the ~0.5 ms dead time of the instrument. A pH jump from 2 to 4.5 did not appear to produce a change in secondary-structure content between the acid-induced unfolded state and the burst-phase intermediate, which seems less consistent with the previous TRCD/ORD results, but an intermediate with native-like CD was detected in the pH-jump experiment with a time constant of formation (0.48 ms) that closely overlapped the mixing dead time. Thus the signal from the late intermediate dominated the CD observed in the first time point measured after mixing and may have obscured detection of the burst phase CD amplitude.

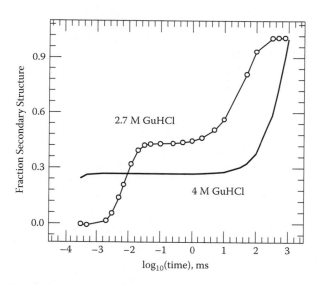

FIGURE 3.3 Comparison of the far-UV TRORD data for redcyt *c* folding in 2.7 and 4 M GuHCl demonstrates the unexpected behavior of the submillisecond kinetics.

The far-UV TRCD/TRORD redcyt c data have been interpreted in terms of the formation of an intermediate on the folding path between the unfolded (U) and natively folded (N) states. Because of the significant formation of secondary structure on that timescale, it was suggested that the intermediate is molten-globule-like (MG) (U → MG → N). Since equilibrium data for a large number of globular proteins in their folded and partially unfolded states show a high correlation between the compactness of a given structure and the amount of ordered secondary-structure formation, it may be that the observed kinetic formation of secondary structure is accompanied by a rapid-collapse phase. Comparison of the relative times of secondary-structure formation and rapid collapse addresses the fundamental question of what comes first—secondary-structure formation, collapse, or tertiary-structure formation? In the framework model, early secondary-structure formation governs later folding events, whereas the hydrophobic collapse model describes an initial collapse as the trigger for secondary-structure formation and full compaction of the protein to the native state. In contrast to the aforementioned two models, the nucleation condensation mechanism suggests that partial elements of tertiary and secondary structures are already present in the transition state. (For a review of these different models, see Daggett and Fersht 2003.) The following comparison of the kinetics of secondary-structure formation, obtained from TRCD/TRORD measurements, with the kinetics of fast collapse that is available in the literature provides insight into the model most appropriate for cyt c folding.

The time constants reported for the very fast collapse phase of oxcyt c range from 16 μs to 90 μs (in 1–2 M GuHCl). One of the first high time-resolution studies to suggest such a rapid collapse into a compact denatured structure ($\tau = 40$ μs) used heme visible band TROD spectroscopy of photoreduction-triggered folding in redcyt c (Pascher et al. 1996). Although Trp59 fluorescence measurements of redcyt c folding that followed CO photodissociation did not detect the 40 μs process (Chan et al. 1997), it was suggested in a response by Winkler and Gray (Chan, Hofrichter, and Eaton 1996) that the comparison was complicated by the different unfolded starting points for the two experiments. Another possible explanation for the discrepancy in the data was that the amplitude of the 40 μs process was not large enough to be easily observed in the Trp fluorescence data. In electron-transfer and fluorescence energy-transfer studies of a variety of modified cyt c systems (Winkler 2004; Pletneva, Gray, and Winkler 2005a, 2005b), it was observed that roughly similar populations of two components, characterized by extended or compact conformations, were established within the stopped-flow dead time of 1 ms. An apparent rapid equilibrium (<1 ms) suggested between the two components implies a cooperative transition between the two subpopulations and would appear to argue that compaction is specific to protein folding and not due to a gradual collapse mechanism general to polymers.

Although the aforementioned studies focused on the collapse process in redcyt c, the most direct observations of the collapse intermediate have been obtained from methods with higher time resolution and from studies on oxcyt c. With submillisecond time resolution, such as that available in rapid-mixing methods, more detailed information about the collapse process was possible. For example, the observation of a collapse process with a simple exponential time course ($\tau \sim 60$ μs) was interpreted as evidence for a discrete free-energy barrier separating a true compacted folding intermediate from the unfolded state. These data were obtained from ultrafast mixing measurements having a 45 μs dead time (Shastry and Roder 1998). A Trp59 fluorescence T-jump study with nanosecond time resolution also reported that microsecond collapse (τ = 90 μs) involved passage over a free-energy barrier (Hagen and Eaton 2000). Small angle x-ray scattering (SAX) monitoring and continuous microfluidic mixing studies demonstrated that the extended, unfolded state collapsed to more compact structure having an apparent R_g of 16 Å within the 150–500 μs time of the first data collection window and then achieved the native R_g value of 14 Å by 10 ms after mixing (Pollack et al. 1999).

The kinetics of the collapse process and of secondary-structure formation suggest that the protein may undergo cooperative collapse to a compacted state in which disordered tertiary interactions stabilize the simultaneous formation of significant amounts of nativelike helical secondary structure. The fluorescence-detected amplitude of the early collapse phase is small, indicating incomplete conversion to the compact form (Shastry and Roder 1998). This amplitude was observed to decrease with increasing GuHCl concentration, whereas the observed folding rate increased. This pattern of amplitudes and observed rates was qualitatively similar to the results of the TRCD/TRORD data. Although there are several reasons why these rates may not be strictly comparable (such as the range of experimental conditions, the different probe methods, the oxidation state of cyt *c* studied, and the absence of studies of the rapid-collapse kinetics that parallels the rate of rapid helix formation in His variants), the available evidence is consistent with the nucleation condensation hypothesis that collapse and secondary-structure formation generally occur simultaneously on the microsecond timescale.

Conformational Diffusion. Returning to the seemingly paradoxical observation mentioned previously of photoreduction-triggered helix formation that preceded the bulk reduction of the sample, one very reactive reductant produced by NADH photoexcitation is the solvated electron. Among the unfolded chains was a conformational subset that was inferred from that observation to react most rapidly with this reductant while remaining kinetically isolated from the bulk conformers because of slow configurational diffusion. The idea that the equilibration rate between unfolded states defines a boundary between the applicable timescales for the energy landscape funnel and classical models of folding (Brooks 1998) was advanced by the results of TRMCD and far-UV TRORD folding studies (Goldbeck et al. 1999; Chen, Goldbeck, and Kliger 2003). In this diffusional viewpoint, the top of the funnel represents the unfolded protein conformations, which travel down the length of the funnel in a free-energy-biased form of configurational diffusion to achieve the lower conformational entropy of the native state. The extent of equilibration of the unfolded conformers at the periphery of the funnel governs the kinetics of protein folding. If the peripheral diffusion time is slower than free-energy-biased, diffusional downhill folding, then the different unfolded conformations can fold along kinetically isolated pathways and exhibit heterogeneous kinetics. In contrast, if interconversion between the unfolded conformers is much faster than the formation of the native state, then such configurational equilibrium leads to homogeneous kinetics that often follow the classical view involving passage over a single transition state. The kinetic heterogeneity that arises from slow equilibration between unfolded polypeptide chains is most likely to affect the earliest events and intermediates in protein folding, but may not necessarily persist long enough to affect the late-time folding processes producing the native state.

Differences between the classical and the landscape approaches to folding may be rather subtle with regard to interpreting experimental measurements that represent ensemble averages over all conformations present, making distinctions between the different models difficult. The CO-photolysis-triggered TROA study of redcyt c by Jones et al., for instance, was interpreted in terms of a classical kinetic model in which CO de-ligation and formation of the immediate five-coordinate heme protein photoproduct were followed by the kinetically homogeneous formation of six-coordinate species involving the binding of various methionine and histidine residues to the heme site (Jones et al. 1993). On the other hand, Goldbeck et al. interpreted the results of their TRMCD study of the same system with a heterogeneous kinetic model that assumed very slow diffusional equilibration between unfolded chain conformers presenting different residues for facile heme binding (Goldbeck et al. 1999). They found that the latter energy landscapelike model produced a better fit to the MCD spectra of the intermediates. This finding was interpreted as providing kinetic evidence for multiple pathways in redcyt c folding in that Met and His binding reactions appeared to proceed separately in kinetically isolated conformational ensembles. This result also implied that conformational diffusion was limited on the timescale of the very fast helix-folding process identified in the far-UV TRCD data.

How rapidly do the unfolded conformations interconvert? This question addresses conformational diffusion from a different perspective than the related question of what is the "speed limit" of folding. Simple considerations of heteropolymer chain diffusion offer estimates of 0.1–1 μs as the fastest time in which a tertiary contact can be formed in a small protein loop ($n = 10$ residues) (Kubelka, Hofrichter, and Eaton 2004). By extension, this is considered to represent the upper limit to the fastest speed that a protein can fold since folding to the native state cannot proceed faster than the rate at which tertiary contacts are formed. The rate of interconversion between all unfolded chain conformations is a more difficult question to answer from a theoretical perspective because the unfolded conformations are highly heterogeneous, involving a wide distribution of possible contacts that can include many elements of native and nonnative structure. However, because the unfolded conformational diffusion time characterizes the loss of the conformer-based kinetic heterogeneity of the energy landscape regime and the arrival at the equilibrated unfolded state that is a prerequisite to the classical transition state theory (TST) folding regime, it too can be considered a type of folding "speed limit."

The TRMCD data offered qualitative evidence that finite intrachain diffusion prevented complete equilibration of the unfolded redcyt c conformers on the timescale of tens of microseconds. Similarly, heterogeneous folding kinetics observed in the photoreduction-triggered folding of redcyt c suggested incomplete equilibration of the unfolded polypeptides on the timescale of the bimolecular reduction process ($\tau \sim 5$ μs) (Chen, Goldbeck, and Kliger 2003). Transient absorption measurements of electron-transfer rates in unfolded cyt c, modified with a Zn-porphyrin heme-substituted excited-state electron donor and an $Ru(NH_3)_5^{3+}$ electron acceptor attached to His33, found that the electron-transfer rates implied a 15-residue closure time of

250 ns (Chang et al. 2003). Extrapolating this result to $n = 10$ residues, which is the size of the most probable tertiary-contact loops expected to form in folding (Camacho and Thirumalai 1995), gave a folding speed limit of $(100 \text{ ns})^{-1}$. Further extrapolation to the $n \sim 62$ residues in the loop formed between Met80 and His18 in the native protein gives 2 μs as a rough estimate for the conformational equilibration time. A more direct measure of the conformational diffusion time (τ_d) for redcyt c was possible when a kinetic model explicitly accounting for the effect of unfolded conformational exchange on the residue-binding kinetics was applied to TRMCD data obtained from CO photolysis-triggered folding studies on several histidine mutants of cyt c (Abel et al. 2007). The resulting τ_d of ~3 μs for unfolded redcyt c was consistent with the extrapolation from the study by Chang et al., with the "molecular timescale" of 2 μs that was reported for a nonheme λ-repressor protein fragment (Yang and Gruebele 2003) and with the time constants for contact formation in nonfolding repeat polypeptides (Krieger et al. 2003). Although many factors differ between the experiments cited earlier, the different measurements of τ_d appear to converge near 1 μs for loop sizes of 60–100 residues. This convergence provides additional support to the suggestion by Gruebele and coworkers that the timescale of conformational diffusion in the unfolded chains represents a more general kinetic limit for classical folding. Folding on slower timescales would imply that the unfolded state(s) and the transition state are in conformation equilibrium, as required by classical TST, whereas faster folding would be better described by the downhill folding scenario of the energy landscape model. In the case of the far-UV TRCD/TRORD studies discussed previously, the 3 μs τ_d reported by TRMCD measurements supports the suggestion that kinetic isolation of the very fast-folding ensemble in redcyt c may be attributed to slow conformational diffusion.

Nonnative Conformational States. The His33 ligand was identified as dominant (80% relative to 20% His26 when both His residues are present) in coordinating the sixth heme axial site in horse oxcyt c (Colon et al. 1997). Thus far-UV TRORD studies were carried out to address how the earliest redcyt c folding events would be affected by changes in the partial constraint on the backbone conformations of the unfolded chains presented by this His ligation. The first of these TRORD studies involved tuna redcyt c, which exhibits markedly different kinetics (in 3.3 and 4 M GuHCl) despite the 80% sequence homology and the high structural similarity with the horse protein (Chen, Goldbeck, and Kliger 2004) (Figure 3.4). That little formation of secondary structure was observed before 5 μs was attributed to the substitution of His33 by a tryptophan residue in the tuna protein. The results of these studies led to several observations about redcyt c folding. First, the data rule out the alternative possibility that the origin of the kinetic heterogeneity was simply due to heme-ligand heterogeneity, a scenario wherein rapid formation of 20% secondary structure would be localized to the minority (20%) His18-Fe-His26 heme-coordinated subpopulation. In such a situation, interconversion of the heme-ligation states may be slow, dependent on the His off rates, and the dominant His26 ligand in tuna redcyt c should then facilitate a significantly larger and faster early folding phase relative to that observed in horse redcyt c. Second, while the data suggest that His33 is in some way important for rapid formation of secondary structure in the

The Mechanism of Cytochrome c Folding

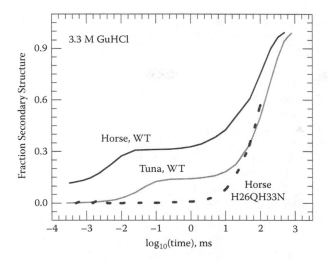

FIGURE 3.4 Sequence dependence of ultrafast helix formation. The kinetic traces for wt horse, wt tuna, and horse H26QH33N are dramatically different, providing insight into the roles of His33 and His26 in redcyt c folding.

fast-folding population, it did not address whether the His18-Fe-His33 configuration facilitates fast folding or whether it slowed folding to a lesser degree than does the His18-Fe-His26 coordination. Finally, the data also indicated that even if His33 heme ligation is a necessary condition for ultrafast folding in a small subset of the unfolded chains, it is not necessarily sufficient because the majority of the unfolded polypeptide chains of the initial oxcyt c sample had this heme ligation.

The dramatic differences in the kinetic traces for fast folding in the tuna and horse wild-type (wt) redcyt c proteins suggest that in the absence of His33, the His26 ligand alone is unable to maintain the structural requirements for fast folding. What happens to the early-phase kinetics in the absence of His26 was explored with far-UV TRORD measurements of redcyt c folding in a horse double mutant (H26QH33N) where His33 and His26 were replaced with low heme affinity residues (Chen et al. 2007). In the presence of GuHCl, the absence of both His33/26 residues leaves the sixth axial site open for other potential nonnative ligands including Met80/65, Lys72/73/79, or the nonacetylated N-terminal group. Using MCD measurements to probe the heme coordination of H26QH33N, a His18-Fe(II)-Met ligation was confirmed for the native redcyt c, whereas His18-Fe(III)-Lys was suggested for unfolded (3.3 M GuHCl) oxcyt c. Thus upon photoreduction of the initial oxidized H26QH33N, folding of redcyt c first passes through the instantaneous His18-Fe(II)-Lys coordinated photoproduct before proceeding to the final redcyt c native protein with the His18-Fe(II)-Met ligation. The secondary structure dynamics that paralleled the heme-ligation changes in the reduced H26QH33N folding reaction do not exhibit the fast phase observed for wt protein folding. With only the slow phase (>1 ms) observed (Figure 3.4), it was concluded that the His18-Fe-His26 heme coordination does not facilitate folding in an absolute sense as does the His18-Fe-His33 ligation.

In cyt c, His33 is associated by van der Waals interactions with a β-turn (residues 21–24) and the carboxy-terminal residue of the protein backbone. In contrast, a triad H-bond network is formed as His26 bridges two Ω-loops (20s and 40s) via two H-bonds to Pro44 and Asn31 and is maintained even with glutamine substitution of His26 and even at 7 M GuHCl (Taler, Navon, and Becker 1998; Yamamoto 1997). With the weaker van der Waals interactions, His33 may be able to facilitate folding because it is not "anchored" by H-bonding. In H26QH33N redcyt c folding, the "stiffness" of His26 induced by the H-bonds may inhibit its ability to form a productive heme-His26 loop to compensate for the absence of His33. This rigidity of the His26 residue, the relative position of His33 and His26 to the heme moiety (Kurchan, Roder, and Bowler 2005), and the presence of surrounding Lys25/Lys27 residues that lead to charge repulsion with the iron (Godbole, Hammack, and Bowler 2000) are all factors that have been proposed to contribute to the lower affinity of His26 for the heme group. The apparently major role of His33 in redcyt c folding and the significant rearrangement of His33 in a membranelike versus aqueous environment, in contrast to the role played by and the movement of His26, suggest more conformational flexibility for His33.

With the current trend toward developing a cellular understanding of protein folding and function, one question of interest in our understanding of redcyt c folding centers on the significance of nonnative conformational states. In this regard, it is interesting to note that it is the reduced state of the heme that is inserted into the protein in the cell (Nicholson and Neupert 1989). In this context, could there be biological significance for nonnative protein tertiary structural states, such as the His18-Fe-His33 heme ligation observed in redcyt c under denaturing conditions? In studies that tailored the experimental conditions to mimic the environment of the mitochondrial membrane, redcyt c exhibited a substantial decrease in the solvent-accessible surface area, as well as a significant conformational difference at His33 (Sivakolundu and Mabrouk 2003). Physiologically, conformational adjustments in redcyt c may help to preserve the specificity of the electron-transfer pathways in cell respiration and to assist cyt c in crossing the mitochondrial membrane into the cytosol in apoptosis. In fact, conformational rearrangements away from the native protein structure can occur on the cellular level, as an increasing number of studies has implicated partially unfolded proteins in cellular functions such as membrane translocation, ligand binding, and signal transduction (see references in Chen, Van Vranken, and Kliger 2008).

Folding Intermediates. For *in vitro* protein folding, His33 was found to have considerable influence on the burst phase dynamics of redcyt c secondary structure formation. However, His33 is also expected to be the dominant His ligand involved in an equilibrium MG-state model of oxcyt c that is formed upon interaction with sodium dodecyl sulfate (SDS) monomers (Das, Mazumdar, and Mitra 1998). The MG state is an important partly unfolded species characterized by near-native secondary structure and fluctuating tertiary contacts. On the cellular level, the MG intermediate has been associated with the chaperone machinery and with human disease (see references in Chen, Van Vranken, and Kliger 2008). In protein folding, the argument of whether the MG state is, first, a true intermediate and second, an obligatory intermediate remains controversial. For redcyt c, the point of contention is whether folding follows a simple two-state mechanism or whether it proceeds through an MG-like intermediate as proposed by the TRCD/TRORD studies (U → MG → N). The adventitious availability of the electron-transfer trigger and the SDS-induced MG model allowed investigation of redcyt c MG at reaction times as fast as hundreds of nanoseconds. This was an unusual opportunity since it is often complicated to make direct comparisons of equilibrium and kinetic intermediates because they are usually not observed under the same experimental conditions. In addition to investigating the dynamics of the interconversion between partly unfolded states, in particular MG, and the more rigid native conformations, this experiment also offered the opportunity to probe the still controversial mechanistic connection between MG folding (MG → N) in more physiological conditions and U → N folding in a harsher, denaturant (GuHCl) environment.

Upon photoreduction of the largely His18-Fe-His ligated oxcyt c MG (0.65 mM SDS), the prompt redcyt c photoproduct, whose protein conformation is presumably unrelaxed from that of the initial oxidized species, folds to the His18-Fe(II)-Met coordinated native state with only a single, slow phase ($\tau \approx 50$ ms, Figure 3.5) (Chen, Van Vranken, and Kliger 2008). This time constant was extrapolated to ~1 ms in zero denaturant by using data from a second concentration of SDS (0.5 mM). A parallel extrapolation of the GuHCl data gave a time constant of ~5.5 ms in zero denaturant for the slow phase (MG → N) of the U → MG → N reaction. The ratio of these observed time constants is consistent with the hypothesis that they arose from the same underlying MG → N rate constant when the 20–30% fractional content of MG intermediate that forms in the fast phase (U → MG) of U → MG → N folding in GuHCl is taken into account. This result supports the hypothesis that the

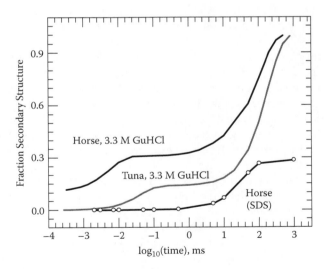

FIGURE 3.5 Comparison of far-UV TRORD kinetic results for U → I → N folding for horse and tuna redcyt c in 3.3 M GuHCl with those for MG → N for horse redcyt c in SDS. These data suggest that I is indeed MG, as proposed by the redcyt c folding studies in GuHCl.

intermediate folding species observed in the photoreduction studies of fully GuHCl-unfolded redcyt c is indeed a productive MG intermediate.

That redcyt c folding deviates from a simple two-state mechanism is additionally supported by the results of equilibrium and kinetic analyses of the acid-induced oxcyt c MG (A state) (Colon and Roder 1996). Stopped-flow fluorescence studies of A-state formation suggested that the U → A reaction proceeds through structural intermediates, similar to those found under native refolding conditions, including a burst-phase compact state and a subsequent intermediate with tertiary contacts between the N- and C-terminal helices. The folding kinetics of a metastable MG state, formed by rapid dilution of GuHCl-unfolded CO-bound redcyt c (M-CO), was probed with absorption methods (Pabit, Roder, and Hagen 2004). The time constants for photolysis-triggered M-CO folding in the presence of viscous cosolutes approached a value of 10 µs at the slowest solvent viscosities. A more detailed study of the CO dissociation and binding kinetics in M-CO under partially and fully denaturing conditions suggested that the M state is structurally similar to M-CO, but higher in energy (Latypov et al. 2008). An on-pathway intermediate assignment for this M species was based on the rollover observed in the chevron plot of the redcyt c unfolding rates. A similar rollover was previously observed in both the folding and unfolding rates of redcyt c measured with stopped-flow fluorescence spectroscopy (Bhuyan and Kumar 2002). The curvature was attributed to denaturant viscosity in the framework of a two-state folding mechanism, even though correction for the effect did not entirely eliminate the nonlinearity in the two limbs of the chevron plot. Support for a two-state model of MG folding was inferred from the observation that the A state of oxcyt c folded within the 3 ms dead time of stopped-flow pH-jump experiments (Sosnick, Mayne, and Englander 1996), too quickly for an MG → N intermediate folding step to be consistent with the ~10 ms barrier observed in U → N folding. However, the partial population of a fast-folding intermediate in rapid equilibrium with U could also account for this observation.

The ongoing nature of the on-pathway/off-pathway argument regarding the kinetic MG intermediate can probably be explained in part by the need for researchers to compare and draw connections between the sometimes daunting variety of folding studies that have been carried out on cytochrome c. For example, a comparison of unfolding free energies for the M state (from the M-CO photolysis studies) that was suggested to be an obligatory unfolding kinetic intermediate, the low-energy foldon state that was posited from the hydrogen exchange evidence to be a folding/unfolding kinetic intermediate with vanishingly small accumulation during folding, and the equilibrium M state detected by heme band CD and MCD spectroscopy, indicates that they may all be essentially the same state (for details see Goldbeck, Chen, and Kliger 2009). Furthermore, in a rough comparison of the $m^{\ddagger}$ and $m^{\ddagger}/m$ values that were calculated to compare the position of the intermediate along the reaction pathway for SDS- and GuHCl-induced redcyt c folding observed with TRORD methods, the values of $m^{\ddagger}/m$ are consistent within the uncertainties of the two measurements, as well as roughly consistent with the values reported by TRFL and TROD studies (Mines et al. 1996; Pascher 2001). Although the debate between on-pathway and off-pathway folding intermediates continues, two lines of evidence from the TRORD studies (the $m^{\ddagger}/m$ value comparison and the coincidence of the signal fraction formed

The Mechanism of Cytochrome c Folding

in the fast phase of redcyt c folding and the ratio of the observed U → N and MG → N rates) suggests that the MG species is a productive intermediate.

Implications for the Redcyt c Folding Mechanism. In placing the results of TRCD/TRORD/TRMCD redcyt c folding studies in the context of the rich array of experiments that have used complementary probes of cyt c folding (Figure 3.6), it is necessary to take into account the wide variety of experimental conditions encountered, including different denaturing conditions, heme oxidation and coordination states, and limitations on triggering and detection time scales. The observed folding dynamics can extend from the hundreds of nanoseconds timescale for initial helix formation (Chen et al. 1998), through the millisecond acquisition of essentially native secondary and tertiary structure, to the final isomerization of proline bonds on the timescale of seconds (Brems and Stellwagen 1983). The folding kinetics of cyt c over this approximately 8 orders of magnitude in time can depend qualitatively (number of intermediate states) and quantitatively (rate constants) on the ligation state of the heme and on its oxidation state, as well as on the denaturant concentration, pH, and temperature. Such differences have contributed to the conflicting (heterogeneous versus homogeneous, two-state versus multistate) descriptions of the redcyt c folding mechanism.

However, with increasing studies the mechanism of redcyt c folding is increasingly being recognized as deviating from a two-state process. A rapidly forming intermediate, attributed to the appearance of an MG-like state, was suggested by these TRCD/TRORD studies to be kinetically isolated from the bulk unfolded protein. This evidence of kinetic heterogeneity implied a finite rate of conformational diffusion in the unfolded protein. To probe the speed with which unfolded proteins exchange between conformational subensembles offering either a Met or His residue to the heme iron for facile binding, TRMCD methods were applied to

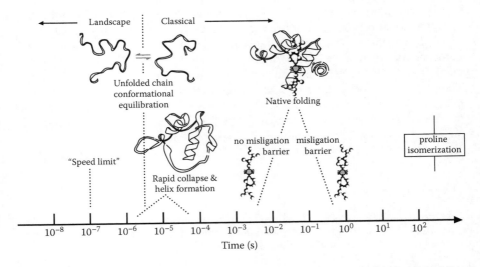

FIGURE 3.6 A summary of the timescales of kinetic processes identified during the folding of cytochrome c.

CO-photolysis-triggered folding in redcyt c proteins from tuna and horse wt and the latter's H33N, H26Q, and H33N/H26Q mutants. The τ_d of ~3 μs provided an explanation for some of the discrepancies observed when comparing various studies of cyt c folding. That is, a classical pathway is indeed appropriate for folding events observed at times much slower than a microsecond when equilibration between unfolded subensembles is achieved. In contrast, the landscape picture more accurately describes folding processes faster than a microsecond. This result may also reconcile the differences in the mechanisms of secondary-structure folding reported for the two oxidation states of cyt c. According to stopped-flow CD (~4 ms dead time) and continuous-flow CD studies (~0.5 ms dead time) of oxcyt c, folding follows a sequential mechanism. Considering that the earliest observation time for either of the two experiments is long past τ_d, a homogeneous model for secondary structure folding accurately describes folding for both proteins on that timescale.

FUTURE STUDIES

Are there differences in the earliest events of secondary structure formation in oxcyt c versus redcyt c? A factor that tends to limit folding studies to either the oxidized or reduced state of the protein is the trigger method. For example, the photoreduction trigger allows only the study of redcyt c folding, although a similar method can, in principle, be implemented with caged oxygen compounds. The CO-photolysis method also has been applied to studies of only the reduced protein. The T-jump TRORD method (Chen et al. 2005) offers the ability to study the secondary-structure dynamics in both oxcyt c and redcyt c because that trigger relies on the thermally induced perturbation of the folding/unfolding equilibrium rather than on the folding free-energy dependence on the oxidation state.

The data obtained from TRORD/TRMCD experiments may compel future studies along an additional avenue. Understanding redcyt c folding is significant because apocytochrome c becomes the holoprotein only after covalent attachment of the reduced state of the heme, which is coupled to the transport of cyt c across the outer mitochondrial membrane (Nicholson and Neupert 1989). *In vivo* then, the attachment of the reduced heme can be considered the trigger for folding of redcyt c. Thus for a better understanding of the applicability of the current *in vitro* kinetic picture of redcyt c folding to those in the cell, we will probe redcyt c folding in the presence of macromolecules such as Ficoll and Dextran in order to mimic the molecular crowding encountered by the protein as it folds in the complex environment of the cell.

REFERENCES

Abbruzzetti, S., C. Viappiani, J. R. Small, L. J. Libertini, and E. W. Small. 2001. Kinetics of histidine deligation from the heme in GuHCl-unfolded Fe(III) cytochrome c studied by a laser-induced pH-jump technique. *J Am Chem Soc* 123 (27):6649–53.

Abel, C. J., R. A. Goldbeck, R. F. Latypov, H. Roder, and D. S. Kliger. 2007. Conformational equilibration time of unfolded protein chains and the folding speed limit. *Biochemistry* 46 (13):4090–99.

Akiyama, S., S. Takahashi, K. Ishimori, and I. Morishima. 2000. Stepwise formation of alpha-helices during cytochrome *c* folding. *Nat Struct Biol* 7 (6):514–20.

Balakrishnan, G., C. L. Weeks, M. Ibrahim, A. V. Soldatova, and T. G. Spiro. 2008. Protein dynamics from time resolved UV Raman spectroscopy. *Curr Opin Struct Biol* 18 (5):623–29.

Bhuyan, A. K., and R. Kumar. 2002. Kinetic barriers to the folding of horse cytochrome *c* in the reduced state. *Biochemistry* 41 (42):12821–34.

Brems, D. N., and E. Stellwagen. 1983. Manipulation of the observed kinetic phases in the refolding of denatured ferricytochromes *c*. *J Biol Chem* 258 (6):3655–60.

Brooks, C. L., III. 1998. Simulations of protein folding and unfolding. *Curr Opin Struct Biol* 8 (2):222–26.

Callender, R., and R. B. Dyer. 2006. Advances in time-resolved approaches to characterize the dynamical nature of enzymatic catalysis. *Chem Rev* 106 (8):3031–42.

Callender, R. H., R. B. Dyer, R. Gilmanshin, and W. H. Woodruff. 1998. Fast events in protein folding: the time evolution of primary processes. *Annu Rev Phys Chem* 49:173–202.

Camacho, C. J., and D. Thirumalai. 1995. Theoretical predictions of folding pathways by using the proximity rule, with applications to bovine pancreatic trypsin inhibitor. *Proc Natl Acad Sci U S A* 92 (5):1277–81.

Chan, C. K., J. Hofrichter, and W. A. Eaton. 1996. Optical triggers of protein folding. *Science* 274 (5287):628–29.

Chan, C. K., Y. Hu, S. Takahashi, D. L. Rousseau, W. A. Eaton, and J. Hofrichter. 1997. Submillisecond protein folding kinetics studied by ultrarapid mixing. *Proc Natl Acad Sci U S A* 94 (5):1779–84.

Chang, I. J., J. C. Lee, J. R. Winkler, and H. B. Gray. 2003. The protein-folding speed limit: intrachain diffusion times set by electron-transfer rates in denatured Ru(NH3)5(His-33)-Zn-cytochrome *c*. *Proc Natl Acad Sci U S A* 100 (7):3838–40.

Chen, E., C. J. Abel, R. A. Goldbeck, and D. S. Kliger. 2007. Non-native heme-histidine ligation promotes microsecond time scale secondary structure formation in reduced horse heart cytochrome *c*. *Biochemistry* 46 (43):12463–72.

Chen, E. F., R. A. Goldbeck, and D. S. Kliger. 2003. Earliest events in protein folding: submicrosecond secondary structure formation in reduced cytochrome *c*. *J Phys Chem* A 107 (40):8149–55.

———. 2004. The earliest events in protein folding: a structural requirement for ultrafast folding in cytochrome *c*. *J Am Chem Soc* 126 (36):11175–81.

Chen, E., V. van Vranken, and D. S. Kliger. 2008. The folding kinetics of the SDS-induced molten globule form of reduced cytochrome *c*. *Biochemistry* 47 (19):5450–59.

Chen, E. F., Y. X. Wen, J. W. Lewis, R. A. Goldbeck, D. S. Kliger, and C. E. M. Strauss. 2005. Nanosecond laser temperature-jump optical rotatory dispersion: Application to early events in protein folding/unfolding. *Rev Sci Instrum* 76 (8):083120–27.

Chen, E. F., P. Wittung-Stafshede, and D. S. Kliger. 1999. Far-UV time-resolved circular dichroism detection of electron-transfer-triggered cytochrome *c* folding. *J Am Chem Soc* 121 (16):3811–17.

Chen, E., M. J. Wood, A. L. Fink, and D. S. Kliger. 1998. Time-resolved circular dichroism studies of protein folding intermediates of cytochrome *c*. *Biochemistry* 37 (16):5589–98.

Chen, E., G. A. Woolley, and D. S. Kliger. 2008. Time-resolved optical rotatory dispersion studies of azobenzene cross-linked peptides. In *Methods in protein structure and stability analysis: NMR and EPR spectroscopies, mass-spectrometry and protein imaging*, ed. V. N. Uversky and E. A. Permyakov, 235–55. Hauppauge, NY: Nova Science.

Colon, W., and H. Roder. 1996. Kinetic intermediates in the formation of the cytochrome *c* molten globule. *Nat Struct Biol* 3 (12):1019–25.

Colon, W., L. P. Wakem, F. Sherman, and H. Roder. 1997. Identification of the predominant non-native histidine ligand in unfolded cytochrome *c*. *Biochemistry* 36 (41):12535–41.

Daggett, V., and A. R. Fersht. 2003. Is there a unifying mechanism for protein folding? *Trends Biochem Sci* 28 (1):18–25.

Das, T. K., S. Mazumdar, and S. Mitra. 1998. Characterization of a partially unfolded structure of cytochrome *c* induced by sodium dodecyl sulphate and the kinetics of its refolding. *Eur J Biochem* 254 (3):662–70.

Elöve, G. A., A. F. Chaffotte, H. Roder, and M. E. Goldberg. 1992. Early steps in cytochrome *c* folding probed by time-resolved circular dichroism and fluorescence spectroscopy. *Biochemistry* 31 (30):6876–83.

Godbole, S., B. Hammack, and B. E. Bowler. 2000. Measuring denatured state energetics: deviations from random coil behavior and implications for the folding of iso-1-cytochrome *c*. *J Mol Biol* 296 (1):217–28.

Goldbeck, R. A., E. Chen, and D. S. Kliger. 2009. Early events, kinetic intermediates and the mechanism of protein folding in cytochrome *c*. *Int J Mol Sci* 10 (4):1476–99.

Goldbeck, R. A., D. B. Kim-Shapiro, and D. S. Kliger. 1997. Fast natural and magnetic circular dichroism spectroscopy. *Annu Rev Phys Chem* 48:453–79.

Goldbeck, R. A., Y. G. Thomas, E. Chen, R. M. Esquerra, and D. S. Kliger. 1999. Multiple pathways on a protein-folding energy landscape: kinetic evidence. *Proc Natl Acad Sci U S A* 96 (6):2782–87.

Gutman, M., and E. Nachliel. 1997. Time-resolved dynamics of proton transfer in proteinous systems. *Annu Rev Phys Chem* 48:329–56.

Hagen, S. J., and W. A. Eaton. 2000. Two-state expansion and collapse of a polypeptide. *J Mol Biol* 301 (4):1019–27.

Hoffman, G. W. 1971. A nanosecond temperature-jump apparatus. *Rev Sci Instrum* 42 (11):1643–47.

Jennings, P. A. 1998. Speeding along the protein folding highway, are we reading the signs correctly? *Nat Struct Biol* 5 (10):846–48.

Jones, C. M., E. R. Henry, Y. Hu, C. K. Chan, S. D. Luck, A. Bhuyan, H. Roder, J. Hofrichter, and W. A. Eaton. 1993. Fast events in protein folding initiated by nanosecond laser photolysis. *Proc Natl Acad Sci U S A* 90 (24):11860–64.

Kliger, D. S., E. Chen, and R. A. Goldbeck. 2004. Kinetic and spectroscopic analysis of early events in protein folding. *Method Enzymol* 380:308–27.

Knight, J. B., A. Vishwanath, J. P. Brody, and R. H. Austin. 1998. Hydrodynamic focusing on a silicon chip: mixing nanoliters in microseconds. *Phys Rev Lett* 80 (17):3863–66.

Krieger, F., B. Fierz, O. Bieri, M. Drewello, and T. Kiefhaber. 2003. Dynamics of unfolded polypeptide chains as model for the earliest steps in protein folding. *J Mol Biol* 332 (1):265–74.

Kubelka, J., J. Hofrichter, and W. A. Eaton. 2004. The protein folding "speed limit." *Curr Opin Struct Biol* 14 (1):76–88.

Kurchan, E., H. Roder, and B. E. Bowler. 2005. Kinetics of loop formation and breakage in the denatured state of iso-1-cytochrome *c*. *J Mol Biol* 353 (3):730–43.

Latypov, R. F., K. Maki, H. Cheng, S. D. Luck, and H. Roder. 2008. Folding mechanism of reduced cytochrome *c*: equilibrium and kinetic properties in the presence of carbon monoxide. *J Mol Biol* 383 (2):437–53.

Mines, G. A., T. Pascher, S. C. Lee, J. R. Winkler, and H. B. Gray. 1996. Cytochrome *c* folding triggered by electron transfer. *Chem Biol* 3 (6):491–97.

Nicholson, D. W., and W. Neupert. 1989. Import of cytochrome *c* into mitochondria: reduction of heme, mediated by NADH and flavin nucleotides, is obligatory for its covalent linkage to apocytochrome *c*. *Proc Natl Acad Sci U S A* 86 (12):4340–44.

Orii, Y. 1993. Immediate reduction of cytochrome *c* by photoexcited NADH: reaction mechanism as revealed by flow-flash and rapid-scan studies. *Biochemistry* 32 (44):11910–14.

Pabit, S. A., H. Roder, and S. J. Hagen. 2004. Internal friction controls the speed of protein folding from a compact configuration. *Biochemistry* 43 (39):12532–38.

Pascher, T. 2001. Temperature and driving force dependence of the folding rate of reduced horse heart cytochrome *c*. *Biochemistry* 40 (19):5812–20.

Pascher, T., J. P. Chesick, J. R. Winkler, and H. B. Gray. 1996. Protein folding triggered by electron transfer. *Science* 271 (5255):1558–60.

Phillips, C. M., Y. Mizutani, and R. M. Hochstrasser. 1995. Ultrafast thermally induced unfolding of RNase A. *Proc Natl Acad Sci U S A* 92 (16):7292–96.

Pletneva, E. V., H. B. Gray, and J. R. Winkler. 2005a. Nature of the cytochrome *c* molten globule. *J Am Chem Soc* 127 (44):15370–71.

Pletneva, E. V., H. B. Gray, and J. R. Winkler. 2005b. Snapshots of cytochrome *c* folding. *Proc Natl Acad Sci U S A* 102 (51):18397–402.

Pollack, L., M. W. Tate, N. C. Darnton, J. B. Knight, S. M. Gruner, W. A. Eaton, and R. H. Austin. 1999. Compactness of the denatured state of a fast-folding protein measured by submillisecond small-angle x-ray scattering. *Proc Natl Acad Sci U S A* 96 (18):10115–17.

Porter, G. 1950. Flash photolysis and spectroscopy: a new method for the study of free radical reactions. *P Roy Soc Lond A Mat* 200 (1061):284–300.

Qi, P. X., T. R. Sosnick, and S. W. Englander. 1998. The burst phase in ribonuclease A folding and solvent dependence of the unfolded state. *Nat Struct Biol* 5 (10):882–84.

Qu, K., J. L. Vaughn, A. Sienkiewicz, C. P. Scholes, and J. S. Fetrow. 1997. Kinetics and motional dynamics of spin-labeled yeast iso-1-cytochrome *c*: 1. Stopped-flow electron paramagnetic resonance as a probe for protein folding/unfolding of the C-terminal helix spin-labeled at cysteine 102. *Biochemistry* 36 (10):2884–97.

Regenfuss, P., R. M. Clegg, M. J. Fulwyler, F. J. Barrantes, and T. M. Jovin. 1985. Mixing liquids in microseconds. *Rev Sci Instrum* 56 (2):283–90.

Shastry, M. C., S. D. Luck, and H. Roder. 1998. A continuous-flow capillary mixing method to monitor reactions on the microsecond time scale. *Biophys J* 74 (5):2714–21.

Shastry, M. C., and H. Roder. 1998. Evidence for barrier-limited protein folding kinetics on the microsecond time scale. *Nat Struct Biol* 5 (5):385–92.

Sivakolundu, S. G., and P. A. Mabrouk. 2003. Structure-function relationship of reduced cytochrome *c* probed by complete solution structure determination in 30% acetonitrile/water solution. *J Biol Inorg Chem* 8 (5):527–39.

Sosnick, T. R., L. Mayne, and S. W. Englander. 1996. Molecular collapse: the rate-limiting step in two-state cytochrome *c* folding. *Proteins* 24 (4):413–26.

Srajer, V., and W. E. Royer Jr. 2008. Time-resolved x-ray crystallography of heme proteins. *Method Enzymol* 437:379–95.

Takahashi, S., S. R. Yeh, T. K. Das, C. K. Chan, D. S. Gottfried, and D. L. Rousseau. 1997. Folding of cytochrome *c* initiated by submillisecond mixing. *Nat Struct Biol* 4 (1):44–50.

Taler, G., G. Navon, and O. M. Becker. 1998. The interaction of borate ions with cytochrome *c* surface sites: a molecular dynamics study. *Biophys J* 75 (5):2461–68.

Telford, J. R., P. Wittung-Stafshede, H. B. Gray, and J. R. Winkler. 1998. Protein folding triggered by electron transfer. *Accounts Chem Res* 31 (11):755–63.

Winkler, J. R. 2004. Cytochrome *c* folding dynamics. *Curr Opin Chem Biol* 8 (2):169–74.

Woody, R. W. 1968. Improved calculation of n-pi* rotational strength in polypeptides. *J Chem Phys* 49 (11):4797–4806.

Yamamoto, Y. 1997. A 1H NMR study of structurally relevant inter-segmental hydrogen bond in cytochrome *c*. *Biochim Biophys Acta* 1343 (2):193–202.

Yang, W. Y., and M. Gruebele. 2003. Folding at the speed limit. *Nature* 423 (6936):193–97.

4 Stability and Folding of Copper-Binding Proteins

Irina Pozdnyakova and Pernilla Wittung-Stafshede

CONTENTS

Distribution of Copper in Living Cells 61
Experimental Approaches to Study Metalloprotein Folding *In Vitro* 63
Blue-Copper Protein as Far-Reaching Model System 65
Proteins Facilitating Cytoplasmic Copper Transfer 68
Multicopper Oxidases: Complex Protein/Metal Systems 71
Studies on Other Copper-Binding Proteins 74
Future Perspective 76
References 76

DISTRIBUTION OF COPPER IN LIVING CELLS

In this chapter we focus on the roles of the cofactors in folding and stability of copper-binding proteins. Copper (Cu) is essential for the survival of living organisms. It is found in the active sites of proteins that participate in cellular respiration, antioxidant defense, neurotransmitter biosynthesis, connective-tissue biosynthesis, and pigment formation (Huffman and O'Halloran 2001; Puig and Thiele 2002; Harris 2003). In Table 4.1 we list the major types of proteins that bind Cu, along with their functional role; we also give some examples of their structures (Figure 4.1). After iron, Cu is the second most common element in biological systems to participate in electron-transfer chains. The thermodynamic and kinetic feasibility of Cu to oxidize/reduce (switching between Cu^{1+} and Cu^{2+}) allows copper-containing proteins to play important roles as electron carriers and redox catalysts in living systems. How and when is the metal inserted into these proteins in the cell? Since non-bonded Cu ions are toxic due to their ability to catalyze the formation of harmful free radicals, the intracellular concentration of Cu is highly regulated (O'Halloran and Culotta 2000). Both prokaryotic and eukaryotic cells regulate Cu homeostasis via dedicated proteins that facilitate its uptake, efflux, as well as distribution to target proteins/enzymes.

In humans, Cu (in the form of Cu^{1+}) is translocated from the plasma to the cytoplasm by high-affinity membrane-spanning proteins of the Ctr1 family (Harris 2003). Cu distribution in the eukaryotic cytoplasm is mediated by several independent Cu chaperones. First, the CCS copper chaperone inserts Cu into the active site of Cu/Zn superoxide dismutase (SOD1) by direct docking with a SOD1 monomer. Second, the COX17 chaperone targets Cu to mitochondrial cytochrome c oxidase (COX), SCOI,

TABLE 4.1
Examples of Copper-Binding Proteins Classified According to Their Cellular Function

Proteins	Function
Electron and Oxygen Management	
Blue-copper proteins	Electron transfer
Hemocyanin	Oxygen management
Enzymes	
Feroxidases (e.g., Ceruloplasmin/Fet3p)	Iron management
Tyrosinase	Production of pigment
Lysyl oxidase	Cross-linking of collagen
Cytochrome c oxidase	Respiration
Superoxide dismutase	Antioxidant defense
Copper Uptake and Delivery	
Cu transporter (Ctr1)	Cu import
P-type ATPases (e.g., ATP7A, ATP7B)	Cu secretion and delivery to secretory pathways
Cu chaperones (e.g., Atox1, Cox17)	Cytoplasmic transport of Cu to targets

Note: We have divided the proteins into three groups: those involved in electron transport and oxygen management, those that are enzymes (catalyzing reactions), and those that facilitate delivery of copper to polypeptides listed in the first two groups (strictly speaking, the last group may not be classified as metalloproteins as they cycle between apo- and holo-forms as part of their function).

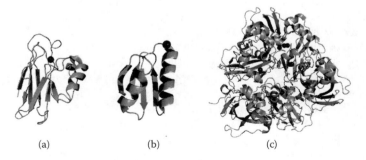

FIGURE 4.1 Structural models of three copper-binding proteins that are discussed in this chapter (i.e., azurin, Atox1, and ceruloplasmin) with the copper ions depicted as black spheres.

and SCOII proteins. Third, in the general path, the human copper chaperone Atox1 mediates Cu delivery to proteins that are made in the secretory pathway. Atox1 interacts directly with the cytosolic metal-binding domains of the copper-transporting P_{1B}-type ATPases ATP7A (i.e., Menkes disease or MNK) and ATP7B (i.e., Wilson disease or WND) proteins localized in the trans-Golgi network. MNK and WND are highly homologous (57% sequence identical) but they are most often expressed in different cell types (Lutsenko, LeShane, and Shinde 2007). Once transferred to MNK/WND, Cu is translocated to the lumen of the Golgi, with ATP hydrolysis, and loaded onto target proteins. We refer the reader to Chapter 8 for more detailed information on WND. Many human copper-dependent enzymes (e.g., blood-clotting factors, tyrosinase, lysyl oxidase, and ceruloplasmin) traverse the secretory pathway and acquire Cu, from either WND or MNK, in the Golgi before reaching their final destination.

In contrast to eukaryotes, prokaryotes lack intracellular compartments; thus organelle-specific carriers of metals may not be essential. However, a homologue of Atox1 (i.e., CopZ) has been described for enteric bacteria (e.g., *Enterococcus hirae*) (O'Halloran and Culotta 2000). In *E. hirae,* which is the best understood prokaryotic Cu homeostasis system, Cu uptake, availability, and export is regulated through a *cop* operon (Solioz and Stoyanov 2003). The *cop* operon is composed of four structural genes that encode for a copper-responsive repressor CopY, the Cu chaperone CopZ, and two Cu ATPases, CopA and CopB. It is quite feasible that other bacteria, including *P. aeruginosa* (from which one of our model proteins comes), may utilize a similar copper regulation system. In fact, structural genomics work has revealed the presence of Atox1 homologues in almost all sequenced genomes (Arnesano et al. 2002).

EXPERIMENTAL APPROACHES TO STUDY METALLOPROTEIN FOLDING *IN VITRO*

To function, most proteins must fold to unique three-dimensional structures. This process occurs spontaneously for most polypeptides once made in the cell. How can we study these reactions? Most *in vitro* studies of protein-folding reactions (with and without metals) involve equilibrium and kinetic experiments with purified proteins using chemical denaturants, such as urea and guanidine hydrochloride (GuHCl), to perturb the structure (most often at pH 7, 20°C) (Maxwell et al. 2005). The conformational state of the protein can be probed by a combination of spectroscopic methods, each method reporting on a different aspect of the protein structure. For example, visible absorption may report on the metal environment, tryptophan fluorescence reports on local environment and solvent exposure of tryptophan residues, and far-UV circular dichroism (CD) reports on the amount of polypeptide secondary structure. To probe un/refolding kinetics, stopped-flow mixing allows for rapid denaturant-concentration jumps, and the triggered reaction can be followed by spectroscopic signal changes. The logarithms of the observed rate constants (k_{obs}; folding-rate constants at low denaturant concentrations; unfolding-rate constants at high denaturant concentrations) are often plotted as a function of denaturant concentration in a so-called chevron plot (Fersht 1999) (Figure 4.2a).

It has been found that small, single-domain proteins (<100 residues), lacking disulfide bonds and proline residues, often fold by two-state equilibrium and kinetic

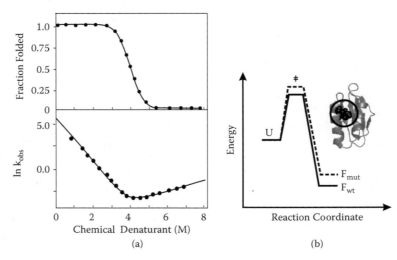

FIGURE 4.2 (a) Schematic plots demonstrating (top) a sigmoid two-state equilibrium unfolding curve (fraction folded on y axis, chemical denaturant on x axis) and (bottom) a semilogarithmic graph, so-called chevron plot, of observed rate constants as a function of chemical denaturant for a two-state kinetic reaction. (b) Energy diagram for a two-state folding protein illustrating φ-value analysis. Solid line corresponds to wild-type (wt) protein; dashed line corresponds to a point mutant with a hydrophobic side chain replaced by Ala (see example in inset; circled side chain removed in the variant). The differences in the transition-state barrier (‡) and the native-state equilibrium stability (difference in energy, $F_{mut} - F_{wt}$) are combined to give the φ-value for the mutated residue. In this example, φ = 1 and the removed side chain makes nativelike interactions in ‡.

mechanisms (Jackson 1998). For such proteins, three states are important for defining the kinetic behavior: the native, transition, and denatured states. The transition state consists of a set of structures (i.e., the transition-state ensemble) whose formation is rate limiting for folding of the native state from the denatured ensemble. The speed of folding for proteins obeying two-state behavior can vary up to a millionfold *in vitro*. One of the fastest-folding proteins is the heme-binding protein cytochrome b_{562}, which folds in only a few μs (Wittung-Stafshede et al. 1999). As the transition state never accumulates, its structural characteristics must be inferred indirectly. The protein-engineering approach has proved to be the most important experimental strategy for obtaining specific information about interactions between residues in the transition state (Matouschek et al. 1990; Matouschek and Fersht 1991). In such experiments structure in the transition state is probed by measuring the kinetic and thermodynamic effects of point mutations (most often hydrophobic-to-alanine substitutions) in different regions of the protein. The results are typically presented as φ-values (de los Rios et al. 2006), which represent the change in stability of the transition state (i.e., effect on folding rate) accompanying the mutation of a residue relative to the effect of the same mutation in the native state (i.e., effect on equilibrium stability) (Figure 4.2b). Therefore φ equal to 1 suggests that the mutated residue is found in a highly structured region of the transition state whereas φ equal to 0 indicates that the altered residue forms very few interactions with other residues in the transition state. Fractional φ-values may be interpreted as possessing different degrees of structure in the folding nucleus (Fersht and Sato 2004; Sanchez and Kiefhaber 2003).

Proteins longer than 100 residues often populate intermediate conformations, either transiently or at equilibrium, and the mechanisms deviate from two-state (Roder and Colon 1997; Matthews 1993). Folding intermediates may be on-pathway (i.e., facilitating the reaction) or off-pathway (i.e., acting as traps that may lead to aggregation) or multiple pathways may be present in which intermediates coexist (Privalov 1996).

Metal-binding proteins fold in a cellular environment where their cognate cofactors are present free in the cytoplasm or bound to delivery proteins. Thus the question arises as to when metals bind (or are delivered) to their corresponding proteins: Specifically, do they coordinate before, during, or after polypeptide folding? It has been demonstrated *in vitro* that many metalloproteins (e.g., cytochrome b_{562}, myoglobin, azurin, and the Cu_A domain) retain strong metal-protein interactions after polypeptide unfolding. This implies that *in vivo* metals may interact with their corresponding proteins before polypeptide folding takes place, which could then impact the folding reaction. Structure in the unfolded protein may form due to specific coordination of the cofactor, and this may decrease the entropy of the unfolded state, limiting the conformational search for the native state. Testing this hypothesis is a major goal in many of our studies. It has been proposed that the ion distribution of the cellular compartment in which a protein folds may control metal specificity in proteins with binding sites suitable for more than one kind of metal (Tottey et al. 2008).

Our approach to reveal folding/binding mechanisms of copper-binding proteins is based on a range of *in vitro* biophysical experiments using strategic proteins (with different Cu sites and varying overall folds) and variants thereof (to probe specific residues). More recently we have used a combination of experimental studies and computational efforts such as molecular dynamics (MD) and quantum-mechanics molecular-mechanics (QM-MM) methods. (Although we describe some of our computational work in this chapter, we direct readers to Chapter 13 for cutting-edge theoretical work on metalloprotein folding.) In the following four sections, we describe our findings of the last 10 years (and work by others as appropriate) on copper-dependent enzymes as well as proteins that facilitate cellular Cu transport. In contrast to the enzymes, which require bound Cu for function, the transport proteins need to constantly cycle between apo- and holo-forms as part of their activity. It is feasible that this difference in how the metal is exploited for function correlates with distinct biophysical features of the two groups of proteins.

BLUE-COPPER PROTEIN AS FAR-REACHING MODEL SYSTEM

Pseudomonas aeruginosa azurin was the first copper-binding protein that we subjected to biophysical folding/binding studies. It is a small (128 residues) blue-copper protein that is believed to facilitate electron transfer in de-nitrification and/or respiration chains (Adman 1991). In recent years, it was proposed that the physiological function of azurin in *P. aeruginosa* involves electron transfer related to the cellular response to oxidative stress (Vijgenboom, Busch, and Canters 1997). Azurin has one α-helix and eight β-strands that fold into a β-barrel structure arranged in a double-wound Greek-key topology (Adman 1991; Nar et al. 1992) (Figure 4.1b).

In *P. aeruginosa* azurin, the redox-active copper (Cu^{1+}/Cu^{2+}) is coordinated in a trigonal bipyramidal geometry by two histidine imidazoles (His46 and His117), one cysteine thiolate (Cys112), and two weaker axial ligands, sulfur of methionine (Met121) and carbonyl of glycine (Gly45). The polypeptide fold appears to define the exact geometry of the metal site, leading to an unusual Cu^{2+} coordination in azurin as well as in other blue-copper proteins (Wittung-Stafshede, Hill et al. 1998). *In vitro, P. aeruginosa* azurin can bind many different metals in the active site (e.g., zinc). Crystal structures have shown that the overall three-dimensional structure of azurin is identical with and without a metal (Cu or zinc) cofactor (Nar et al. 1991, 1992). *P. aeruginosa* azurin has been found to interact with the tumor-suppressor gene product p53 and act as an anticancer agent in cell-culture studies (Yamada et al. 2002; Apiyo and Wittung-Stafshede 2005). Thus not only is *P. aeruginosa* azurin an excellent model system, being a small single-domain protein with a common fold and one Cu ion, it is also a putative cancer-drug candidate.

Early equilibrium-unfolding experiments of oxidized (Cu^{2+}) and reduced (Cu^{1+}) azurin showed that the oxidized form is more stable than the reduced form. The unfolding reactions were reversible and exhibited no protein-concentration dependence. This implied that the metal remained bound to the unfolded state since, otherwise, higher protein concentrations should have resulted in unfolding curves shifted to higher GuHCl concentrations (Leckner, Wittung et al. 1997). Detection with different spectroscopic probes yielded identical unfolding curves, in accord with two-state equilibrium-unfolding transitions. Based on the difference in thermodynamic stability for the two redox forms, we predicted that the Cu in unfolded azurin has a reduction potential 0.13 V higher than that in folded azurin (Leckner, Wittung et al. 1997); subsequent cyclic-voltammetry experiments confirmed this prediction (20°C, pH 7) (Wittung-Stafshede, Hill et al. 1998). The high Cu reduction potential in the unfolded state was rationalized by a trigonal metal coordination that favored the Cu^{1+} form. This observation also implied that Cys112 remained a Cu ligand in the unfolded-state complex (Wittung-Stafshede, Hill et al. 1998). Equilibrium-unfolding studies of zinc-loaded azurin revealed that zinc, like Cu, stayed bound to the polypeptide upon unfolding (Leckner, Bonander et al. 1997; Marks et al. 2004).

Extended X-ray absorption fine structure (EXAFS) experiments in high GuHCl concentrations established that Cu in unfolded azurin is coordinated in a trigonal geometry to one thiolate (i.e., Cys112), one histidine imidazole, and a third unknown ligand (DeBeer et al. 2000). Using various single-site mutants, we then elucidated that His117 and Met121 are the second and third Cu ligands in unfolded azurin (Pozdnyakova, Guidry, and Wittung-Stafshede 2001b; Marks et al. 2004). Complementary evidence for the Cu ligands in unfolded azurin came from model-peptide studies using a 13-residue peptide corresponding to residues 111–123 (Pozdnyakova, Guidry, and Wittung-Stafshede 2000). Upon Cu^{2+} binding to this peptide, β-like secondary structure formed and we proposed that this may be the nucleation site for folding of the full-length protein. Since Cu and zinc stays bound upon unfolding, the net effect on azurin's stability as compared to the stability of apo-azurin corresponds to the difference in metal affinity for the folded and unfolded polypeptide. Thermodynamic cycles demonstrated that the metals greatly stabilize native azurin; zinc-, Cu^{1+}-, and Cu^{2+}-forms of azurin have thermodynamic stabilities of 39, 40, and 52 kJ/mol, respectively, whereas the stability of apo-azurin is 29 kJ/mol (pH 7, 20°C) (Leckner, Wittung et al. 1997; Pozdnyakova, Guidry, and Wittung-Stafshede 2002; Pozdnyakova and Wittung-Stafshede 2003).

To address possible pathways for formation of *active* azurin (i.e., folded protein with Cu in the active site) *in vivo*, we investigated the timescales for the two extreme scenarios (Figure 4.3a), Cu binding before polypeptide folding (Pathway 1) and copper binding after polypeptide folding (Pathway 2). The folding and unfolding kinetics for apo-azurin follows two-state behavior (Pozdnyakova, Guidry, and Wittung-Stafshede 2001a, 2002; Pozdnyakova and Wittung-Stafshede 2001a; Pozdnyakova and Wittung-Stafshede 2001b; Pozdnyakova and Wittung-Stafshede 2003). The extrapolated folding time in water ($\tau \sim 7$ ms) is fast whereas Cu uptake by folded apo-azurin, which results in active protein through Pathway 2, is slow (i.e., $\tau \sim$ min-to-hours depending on protein-to-copper excess) (Pozdnyakova and Wittung-Stafshede 2001a, 2001b). In contrast, we found the formation of active (holo) azurin to be much faster when the polypeptide folds in the presence of Cu (Pozdnyakova and Wittung-Stafshede 2001a, 2001b). We concluded that in these latter experiments, active azurin formation follows Pathway 1, with rapid Cu uptake before polypeptide folding. Thus introducing Cu prior to protein folding results in more than 1000-fold faster formation of *active* azurin (i.e., the folded holo-form) (Pozdnyakova and Wittung-Stafshede 2001a, 2001b). Thus if time of biosynthesis should be minimized in the cell, Cu coordination before folding may be biologically relevant. In parenthesis, we found that the polypeptide folding speeds for apo-, copper- and zinc-loaded forms of azurin were all rather similar. Based on this, as well as reported folding speeds for other β-barrel proteins, we proposed that there is a speed limit for β-barrel formation on the order of a few ms (Pozdnyakova and Wittung-Stafshede 2003).

To compare the folding-transition states of azurin without and with metal, we turned to φ-value analysis. Here we used zinc instead of Cu as the holo-form; as zinc is inert, unwanted Cu redox reactions are avoided, which otherwise hamper the measurements. Our initial focus was on eight core residues found by others to be so-called "structural determinants" in 94% of all sandwichlike proteins, including blue-copper proteins like azurin (Kister, Finkelstein, and Gelfand 2002). Interestingly, in the apo-form of azurin, we found that half of the structural determinant residues were important for folding mechanism (i.e., high φ-values) whereas the other half instead contributed largely to native-state stability (Wilson and Wittung-Stafshede 2005a; Zong et al. 2006). This finding indicated that residues may be conserved for mechanistic reasons, a topic of some controversy. To get a more complete picture of apo-azurin's transition state structure, we extended the φ–value analysis to 18 positions covering all secondary structure elements. Based on our data, apo-azurin's folding nucleus appears highly polarized to a few core residues (e.g., Val31, Leu33 and Leu50) that have φ-values of 1 (Wilson and Wittung-Stafshede 2005a; Chen et al. 2006). In contrast to apo-azurin, the semilogarithmic plot of zinc-substituted azurin folding and unfolding rate constants versus denaturant concentration exhibits pronounced curvature in both folding and unfolding arms (Pozdnyakova and Wittung-Stafshede 2003). In general, such behavior can be caused by transient aggregation, burst-phase intermediates, or movement of the transition-state placement. In the case of zinc-substituted azurin, we assigned the curvature to movement of the transition

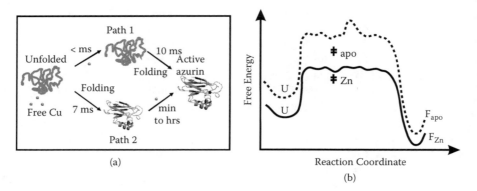

FIGURE 4.3 (a) The two extreme paths to active holoazurin: first polypeptide folding, then copper binding to the folded apo-protein (path 2), and copper binding to the unfolded protein followed by polypeptide folding (path 1). Timescales for the different steps are indicated. (b) Proposed energy profiles for folding of apo (top) and holo (bottom) azurin that is in agreement with a fixed transition state for apo (elevated feature is due to water removal) but a moving, diffuse transition state for the zinc-form (broad energy barrier that includes water molecules).

state as a function of denaturant concentration. At low concentrations of denaturant the transition state occurs early in the folding reaction, whereas at high denaturant concentration it moves closer to the native structure. Using φ-value analysis on the data for the zinc-substituted variants, we could therefore probe the growth of the transition-state structure with residue-specific resolution. We found delocalized interactions taking place throughout the protein that gradually grow more native-like around a few residues centered in the core when the transition state becomes more and more organized (Wilson and Wittung-Stafshede 2005b).

The dramatic difference in apparent kinetic behavior for the two forms of azurin (fixed, polarized transition state for apo-form; moving, diffuse transition state for zinc-form) was rationalized as a minor alteration on a *common* free energy profile that exhibits a broad activation barrier (Figure 4.3b). We speculated that the fixed transition state for apo-azurin may be a result of a small pointed feature projecting from the top of an otherwise broad free-energy profile. The presence of zinc suppresses this local bump, and the activation barrier becomes broad. The reason for the energetic bump in the apo-protein's energy landscape was later explained by complete water exclusion in the apo-protein folding-transition state, whereas zinc-substituted azurin's folding-transition state contained weakly bound water molecules (Wilson, Apiyo, and Wittung-Stafshede 2006).

PROTEINS FACILITATING CYTOPLASMIC COPPER TRANSFER

The reoccurring question of how Cu is presented to azurin in the cell prompted us to study proteins involved in Cu transport in bacteria as well as humans. The human Cu chaperone Atox1 is a 68-residue protein that, like the cytosolic metal-binding domains in WND/MNK, has a ferredoxinlike fold and a single MxCxxC copper-binding

Stability and Folding of Copper-Binding Proteins

motif (Huffman and O'Halloran 2001; Arnesano et al. 2002). Structural work has demonstrated that copper chaperones and target metal-binding domains from many different organisms possess the same ferredoxinlike fold and copper-binding motif (Arnesano et al. 2002). Atox1 binds Cu^{1+} via the two conserved Cys positioned in a surface-exposed loop. A transfer mechanism where Cu moves from Atox1 to one of the six metal-binding domains in WND/MNK through a transient copper-bridged hetero-dimer has been proposed (Banci et al. 2008). Why MNK and WND have six similar metal-binding domains (like pearls on a string) is not clear: It appears that some are involved in interactions with Atox1 while others may play regulatory roles. A scheme of WND's multidomain structure is shown in Figure 4.4a.

Cycling between apo- and holo-forms as part of activity may put constraints on how tight Cu could bind and require built-in structural dynamics. To explore these questions, we first compared *in vitro* unfolding and stability of Atox1 with the same properties of the homologue *Bacillus subtilis* CopZ as a function of Cu (Hussain and Wittung-Stafshede 2007). We discovered that unfolding of both apo- and Cu^{1+}-forms of CopZ and Atox1, induced by the chemical denaturant GuHCl and by thermal perturbation, are reversible two-state reactions. Despite the same overall structure,

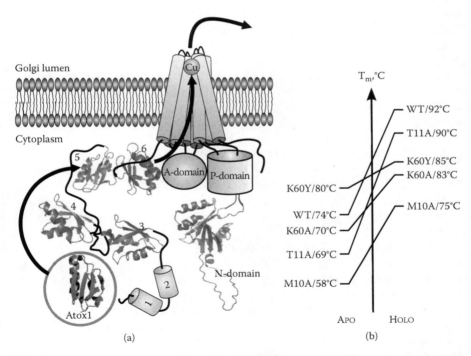

FIGURE 4.4 (a) Schematic diagram of WND (ATP7B) with all its domains. For domains that have been structurally determined (i.e., metal-binding domains 3–6, ATP-binding N-subdomain), the reported structure is shown. The structure of Atox1 is also shown (circled). Arrows show possible flow of copper from Atox1 toward the Golgi. (b) Temperature axis with thermal midpoints (pH 7 *in vitro*) indicated for various Atox1 variants in apo- (left) and holo- (right) forms.

the apo-form of CopZ has more than 20°C lower thermal stability as compared to the apo-form of Atox1. Variations in surface charges may explain this difference between the proteins on a molecular level (Arnesano et al. 2002). From a biological standpoint, the presence of competing Cu chaperones in mammalian cells may require a more stable apo-form in eukaryotes as compared to in prokaryotes.

Although the copper site is located in a loop at the protein surface, Cu binding favorably affects both thermal and chemical stability of the two proteins. This suggests that copper binding induces rearrangements throughout the protein structure that improves overall stability. Based on the absence of protein-concentration dependence and prompt reversibility of the unfolding reactions, we speculate that Cu remains coordinated to the two Cys after protein unfolding (Hussain and Wittung-Stafshede 2007). This appears logical since the two Cys side chains coordinating the Cu are close in amino acid sequence.

To assess the molecular origin of the stability differences between Atox1 and CopZ, we used a combination of QM-MM and MD computational methods (Rodriguez-Granillo and Wittung-Stafshede 2008). Simulations are a useful complement to *in vitro* data that may give insights not accessible by means of experimental techniques. We discovered that the QM-MM-optimized Cu^{1+} geometries differ between the two holo-proteins despite the identical nature of the Cu ligands. Cu^{1+} in Atox1 favors a linear Cys(S)-Cu-Cys(S) arrangement whereas this angle is close to 150 degrees in CopZ. Both proteins become less dynamic in the presence of Cu: Cu binding to CopZ has the largest effect, which thus matches our *in vitro* stability data. Both average fluctuation (i.e., root mean square deviation or RMSD) and radius of gyration (i.e., compaction) data demonstrate that the effects of Cu coordination extend throughout the proteins. Albeit distinct deviations between the two homologues are found in protein-solvent interactions, we discovered for both a biphasic distribution of Cys(S)-Cys(S) distances in the apo-forms that are separated by 0.5–1 kcal/mol barriers. We propose that the conformations with long Cys(S)-Cys(S) distances play key roles in Cu uptake and release in this group of proteins (Rodriguez-Granillo and Wittung-Stafshede 2008). Taken together, differential stability of apo- and holo-forms seems linked to conformational/dynamic differences in the whole protein (but specifically in the copper-binding loop and nearby peptide regions) that, in turn, offer a mechanism for *in vivo* target recognition.

Several residues that are not interacting directly with the Cu are conserved in this group of proteins. For example, Met10 in the Cu-binding loop is completely conserved and Lys60 in Atox1 is always a Lys in eukaryotes but a Tyr in prokaryotes. What roles do these residues play in chaperone biophysics and function? We obtained some answers to this from characterization of Cu release from wild-type Atox1 and two point mutants (Met10Ala and Lys60Ala Atox1). The dynamics of Cu displacement from holo-Atox1 were measured using the Cu^{1+} chelator BCA (bicinchonic acid) as the metal acceptor. We first deduced that BCA removes Cu from Atox1 in a three-step process involving bimolecular formation of an initial Atox1-Cu-BCA complex followed by dissociation of Atox1 and the binding of a second BCA to generate apo-Atox1 and $Cu-BCA_2$ (Hussain, Olson, and Wittung-Stafshede 2008). This copper-bridged protein-chelator complex may kinetically mimic the ternary chaperone-target complex involved in real metal transfer *in vivo*. Both mutants lose

Stability and Folding of Copper-Binding Proteins

Cu more readily than wild-type Atox1 due to more rapid and facile displacement of the protein from the Atox1-Cu-BCA intermediate by the second BCA. Remarkably, we found that Cu^{1+} uptake from solution by BCA is much slower than the transfer from holo-Atox1, presumably due to slow dissociation of DTT-Cu complexes formed in DTT containing buffer (Hussain, Olson, and Wittung-Stafshede 2008). This result suggests that Cu chaperones play a role in making Cu rapidly accessible to substrates, in addition to providing transfer specificity. We also subjected the mutated Atox1 variants (adding Thr11Ala since position 11 is conserved as Thr in all eukaryotes and Lys60Tyr since this position is a Tyr in prokaryotes) to MD simulations (Rodriguez-Granillo and Wittung-Stafshede 2009a). Surprisingly, both apo and holo Atox1 become more rigid in the absence of either Thr11 or Lys60, suggesting that these residues introduce protein *flexibility*. Moreover, Lys60 and Thr11 were found to participate in electrostatic networks that stabilize the Cu-bound form and, in the apo-form, determine the solvent exposure of the two Cys. In contrast, a substitution of Met10 buried in the hydrophobic core of Atox1 results in a protein with more structural dynamics. Similar trends in rigidity, structural dynamics, and interaction networks were found in corresponding point-mutated CopZ variants (Rodriguez-Granillo and Wittung-Stafshede 2009b). *In silico* structural flexibility correlates with protein stability as demonstrated by *in vitro* thermal unfolding experiments on the purified Atox1 variants. As expected, Lys60Ala, Lys60Tyr, and Thr11Ala Atox1 variants were similar in thermal stability as wild-type Atox1 in both apo- and holo-forms whereas Met10Ala Atox1 was less resistant to heat than wild-type Atox1 (data summarized in Figure 4.4b). In all cases the holo-form is more stable than the apo-form. The discovered built-in flexibility supports the idea of structural changes being required for the formation of transient Atox1-Cu-target complexes *in vivo*. The importance of electrostatic attraction provided by Lys60 for complex formation (and Cu transfer to the target) between Cu-Atox1 and the fourth WND metal-binding domain could be demonstrated by a combination of *in vitro* and *in silico* experiments (Hussain, Rodriguez-Granillo, and Wittung-Stafshede 2009).

As a side step, we studied unfolding of the WND N-subdomain (Figure 4.4a) as a function of ATP and the most common Wilson's disease mutation, H1069Q (which is found in this domain), by spectroscopic and calorimetric methods. To our initial surprise, we found that the mutation has little effect on either protein stability or ATP affinity (Rodriguez-Granillo, Sedlak, and Wittung-Stafshede 2008). Nonetheless, computational work revealed that the mutant, in contrast to wild type, coordinated ATP in an orientation that hinders phosphorylation. Lack of phosphorylation has been reported for this variant (Tsivkovskii, Efremov, and Lutsenko 2003). This provides a molecular explanation for the observed disease phenotype for this mutation.

MULTICOPPER OXIDASES: COMPLEX PROTEIN/METAL SYSTEMS

To address the role of Cu in folding and stability of more complex systems (i.e., with multi-domain structures and multiple Cu sites), we characterized two multi-copper oxidases (MCOs): human ceruloplasmin and yeast Fet3p (Figure 4.5a). Despite the fact that unfolding is irreversible for both proteins, the work on these systems demonstrates that one may still extract a great deal of information about

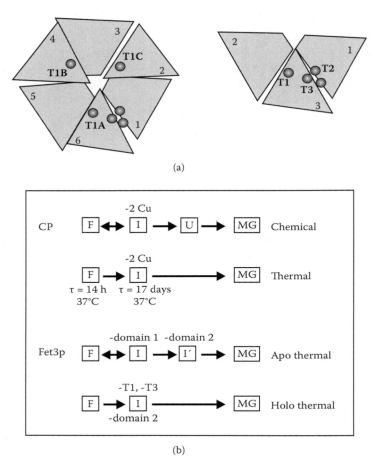

FIGURE 4.5 (a) Schematic structures of a three-domain MCO (Fet3p, right) and a six-domain MCO (ceruloplasmin, left) with copper sites indicated (T1, T2, and T3). (b) Key aspects of the unfolding reactions of holo ceruloplasmin (CP; thermal and chemical perturbations) and holo and apo wild-type forms of Fet3p (F = folded; I = intermediates; U = unfolded; MG = molten globule).

folding/unfolding/binding mechanisms. More than 1000 proteins have been identified as MCOs based on the multiples of the cupredoxin motifs they contain (Kosman 2002). MCO proteins are found in all kingdoms of life; the majority of the known MCOs contain three cupredoxin domains and four Cu ions distributed in three distinct sites (Kosman 2002) (see Fet3p, Figure 4.5a). Apart from their biological relevance, MCOs are highly interesting as catalysts in biofuel cells and in other biotechnological applications.

Human ceruloplasmin (CP) plays an important role in iron metabolism due to its ability to oxidize Fe^{2+} to Fe^{3+}, which allows for subsequent incorporation of Fe^{3+} into apotransferrin (Hellman and Gitlin 2002). The protein has six β-barrel domains with one type-1 (T1) Cu in each of domains 2, 4, and 6; the remaining Cu ions form a catalytic trinuclear cluster, one type-2 (T2) and two type-3 (T3) coppers, at the interface between domains 1 and 6. First, we characterized urea-induced unfolding of holo- and apo-forms of CP by a range of probes that included: far-UV CD, intrinsic fluorescence, ANS (8-*anilino*-1-naphthalene sulfonic acid) binding, visible absorption, Cu content, and oxidase activity (pH 7, 20°C) (Sedlak and Wittung-Stafshede 2007). We found that holo-CP unfolds in a complex reaction at these conditions with at least one equilibrium intermediate. Formation of the intermediate correlated with decreased secondary structure, exposure of aromatics, loss of two coppers, and reduced oxidase activity. This step was reversible, indicating that the trinuclear cluster remained intact. Further additions of urea triggered a complete unfolding of the protein and loss of all coppers. Attempts to refold this species resulted in an inactive apoprotein with molten-globule characteristics (Figure 4.5b). The apo-form of CP also unfolds in a multi-step reaction albeit the intermediate appears at slightly lower urea concentration. Again, correct refolding was possible from the intermediate but not the unfolded state. These observations demonstrate that *in vitro* equilibrium unfolding of CP involves intermediates and that the copper ions are removed in stages (Sedlak and Wittung-Stafshede 2007). When the catalytic Cu site is finally destroyed, refolding is not possible at neutral pH. Thus timely Cu incorporation, before apo-protein misfolding to the molten globule, appears crucial for successful *in vivo* biosynthesis of functional CP. This implies a mechanistic role for the trinuclear Cu cluster as a nucleation point, aligning domains 1 and 6, during CP folding *in vivo*.

We also assessed the role of the coppers in CP thermal stability. For this, we probed the thermal unfolding process as a function of scan rate of holo- and apo-forms again with a combination of detection methods (CD, aromatic and ANS fluorescence, visible absorption, activity, and differential scanning calorimetry [DSC]) (Sedlak, Zoldak, and Wittung-Stafshede 2008). Both apo- and holo-forms of CP undergo irreversible thermal reactions to denatured states with significant residual structure. The spectroscopic signals for these species are similar to the dead-end molten globule found in the urea experiments; thus heating of CP results in molten globule formation without any detectable population of fully unfolded structures (Figure 4.5b). For identical scan rates, the thermal midpoint appears at 15–20°C higher temperatures for the holo- as compared to the apo-form. The thermal data for both CP forms were fit by a mechanistic model involving two consecutive, irreversible steps ($N \rightarrow I \rightarrow D$) (Sedlak, Zoldak, and Wittung-Stafshede 2008). The thermal holo-intermediate, I, has lost two Cu ions (one blue T1) and secondary structure in at least one domain; however, the trinuclear Cu cluster remains intact as it is functional in oxidase activity. Thus the urea-induced I (pH 7, 20°C) and the thermal I are similar in many aspects.

The activation parameters obtained from the fits to the thermal transitions were used to assess the kinetic stability of apo- and holo-forms of CP at a physiological temperature of 37°C (Sedlak, Zoldak, and Wittung-Stafshede 2008). From the analysis, it emerged that native CP (i.e., with six coppers), despite the apparent high thermal stability, is rather unstable and converts to I in less than a day at 37°C. Nonetheless, this form remains intact for over 2 weeks and may be a biologically relevant state of CP *in vivo*. In the absence of all coppers, CP is even more unstable;

apo-CP is completely unfolded in less than 2 days at 37°C. The low kinetic stability of the apo-form of CP may correlate with its rapid degradation *in vivo* in various disease conditions.

To specifically probe the role of each Cu site in unfolding/stability of an MCO system, we turned to *Saccharomyces cerevisiae* Fet3p that contains three β-barrel domains and four Cu ions located in three metal sites (T1 in domain 3; T2 and the binuclear T3 at the interface between domains 1 and 3) (Figure 4.5a). In contrast to CP, a collection of de-metallated forms of Fet3p has been prepared, and electronic and redox properties of each Cu site have been characterized (Kosman 2002; Palmer et al. 2002). The partially metallated Fet3p forms are designated T1D, T2D, and T1D/T2D (D, depleted), in which the T1, T2, or both the T1 and T2 Cu ions are absent. Upon assessing thermal unfolding of the various de-metallated forms of Fet3p using spectroscopic and calorimetric methods *in vitro* (pH 7), we could reveal how stability is defined by each Cu site (Sedlak et al. 2008).

Like apo- and holo-forms of CP, the thermal unfolding reactions of apo- and the different holo-forms of Fet3p are irreversible reactions that depend on temperature scan rate. The domains in apo-Fet3p unfold sequentially (T_m of 45°C, 62°C, and 72°C; 1 K/min) (Figure 4.5b). Addition of T3 (i.e., the T1D/T2D variant) imposes strain in the apo-structure that results in coupled domain unfolding and low stability (T_m of 50°C; 1 K/min). Further inclusion of T2 (i.e., the T1D variant) increases overall stability by ~5°C but unfolding remains coupled in one step. Introduction of T1, producing fully-loaded holo-Fet3p (or in the absence of T2, i.e., the T2D variant), results in stabilization of domain 3, which uncouples unfolding of the domains. For the wild-type holo-form, unfolding of domain 2 occurs first along with Cu-site perturbations (T_m 50–55°C; 1 K/min), followed by unfolding of domains 1 and 3 (~65–70°C; 1 K/min) (Figure 4.5b). That unfolding of domains 1 and 3 is coupled is likely due to the T3 coppers "stitching" these domains together. Our data articulate the importance of T3 in holding the Fet3p trimeric structure together (Sedlak et al. 2008), which may be a general feature of MCO proteins as the same conclusion was made for CP. The closed structural arrangement is required for Fet3p function but, nonetheless, results in loss of overall protein stability as compared to the nonfunctional apo-form of Fet3p. Fet3p, thus, is an example of a biological situation in which trade-offs between stability and structure are mediated by metal prosthetic groups to obtain a desired function.

STUDIES ON OTHER COPPER-BINDING PROTEINS

In addition to the already discussed groups of proteins, there are limited folding/stability data on some other copper-binding proteins, some of which we will briefly mention here.

The superoxide dismutase (SOD) family of protein is a large class of metalloproteins that catalyze the conversion of superoxide to hydrogen peroxide and are an important component of the antioxidant defense in the cell. SOD is made up from two identical subunits (each monomer is approximately 16 kDa) held together by hydrophobic interactions and an inter-protein disulfide bridge. In Cu/Zn SOD (which contains one Cu and one Zn), Cu is involved in the catalytic reaction whereas Zn plays a structural role (Battistoni et al. 1998). Although *in vitro* equilibrium unfolding of SOD is not two-state, it is clear that the holo-form is more stable than the apo-form toward chemical perturbation and toward acid-induced denaturation (Lynch and Colon 2006). That suggested that Cu and Zn stabilize the β-barrel fold of SOD, which is explained by nuclear magnetic resonance (NMR) studies revealing that the metals limit local unfolding (Assfalg et al. 2003).

It is important to point out that *in vitro* the denatured state of SOD in the presence of metal ions was more compact than in the absence of the cofactors. Although the nativelike coordination geometry of both Cu and Zn was lost in high GuHCl concentrations, it is possible that the metal ions still interact with the unfolded polypeptide (but in a different mode), which results in local ordering and compaction (Assfalg et al. 2002; Libralesso et al. 2005).

As part of our own Cu chaperone work, we investigated CopC from *Pseudomonas syringae pathovar* tomato (Hussain, Sedlak, and Wittung-Stafshede 2007). This protein is proposed to function as a Cu chaperone in the periplasm (Arnesano et al. 2003). In contrast to the cytoplasmic Cu chaperones, CopC has two Cu sites: one specific for Cu^{1+} and the other for Cu^{2+}. CopC is characterized by a β-barrel structure similar to azurin and the individual domains of the MCO proteins. Through a range of chemical and thermal unfolding experiments, we revealed that Cu^{2+} coordination affects both CopC stability and unfolding pathway, whereas the presence of Cu^{1+} has only a small effect on stability (Hussain, Sedlak, and Wittung-Stafshede 2007). Similarly, the thermodynamic stability of the Cu_A domain of cytochrome c oxidase (with a binuclear purple Cu-Cu center) from *Thermus thermophilus* is affected by the oxidation state of the Cu cofactors. The oxidized Cu^{2+}/Cu^{2+} form remains folded up to 8 M GuHCl (at 20°C, pH 7) while unfolding of the reduced (Cu^{1+}/Cu^{1+}) CuA domain is completed already in 7 M GuHCl. The difference in thermodynamic stability between oxidized and reduced forms was estimated to ~20 kJ/mol and explained by the geometry of the Cu site being optimal for the coordination of Cu^{2+} but not for Cu^{1+} (Wittung-Stafshede, Malmstrom et al. 1998).

Another member of the MCO family, the laccase protein CotA from *B. subtilis*, has been subjected to quite extensive folding and stability studies. Similar to the other MCO proteins (e.g., Fet3p), the overall fold of CotA comprises three cupredoxinlike domains and accommodates four Cu ions in T1 and T2/T3 sites (Enguita et al. 2003). Recombinant CotA can be produced in *Escherichia coli* in copper-supplemented media under microaerobic conditions. When the protein was expressed at aerobic conditions (resulting in lower cellular Cu levels), the catalytic efficiency of purified CotA was markedly reduced and some alterations of the T2/T3 center geometry were detected by electron paramagnetic resonance (EPR). The authors hypothesized that at conditions when intracellular levels of Cu are limited, CotA fails to fold into its native conformation, indicating that Cu might be required for proper folding of CotA *in vivo* (Durao et al. 2008). This correlates with our finding on ceruloplasmin and Fet3p described earlier in this chapter.

The importance of the T1 Cu site for CotA enzymatic activity and thermodynamic stability was interrogated using mutational analysis (Durao et al. 2006). Replacement of the axial Cu ligand (i.e., Met50) by nonligating residues (i.e., Leu and Phe) leads to an increase in the Cu^{2+} reduction potential by approximately 100 mV relative to that of the wild-type protein. Although the T1 mutations did not perturb either the overall geometry of the T1 Cu site or global protein structure, the catalytic activity (using a range of substrates) of the T1-site mutants was severely compromised. Equilibrium unfolding of CotA was also affected by the mutations: In contrast to the wild-type CotA, unfolding of the T1-site mutants appeared non-two state involving an equilibrium intermediate.

The MCO, McoA from *Aquifex aeolicus*, is closely related to CotA and, as expected for a thermophile, it withstands high temperatures: unfolded of partially Cu-loaded McoA does not occurs until 110°C (Fernandes, Martins, and Melo 2009). In the fully Cu-loaded state, McoA resists unfolding even at 130°C, which indicates a strong stabilizing effect of the Cu ions. Despite its high thermal stability, McoA appears sensitive to the presence of the chemical denaturant, GuHCl. Even low concentrations (less than 0.2 M) of GuHCl cause bleaching of Cu from the trinuclear center and, as a result, a loss of enzymatic activity occurs at lower denaturant concentrations than the global protein unfolding. Despite this, kinetic measurements indicated that presence of Cu has a stabilizing effect on McoA via Cu-induced reduction of the unfolding speed.

FUTURE PERSPECTIVE

Copper-binding proteins are essential for living organisms. At the same time, an increasing amount of evidence suggests that Cu ions also participate in disease progression. The effects of Cu binding on protein stability and folding reactions vary dramatically. For example, the Cu in *P. aeruginosa* azurin stabilizes the protein dramatically whereas in beta-2-microglobulin Cu binding results in native-state destabilization (Eakin et al. 2002). In several well-behaved protein systems, we have found that Cu is able to bind to unfolded as well as partially folded structures *in vitro;* this may be linked to how Cu is involved in, or even triggers, diseases that involve misfolded and aggregated proteins and/or peptides. See Chapters 9 and 11, for example, for the role of metals (including Cu) in Parkinson's and prion-protein diseases. Better mechanistic understanding of the proteins that facilitate Cu transport in living systems may lead to the development of new drugs that modulate cellular Cu levels. We want to emphasize that for all future work, one ought to take into account the *in vivo* environment where the proteins are found. It has been reported that the presence of macromolecular crowding agents *in vitro*, which create a crowded scenario similar to that in the cytoplasm, can affect properties such as protein stability, folding speed, secondary-structure content, as well as overall protein shape (Homouz et al. 2008). Thus the cellular environment may also affect metal-protein interactions/affinities and thereby tune metalloprotein function.

REFERENCES

Adman, E. T. 1991. Copper protein structures. *Adv Protein Chem* 42:145–97.
Apiyo, D., and P. Wittung-Stafshede. 2005. Unique complex between bacterial azurin and tumor-suppressor protein p53. *Biochem Biophys Res Commun* 332 (4):965–68.

Arnesano, F., L. Banci, I. Bertini, S. Ciofi-Baffoni, E. Molteni, D. L. Huffman, and T. V. O'Halloran. 2002. Metallochaperones and metal-transporting ATPases: a comparative analysis of sequences and structures. *Genome Res* 12 (2):255–71.

Assfalg, M., L. Banci, I. Bertini, P. Turano, and P. R. Vasos. 2003. Superoxide dismutase folding/unfolding pathway: role of the metal ions in modulating structural and dynamical features. *J Mol Biol* 330 (1):145–58.

Banci, L., I. Bertini, F. Cantini, A. C. Rosenzweig, and L. A. Yatsunyk. 2008. Metal binding domains 3 and 4 of the Wilson disease protein: solution structure and interaction with the copper(I) chaperone HAH1. *Biochemistry* 47 (28):7423–29.

Battistoni, A., S. Folcarelli, L. Cervoni, F. Polizio, A. Desideri, A. Giartosio, and G. Rotilio. 1998. Role of the dimeric structure in Cu, Zn superoxide dismutase. pH-dependent, reversible denaturation of the monomeric enzyme from *Escherichia coli*. *J Biol Chem* 273 (10):5655–61.

Chen, M., C. J. Wilson, Y. Wu, P. Wittung-Stafshede, and J. Ma. 2006. Correlation between protein stability cores and protein folding kinetics: a case study on *Pseudomonas aeruginosa* apo-azurin. *Structure* 14 (9):1401–10.

DeBeer, S., P. Wittung-Stafshede, J. Leckner, G. Karlsson, J. R. Winkler, H. B. Gray, B. Malmstrom, E. I. Solomon, B. Hedman, and K. Hodgson. 2000. X-ray absorption spectroscopy of unfolded copper(I) azurin. *Inorg Chim Acta* 297:278–82.

de los Rios, M. A., B. K. Muralidhara, D. Wildes, T. R. Sosnick, S. Marqusee, P. Wittung-Stafshede, K. W. Plaxco, and I. Ruczinski. 2006. On the precision of experimentally determined protein folding rates and phi-values. *Protein Sci* 15 (3):553–63.

Durao, P., I. Bento, A. T. Fernandes, E. P. Melo, P. F. Lindley, and L. O. Martins. 2006. Perturbations of the T1 copper site in the CotA laccase from *Bacillus subtilis:* structural, biochemical, enzymatic and stability studies. *J Biol Inorg Chem* 11 (4):514–26.

Durao, P., Z. Chen, C. S. Silva, C. M. Soares, M. M. Pereira, S. Todorovic, P. Hildebrandt, I. Bento, P. F. Lindley, and L. O. Martins. 2008. Proximal mutations at the type 1 copper site of CotA laccase: spectroscopic, redox, kinetic and structural characterization of I494A and L386A mutants. *Biochem J* 412 (2):339–46.

Eakin, C. M., J. D. Knight, C. J. Morgan, M. A. Gelfand, and A. D. Miranker. 2002. Formation of a copper specific binding site in non-native states of beta-2-microglobulin. *Biochemistry* 41 (34):10646–56.

Enguita, F. J., L. O. Martins, A. O. Henriques, and M. A. Carrondo. 2003. Crystal structure of a bacterial endospore coat component. A laccase with enhanced thermostability properties. *J Biol Chem* 278 (21):19416–25.

Fernandes, A. T., L. O. Martins, and E. P. Melo. 2009. The hyperthermophilic nature of the metallo-oxidase from *Aquifex aeolicus*. *Biochim Biophys Acta* 1794 (1):75–83.

Fersht, A. 1999. *Structure and mechanism in protein science*. New York: Freeman.

Fersht, A. R., and S. Sato. 2004. Phi-value analysis and the nature of protein-folding transition states. *Proc Natl Acad Sci U S A* 101 (21):7976–81.

Harris, E. D. 2003. Basic and clinical aspects of copper. *Crit Rev Clin Lab Sci* 40 (5):547–86.

Hellman, N. E., and J. D. Gitlin. 2002. Ceruloplasmin metabolism and function. *Annu Rev Nutr* 22:439–58.

Homouz, D., M. Perham, A. Samiotakis, M. S. Cheung, and P. Wittung-Stafshede. 2008. Crowded, cell-like environment induces shape changes in aspherical protein. *Proc Natl Acad Sci U S A* 105 (33):11754–59.

Huffman, D. L., and T. V. O'Halloran. 2001. Function, structure, and mechanism of intracellular copper trafficking proteins. *Annu Rev Biochem* 70:677–701.

Hussain, F., J. S. Olson, and P. Wittung-Stafshede. 2008. Conserved residues modulate copper release in human copper chaperone Atox1. *Proc Natl Acad Sci U S A* 105 (32):11158–63.

Hussain, F., A. Rodriguez-Granillo, and P. Wittung-Stafshede. 2009. Lysine-60 in copper chaperone atox1 plays an essential role in adduct formation with a target Wilson disease domain. *J Am Chem Soc* 131 (45):16371–73.

Hussain, F., E. Sedlak, and P. Wittung-Stafshede. 2007. Role of copper in folding and stability of cupredoxin-like copper-carrier protein CopC. *Arch Biochem Biophys* 467 (1):58–66.

Hussain, F., and P. Wittung-Stafshede. 2007. Impact of cofactor on stability of bacterial (CopZ) and human (Atox1) copper chaperones. *Biochim Biophys Acta* 1774 (10):1316–22.

Jackson, S. E. 1998. How do small single-domain proteins fold? *Fold Des* 3 (4):R81–91.

Kister, A. E., A. V. Finkelstein, and I. M. Gelfand. 2002. Common features in structures and sequences of sandwich-like proteins. *Proc Natl Acad Sci U S A* 99 (22):14137–41.

Kosman, D. J. 2002. FET3P, ceruloplasmin, and the role of copper in iron metabolism. *Adv Protein Chem* 60:221–69.

Leckner, J., N. Bonander, P. Wittung-Stafshede, B. G. Malmstrom, and B. G. Karlsson. 1997. The effect of the metal ion on the folding energetics of azurin: a comparison of the native, zinc and apoprotein. *Biochim Biophys Acta* 1342 (1):19–27.

Leckner, J., P. Wittung, N. Bonander, G. Karlsson, and B. Malmstrom. 1997. The effect of redox state on the folding free energy of azurin. *J Biol Inorg Chem* 2:368–71.

Libralesso, E., K. Nerinovski, G. Parigi, and P. Turano. 2005. 1H nuclear magnetic relaxation dispersion of Cu,Zn superoxide dismutase in the native and guanidinium-induced unfolded forms. *Biochem Biophys Res Commun* 328 (2):633–39.

Lutsenko, S., E. S. LeShane, and U. Shinde. 2007. Biochemical basis of regulation of human copper-transporting ATPases. *Arch Biochem Biophys* 463 (2):134–48.

Lynch, S. M., and W. Colon. 2006. Dominant role of copper in the kinetic stability of Cu/Zn superoxide dismutase. *Biochem Biophys Res Commun* 340 (2):457–61.

Marks, J., I. Pozdnyakova, J. Guidry, and P. Wittung-Stafshede. 2004. Methionine-121 coordination determines metal specificity in unfolded *Pseudomonas aeruginosa* azurin. *J Biol Inorg Chem* 9 (3):281–88.

Matouschek, A., and A. R. Fersht. 1991. Protein engineering in analysis of protein folding pathways and stability. *Methods Enzymol* 202:82–112.

Matouschek, A., J. T. Kellis Jr., L. Serrano, M. Bycroft, and A. R. Fersht. 1990. Transient folding intermediates characterized by protein engineering. *Nature* 346 (6283):44045.

Matthews, C. R. 1993. Pathways of protein folding. *Annu Rev Biochem* 62:653–83.

Maxwell, K. L., D. Wildes, A. Zarrine-Afsar, M. A. De Los Rios, A. G. Brown, C. T. Friel, L. Hedberg, J. C. Horng, D. Bona, E. J. Miller, A. Vallee-Belisle, E. R. Main, F. Bemporad, L. Qiu, K. Teilum, N. D. Vu, A. M. Edwards, I. Ruczinski, F. M. Poulsen, B. B. Kragelund, S. W. Michnick, F. Chiti, Y. Bai, S. J. Hagen, L. Serrano, M. Oliveberg, D. P. Raleigh, P. Wittung-Stafshede, S. E. Radford, S. E. Jackson, T. R. Sosnick, S. Marqusee, A. R. Davidson, and K. W. Plaxco. 2005. Protein folding: defining a "standard" set of experimental conditions and a preliminary kinetic data set of two-state proteins. *Protein Sci* 14 (3):602–16.

Nar, H., A. Messerschmidt, R. Huber, M. van de Kamp, and G. W. Canters. 1991. Crystal structure analysis of oxidized *Pseudomonas aeruginosa* azurin at pH 5.5 and pH 9.0. A pH-induced conformational transition involves a peptide bond flip. *J Mol Biol* 221 (3):765–72.

———. 1992. Crystal structure of *Pseudomonas aeruginosa* apo-azurin at 1.85 A resolution. *FEBS Lett* 306 (2–3):119–24.

O'Halloran, T. V., and V. C. Culotta. 2000. Metallochaperones, an intracellular shuttle service for metal ions. *J Biol Chem* 275 (33):25057–60.

Palmer, A. E., L. Quintanar, S. Severance, T. P. Wang, D. J. Kosman, and E. I. Solomon. 2002. Spectroscopic characterization and O2 reactivity of the trinuclear Cu cluster of mutants of the multicopper oxidase Fet3p. *Biochemistry* 41 (20):6438–48.

Pozdnyakova, I., J. Guidry, and P. Wittung-Stafshede. 2000. Copper triggered b-hairpin formation. Initiation site for azurin folding? *J AM Chem Soc* 122:6337–38.

———. 2001a. Copper stabilizes azurin by decreasing the unfolding rate. *Arch Biochem Biophys* 390:146–48.

———. 2001b. Probing copper ligands in denatured *Pseudomonas aeruginosa* azurin: unfolding His117Gly and His46Gly mutants. *J Biol Inorg Chem* 6:182–88.

———. 2002. Studies of *Pseudomonas aeruginosa* azurin mutants: cavities in beta-barrel do not affect refolding speed. *Biophys J* 82 (5):2645–51.

Pozdnyakova, I., and P. Wittung-Stafshede. 2001a. Biological relevance of metal binding before protein folding. *J Am Chem Soc* 123 (41):10135–36.

———. 2001b. Copper binding before polypeptide folding speeds up formation of active (holo) *Pseudomonas aeruginosa* azurin. *Biochemistry* 40 (45):13728–33.

———. 2003. Approaching the speed limit for Greek Key beta-barrel formation: transition-state movement tunes folding rate of zinc-substituted azurin. *Biochim Biophys Acta* 1651 (1–2):1–4.

Privalov, P. L. 1996. Intermediate states in protein folding. *J Mol Biol* 258 (5):707–25.

Puig, S., and D. J. Thiele. 2002. Molecular mechanisms of copper uptake and distribution. *Curr Opin Chem Biol* 6 (2):171–80.

Roder, H., and W. Colon. 1997. Kinetic role of early intermediates in protein folding. *Curr Opin Struct Biol* 7 (1):15–28.

Rodriguez-Granillo, A., E. Sedlak, and P. Wittung-Stafshede. 2008. Stability and ATP binding of the nucleotide-binding domain of the Wilson disease protein: effect of the common H1069Q mutation. *J Mol Biol* 383 (5):1097–1111.

Rodriguez-Granillo, A., and P. Wittung-Stafshede. 2008. Structure and dynamics of Cu(I) binding in copper chaperones Atox1 and CopZ: a computer simulation study. *J Phys Chem B* 112 (15):4583–93.

———. 2009a. Differential roles of Met10, Thr11, and Lys60 in structural dynamics of human copper chaperone Atox1. *Biochemistry* 48 (5):960–72.

———. 2009b. Tuning of copper-loop flexibility in *Bacillus subtilis* CopZ copper chaperone: role of conserved residues. *J Phys Chem B* 113 (7):1919–32.

Sanchez, I. E., and T. Kiefhaber. 2003. Origin of unusual phi-values in protein folding: evidence against specific nucleation sites. *J Mol Biol* 334 (5):1077–85.

Sedlak, E., and P. Wittung-Stafshede. 2007. Discrete roles of copper ions in chemical unfolding of human ceruloplasmin. *Biochemistry* 46 (33):9638–44.

Sedlak, E., L. Ziegler, D. J. Kosman, and P. Wittung-Stafshede. 2008. *In vitro* unfolding of yeast multicopper oxidase Fet3p variants reveals unique role of each metal site. *Proc Natl Acad Sci U S A* 105 (49):19258–63.

Sedlak, E., G. Zoldak, and P. Wittung-Stafshede. 2008. Role of copper in thermal stability of human ceruloplasmin. *Biophys J* 94 (4):1384–91.

Solioz, M., and J. V. Stoyanov. 2003. Copper homeostasis in *Enterococcus hirae*. *FEMS Microbiol Rev* 27 (2–3):183–95.

Tottey, S., K. J. Waldron, S. J. Firbank, B. Reale, C. Bessant, K. Sato, T. R. Cheek, J. Gray, M. J. Banfield, C. Dennison, and N. J. Robinson. 2008. Protein-folding location can regulate manganese-binding versus copper- or zinc-binding. *Nature* 455 (7216):1138–42.

Tsivkovskii, R., R. G. Efremov, and S. Lutsenko. 2003. The role of the invariant His-1069 in folding and function of the Wilson's disease protein, the human copper-transporting ATPase ATP7B. *J Biol Chem* 278 (15):13302–8.

Vijgenboom, E., J. E. Busch, and G. W. Canters. 1997. *In vivo* studies disprove an obligatory role of azurin in denitrification in *Pseudomonas aeruginosa* and show that azu expression is under control of rpoS and ANR. *Microbiology* 143 (Pt 9):2853–63.

Wilson, C. J., D. Apiyo, and P. Wittung-Stafshede. 2006. Solvation of the folding-transition state in *Pseudomonas aeruginosa* azurin is modulated by metal: solvation of azurin's folding nucleus. *Protein Sci* 15 (4):843–52.

Wilson, C. J., and P. Wittung-Stafshede. 2005a. Role of structural determinants in folding of the sandwich-like protein *Pseudomonas aeruginosa* azurin. *Proc Natl Acad Sci U S A* 102 (11):3984–87.

———. 2005b. Snapshots of a dynamic folding nucleus in zinc-substituted *Pseudomonas aeruginosa* azurin. *Biochemistry* 44 (30):10054–62.

Wittung-Stafshede, P. , M. G. Hill, E. Gomez, A. Di Bilio, G. Karlsson, J. Leckner, J. R. Winkler, H. B. Gray, and B. G. Malmstrom. 1998. Reduction potentials of blue and purple copper proteins in their unfolded states: a closer lock at rack-induced coordination. *J Biol Inorg Chem* 3:367–70.

Wittung-Stafshede, P., J. C. Lee, J. R. Winkler, and H. B. Gray. 1999. Cytochrome b562 folding triggered by electron transfer: approaching the speed limit for formation of a four-helix-bundle protein. *Proc Natl Acad Sci U S A* 96 (12):6587–90.

Wittung-Stafshede, P., B. G. Malmstrom, D. Sanders, J. A. Fee, J. R. Winkler, and H. B. Gray. 1998. Effect of redox state on the folding free energy of a thermostable electron-transfer metalloprotein: the CuA domain of cytochrome oxidase from *Thermus thermophilus*. *Biochemistry* 37 (9):3172–77.

Yamada, T., M. Goto, V. Punj, O. Zaborina, M. L. Chen, K. Kimbara, D. Majumdar, E. Cunningham, T. K. Das Gupta, and A. M. Chakrabarty. 2002. Bacterial redox protein azurin, tumor suppressor protein p53, and regression of cancer. *Proc Natl Acad Sci U S A* 99 (22):14098–103.

Zong, C., C. J. Wilson, T. Shen, P. G. Wolynes, and P. Wittung-Stafshede. 2006. Phi-value analysis of apo-azurin folding: comparison between experiment and theory. *Biochemistry* 45 (20):6458–66.

5 Iron-Sulfur Clusters, Protein Folds, and Ferredoxin Stability

Sónia S. Leal and Cláudio M. Gomes

CONTENTS

Introduction 81
Biogenesis of Iron-Sulfur Proteins: An Overview 82
The Interplay between Protein Folding and Iron-Sulfur Cluster Binding 83
Folding and Stability of Small Iron-Sulfur Proteins 83
Di-Cluster Ferredoxins as a Case Study 86
    Ferredoxin Hyperstability and Unfolding Pathways 86
    Effects of Electrostatic Interactions and Metal Centers
    on Ferredoxin Stability 88
    Role of the His/Asp Zinc Center in Ferredoxin Stability 90
    Monitoring Ferredoxin Unfolding at Different Levels
    of Metalloprotein Organization 91
    The Molten Globule State of Ferredoxin 92
Future Perspective 94
References 94

INTRODUCTION

Iron-sulfur proteins are a multifaceted class of proteins containing iron-sulfur clusters (Fe-S) as a prosthetic group. These proteins are ubiquitously found within all life domains and are involved in a plethora of essential biological processes and cellular pathways, such as respiration and photosynthesis, DNA and RNA metabolism, and intervening in the regulation of iron homeostasis and gene expression. This functional versatility can be well correlated to the structural diversity and chemical properties of the clusters. In fact, these amazing complexes of iron and cysteinate sulfur can be found in structures with different nuclearities and have the ability to undergo interconversions and ligand exchange reactions (Figure 5.1). Fe-S clusters are in great majority bound to proteins by thiolate ligation, and accordingly cysteinyl sulfur is the frequent ligand of Fe-S active sites. The evolution of multiple folds suitable to host Fe-S clusters has also expanded and fine-tuned the biological functions of these proteins. However, in spite of the importance of Fe-S proteins in life, much remains to be understood about the molecular mechanisms of biological insertion of Fe-S clusters and their influence on the folding of the recipient apoproteins.

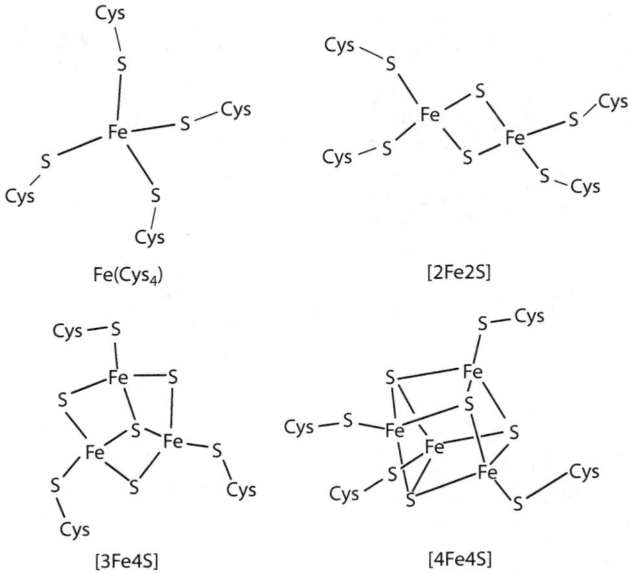

FIGURE 5.1 Structures of Fe-S clusters.

BIOGENESIS OF IRON-SULFUR PROTEINS: AN OVERVIEW

Iron-sulfur clusters are inorganic structures formed from the chemical assembly of ferrous iron, thiol, and sulfide, under reducing conditions. Although many apo-forms of Fe-S proteins may reincorporate *in vitro* their clusters (FeCys$_4$, [2Fe-2S], [4Fe-4S]) upon addition of the aforementioned components, the fact is that *in vivo* the biosynthesis of Fe-S proteins is far more complex (reviewed in Johnson et al. 2005; Lill and Muhlenhoff 2006). The need for assisted, rather than spontaneous, Fe-S cluster assembly and insertion into apoproteins is most likely the result of exigent anaerobic conditions, and of the toxicity that would arise from the high concentrations of iron and sulfide ions required for efficient chemical reconstitution inside the cell. Furthermore, protein-mediated assembly of Fe-S clusters is likely to be more tightly regulated and efficient under biological control. Different systems for the biosynthesis of Fe-S clusters operate in eukaryotes and prokaryotes: ISC (iron-sulfur cluster) for mitochondrial iron-sulfur clusters; SUF (sulfur mobilization) machinery; Nif, NIF (nitrogen fixation) system and the CIA (cytosolic iron-sulfur protein assembly) machinery (Lill and Muhlenhoff 2006). Independently of the system involved, the biogenesis process relies on the concerted action of a cysteine desulfurase and an Fe-S cluster scaffolding protein. The first steps of the process are believed to involve the formation of a transient complex between these two proteins, which results in sulfur transfer to the scaffold protein (Mihara and Esaki 2002; Smith et al. 2001; Urbina et al. 2001). The source of Fe is not consensual, although frataxin and homologues have been shown to function as iron donors (Lill and Muhlenhoff 2006). Presently, the molecular biochemistry of Fe-S cluster assembly in scaffold proteins remains poorly understood, as do the number and the precise type of clusters that can be formed in each class of scaffold proteins. In addition it is also suggested that cluster assembly is likely to initially form [2Fe-2S] clusters as the basic building block for clusters with other stoichiometries. This process is likely to engage molecular chaperones but it is certainly driven by relatively unspecific protein-protein interactions. One may speculate that the abundant hydrophobic residues that frequently surround the cluster-binding sites in the Fe-S protein folds may play a role. Clearly, although the major players in biological assembly of Fe-S clusters have been identified, little is known about the molecular mechanisms that determine these processes, and this remains a challenge for the future.

THE INTERPLAY BETWEEN PROTEIN FOLDING AND IRON-SULFUR CLUSTER BINDING

A large diversity of proteins bind Fe-S clusters: A recent analysis has identified nearly 50 distinct structural folds that are able to accommodate Fe-S clusters (Meyer 2008). These folds are found in a variety of Fe-S proteins, from large multidomain complexes to small proteins. To some extent, in all cases the protein conformation and stability is affected by binding of the Fe-S cluster. However, the cross talk between the folding of an Fe-S protein and binding of its cluster(s) may be rather diverse. Binding of the Fe-S cluster to its recipient apoprotein will shape the protein structure and local conformation, stabilizing the protein as a result of the metal-protein interactions. In return the polypeptide chain accommodates the cluster and generates a protective ligand framework against possible oxidative degradation of the oxygen-sensitive center. However, there are also many cases of Fe-S folds in which the clusters are harbored in less protective environments. What then could determine these differences?

Evolutionarily, one can speculate that nature may have taken advantage of this successful interplay, which has led to the ferredoxin-type proteins, whose fold is considered to be primordial. In fact, for small Fe-S proteins preferentially the stability of the Fe-S clusters is essential for biological function and therefore the clusters are generally hosted in a relatively rigid and protective fold. This is well illustrated in several studies in which sets of hydrophobic residues have been shown to be critical for Fe-S cluster stability and function in rubredoxins, HiPIP, and ferredoxins. However, throughout evolution the functions of Fe-S proteins have expanded as a result of nature having recruited or generated protein folds with different characteristics, where the Fe-S sites are either labile or located at rather exposed domains or within subunit interfaces. This is the case, for example, in the structure of SoxR (Watanabe et al. 2008) and CnfU (Yabe et al. 2008) proteins, which evidence a completely surface-exposed [2Fe-2S] cluster, or of poplar glutaredoxin C1 protein, which showed a bridging iron-sulfur cluster at the active site (Feng et al. 2006). Interestingly, proteins with unwrapped clusters seem to be mostly related with regulatory or scaffolding functions.

FOLDING AND STABILITY OF SMALL IRON-SULFUR PROTEINS

Small iron-sulfur proteins such as rubredoxins and ferredoxins are a priori good models for folding and stability studies considering their sizes, available structural

84   Protein Folding and Metal Ions: Mechanisms, Biology and Disease

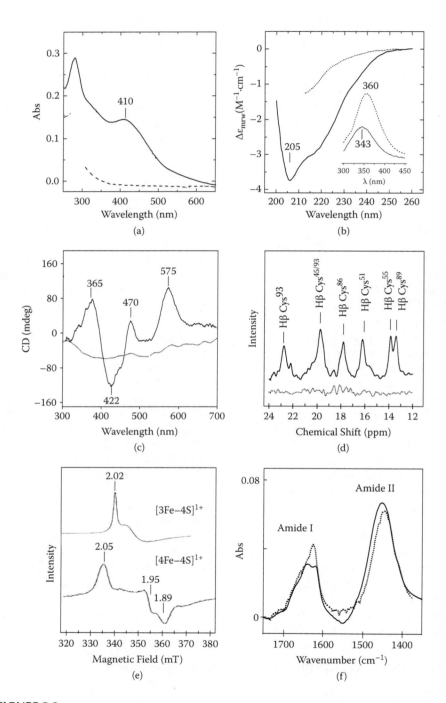

FIGURE 5.2

information, and the fact that they are usually very well characterized proteins, as well as in respect to the properties of the Fe-S clusters themselves, which are amenable for different spectroscopic methods. This is rather advantageous in experimental design as it allows combining methodologies that monitor changes in protein structure (such as far-ultraviolet [far-UV] circular dichroism [CD], Trp emission, and Fourier transform infrared spectroscopy [FTIR] with those that report modifications at the Fe-S cluster itself (such as electron paramagnetic resonance spectroscopy [EPR], resonance Raman [RR], and visible absorption or CD). This allows for a broad coverage of protein folding and unfolding events, which can be monitored under different perspectives, as illustrated in Figure 5.2, which shows a number of different spectroscopic fingerprints of a di-cluster ferredoxin that can effectively be used to monitor the folding status of the protein. In recent years we have been implementing this multitechnique approach to study the conformational properties and folding of rubredoxin (Henriques, Saraiva, and Gomes 2006), Rieske proteins (Botelho et al. 2010; Kletzin et al. 2005), and di-cluster ferredoxins (Leal and Gomes 2007; Todorovic et al. 2007).

One major pitfall in folding studies using Fe-S proteins arises from the fact that unfolding is in general irreversible, as a result of the disintegration of the Fe-S centers, which are required for the protein to refold into its native conformation, and are unable to self-assemble from the individual components in solution without the assistance of the Fe-S biogenesis proteins. By itself this is indicative that Fe-S clusters are key structural and stabilizing elements that act as relevant nucleation points for the folded protein state. Nevertheless, thermodynamic information can be extracted from thermal transitions by studying the dependence of the apparent melting temperature (T_m) with the heating rate (v), which allows for elimination of the artifacts of the kinetics of the irreversible transition, which become negligible at high heating rates. Therefore the y-axis intersect of a T_m vs. $1/v$ plot gives the melting temperature at an infinite heating rate, i.e., at equilibrium. This formalism was originally developed to deal with protein aggregation events during calorimetric analysis (Sanchez-Ruiz et al. 1988) and has subsequently been effectively used to analyze the thermodynamics of thermal unfolding of several Fe-S proteins (Henriques, Saraiva, and Gomes 2006; Mitou et al. 2003). In fact, this methodology also allows one to infer what

FIGURE 5.2 (Opposite) Spectroscopic tools to analyze native and unfolded states of Fe-S proteins. The spectroscopic fingerprints of native (solid lines) and unfolded (dotted/dashed lines) of the *Acididanus ambivalens* [3Fe-4S][4Fe-4S] ferredoxin (*Aa*Fd) are shown, using techniques that monitor differences on the protein structure (far-UV CD, FTIR, aromatic emission) and on the status of Fe-S clusters (UV-Vis spectroscopy, visible-CD, ^{1}H-NMR, EPR). A. Visible absorption; B. Far-UV CD, inset aromatic fluorescence emission; C. Visible CD; D. ^{1}H-NMR of beta protons of Fe-S cluster ligand cysteine residues; E. EPR of the [3Fe-4S] cluster (oxidized) and [4Fe-4S] cluster (dithionite reduced); F. FT-infrared spectroscopy. For further details, see Gomes (1998), Leal and Gomes (2008), and Todorovic et al. (2007). (Gomes, C. M. et al. 1998. Di-cluster, seven-iron ferredoxins from hyperthermophilic *Sulfolobales*. *JBIC* 3 (1):499–507. Leal, S. S., and C. M. Gomes. 2008. On the relative contribution of ionic interactions over iron-sulfur clusters to ferredoxin stability. *Biochim Biophys Acta* 1784 (11):1596–1600. Todorovic, S. et al. 2007. A spectroscopic study of the temperature induced modifications on ferredoxin folding and iron-sulfur moieties. *Biochemistry* 46 (37):10733–38. With permission.)

might be the source of irreversibility, when the effect of varying the heating rate is monitored using a combination of different techniques. This is exemplified by studies on the thermal unfolding transition of a rubredoxin, which was monitored by differential scanning calorimetry and visible spectroscopy at increasing heating rates. In this case, while the Tm values determined by calorimetry almost did not change, those determined by visible absorption, which monitored directly the iron site degradation, were very affected, clearly suggesting that the Fe-S site is the source of the irreversibility of the reaction (Henriques, Saraiva, and Gomes 2006). One other approach to overcoming this limitation is the characterization of unfolding pathways, which may elicit the involvement of intermediates or highlight the influence of a particular interaction in the protein or Fe-S cluster, therefore providing an insight into the reverse process. Also, a direct comparison between related proteins may disclose different unfolding mechanisms, such as those observed between the rubredoxin from *Methanocaldococcus jannaschii*, which undergoes thermal denaturation via a simple two-step mechanism with concomitant loss of tertiary structure, hydrophobic collapse, and disintegration of the iron-sulfur center (Henriques, Saraiva, and Gomes 2005), and that of *Pyrococcus furiosus*, which has a complex kinetic behavior comprising at least three intermediate steps (Cavagnero et al. 1998).

DI-CLUSTER FERREDOXINS AS A CASE STUDY

In recent years we have been extensively studying the family of seven iron di-cluster ferredoxins from thermoacidophilic archaea as working models for folding and stability studies in Fe-S proteins. These are small (~11.6 kDa) acidic (pI ~3.5) proteins that contain a $[3Fe-4S]^{1+/0}$ (cluster I) and a $[4Fe-4S]^{2+/1+}$ (cluster II) centers within a $(\beta\alpha\beta)_2$ core fold and an N-terminal extension (~30 amino acids) that comprise a His/Asp Zn^{2+} site (Gomes et al. 1998) (Figure 5.3). These are highly abundant cytosolic proteins, which has allowed *in-cell* EPR studies to establish the biological relevance of $[3Fe-4S]^{1+/0}$ centers (Teixeira et al. 1995), in opposition to the possibility that they could have arisen from oxidative damage of a $[4Fe-4S]^{2+/1+}$ cluster (Moura et al. 1982; Bell et al. 1982). The crystal structure of the *Sulfolobus tokodaii* ferredoxin elicited the presence of an as yet unknown zinc site (Fujii et al. 1997). However, this structure was obtained from an oxidative corruption of the native form, with the tetranuclear center converted into a trinuclear form originating two $[3Fe-4S]^{1+/0}$ centers (Fujii et al. 1997), which prompted several studies attempting to explain the discrepancy between the native and the oxidative damaged form (Iwasaki et al. 2000; Iwasaki and Oshima 1997). This aspect has been recently clarified with the determination of the crystal structure of the *Acidianus ambivalens* ferredoxin (*Aa*Fd) at 2.0 Å resolution, in its physiological form harboring the intact clusters (Frazao et al. 2008). Unless otherwise noted, the latter has been our main working model, which is described in the following sections.

FERREDOXIN HYPERSTABILITY AND UNFOLDING PATHWAYS

The proteomes from thermophiles consist of naturally thermal-resistant proteins, but early studies have shown that thermophilic di-cluster ferredoxins proteins were amazingly resistant to denaturation: Incubation at 70°C for 72 h has no denaturing effect

Iron-Sulfur Clusters, Protein Folds, and Ferredoxin Stability

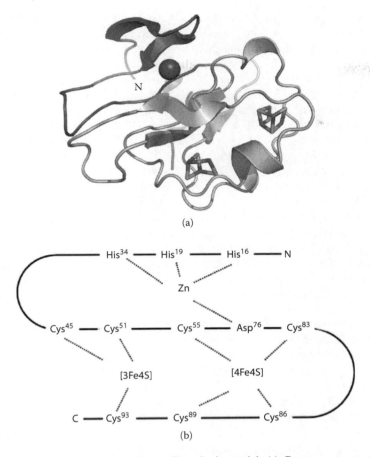

FIGURE 5.3 Structure of the di-cluster Ferredoxin model. (a) Cartoon representation of the AaFd (PBD 2vkr) representing the Fe-S clusters, the zinc ion, and the N-terminal 36 amino acid extension (darker chain). (b) Scheme depicting the metal ion and Fe-S coordination. More details on the protein structure at Frazao et al. (2008). (Frazao, C. et al. 2008. Crystallographic analysis of the intact metal centres [3Fe-4S](1+/0) and [4Fe-4S](2+/1+) in a Zn(2+)-containing ferredoxin. *FEBS Lett* 582 (5):763–67. With permission.)

(Gomes et al. 1998), and in particular for AaFd incubation at neutral pH with 8 M GuHCl also does not result in protein unfolding (Wittung-Stafshede, Gomes, and Teixeira 2000). These are unique properties considering the fact that these proteins harbor cuboid inorganic structures, which may undergo degradative conversions, as in many other Fe-S proteins (Beinert, Holm, and Munck 1997). Therefore the AaFd was a challenging model to use as a tool to characterize the source of protein hyperstability and to discriminate between the relative contributions arising from the chemical properties of the polypeptide chain and those of the Fe-S clusters and zinc site. Since AaFd was not unfolding at pH 7.0, even in the presence of 8 M GuHCl, more extreme pH values (4 < pH > 10) had to be screened in order to perturb electrostatic interactions and therefore decrease the melting transition to below the boiling point of water (Moczygemba et al. 2001; Wittung-Stafshede, Gomes, and Teixeira 2000).

For these reasons, the initial mechanistic studies on *Aa*Fd unfolding were carried out at pH 10, and it has been suggested that ferredoxin unfolding would involve the degradation of the Fe-S clusters via a linear three-iron-sulfur center intermediate, on the basis of the appearance of visible bands at 520 nm and 610 nm (Jones et al. 2002), identical to the one observed when this cluster is formed in purple aconitase (Kennedy et al. 1984). The latter is produced when inactive [3Fe4S] aconitase is exposed to pH >9 or is partly denatured with urea (Kennedy et al. 1984). A purple compound, for which there is no known biological function, is then generated with a characteristic set of spectroscopic signatures at 520 and 610 nm, which can be reconverted back to the cubic [4Fe4S] active form, upon lowering the pH and incubation with an iron salt and a thiol. This observation in *Aa*Fd was later generalized to other iron-sulfur proteins (Griffin et al. 2003; Pereira et al. 2002), even those containing Fe-S centers with lower nuclearities (Higgins, Meyer, and Wittung-Stafshede 2002; Higgins and Wittung-Stafshede 2004); unlike in aconitase, the reconversion back to the original clusters was not taking place. However, subsequent kinetic and spectroscopic analysis aimed at analyzing if one of the clusters had a preponderant role in the formation of this hypothetical intermediate have established that shortly after initial protein unfolding, iron release proceeds monophasically at a rate comparable to that of cluster degradation, and that no typical EPR features of linear three-iron-sulfur centers are observed (Leal, Teixeira, and Gomes 2004). Further, it was observed that EDTA prevents the formation of the transient bands and that sulfide significantly enhances its intensity and lifetime, even after protein unfolding. Altogether these observations showed that such intermediate was instead a soluble iron sulfide that was produced when the Fe-S clusters were disassembled at pH 10; in fact, the misleading spectroscopic fingerprint was even shown to be generated upon alkaline unfolding of other Fe-containing proteins, provided that sulfide was added when iron was released (Leal and Gomes 2005). On the other hand, ferredoxin unfolding under acid conditions (pH 2) released sulfides converted to the volatile hydrogen sulfide forms, the unfolding kinetics is monophasic, and cluster disruption and protein unfolding are simultaneous events.

Effects of Electrostatic Interactions and Metal Centers on Ferredoxin Stability

What are then the factors underlying the extraordinary stability of this ferredoxin? Clearly electrostatic interactions are playing a role, and so are the Fe-S clusters and eventually the His/Asp zinc site as well, but can we determine the relative contributions of ionic interactions over iron-sulfur clusters to the stability of this ferredoxin? With this goal in mind, we have carried out a study in which the protein conformational changes and stability of Fe-S clusters were evaluated upon poising the protein at different pH values, within the 2 < pH < 11.5 range. In the pH 5–8 interval, the protein has a very high apparent melting temperature (T_m ~ 120°C), which nevertheless decreases toward pH extremes, and acidification is more destabilizing than alkalinization. The structural rationale for this observation is that on the one hand the protein has no notorious superficial basic residues near the Fe-S clusters, whose deprotonation would be likely to influence the stability of the centers. On the other hand, a set of five ion pairs could be disrupted upon acidification, thus contributing more to destabilization.

Acidification below pH 5 triggers events in two steps: Down to the isoelectric point (pH 3.5) the Fe-S clusters remain unchanged, but there is a rearrangement of the secondary structure, and the single Trp becomes more solvent shielded, denoting structural packing resulting from protonation, presumably of Asp and Glu residues (Figure 5.4). Interestingly, the pH change within the interval 3.5–2.5 has a minor effect on the Fe-S

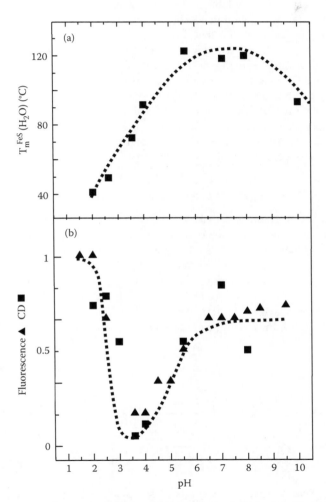

FIGURE 5.4 Effect of the pH on the stability and conformation of AaFd. (a) Effect on the thermal stability of the Fe-S clusters as determined from monitoring cluster disassembly at 410 nm. (b) Effects on the protein conformation as evaluated by the normalized variations on the 222 nm CD signal (■) and from the shift in the Trp-emission maximum (▲). Further details in Leal and Gomes (2008). (Leal, S. S., and C. M. Gomes. 2008. On the relative contribution of ionic interactions over iron-sulfur clusters to ferredoxin stability. *Biochim Biophys Acta* 1784 (11):1596–1600. With permission.)

clusters, as shown by the fact that the 410 nm absorption band and chemical shifts of the β-CH$_2$ protons from the clusters ligand cysteines remain essentially unaffected (Leal and Gomes 2008). The relative stabilizing contribution of the clusters becomes evident when stabilizing ionic interactions are switched off as a result of poising the protein at pH 3.5, at an overall null charge: Under these conditions, the Fe-S clusters disassemble at $T_m = 72°C$ (Figure 5.4), whereas the protein unfolds at $T_m = 52°C$. This stabilizing effect is also evident under other buffer conditions as the presence of EDTA lowers the observed melting temperature in thermal ramp experiments and the midpoint denaturant concentration in equilibrium chemical unfolding experiments (Leal, Teixeira, and Gomes 2004). Further drop to pH 1.5 resulted in a marked increase in the 1-anilino naphthalene-8-sulfonic acid (ANS) fluorescence emission that is concomitant with the disintegration of Fe-S clusters. In fact, at pH 1.5 the spectroscopic signature of the Fe-S clusters is absent and a ~20-fold increase in the ANS emission is observed, in comparison to that of the native state at pH 7 (Figure 5.5). These results clearly indicate that acidification below pH 3.5 will disintegrate the Fe-S clusters, making hydrophobic regions simultaneously accessible. Figure 5.5 summarizes schematically these modifications as a function of pH, and it is clear that the Fe-S clusters play a key role in the high stability of this ferredoxin.

ROLE OF THE HIS/ASP ZINC CENTER IN FERREDOXIN STABILITY

At this point we hypothesized that the zinc center could as well play a role in this process, as protonation of the His/Asp zinc ligands could induce a destabilizing effect. In agreement, a thermal stability study carried out on the related ferredoxin isoforms from *Sulfolobus metallicus* (which differ almost exclusively on the presence or absence of the zinc center ligands and zinc ion) showed that the isoform carrying the zinc center (FdA) was considerably more destabilized by acidic conditions ($T_m \approx 41°C$ at pH 2) than the ferredoxin lacking the zinc center (FdB), which was more stable ($T_m \approx 64°C$) (Rocha et al. 2006). Interestingly, in these two ferredoxin isoforms the Zn^{2+}-lacking

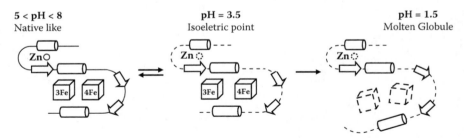

FIGURE 5.5 Scheme depicting the effects of electrostatic interactions on ferredoxin structure and metal centers stability. Protein destabilization is illustrated by the dashed line, and the dashed cubes represent the disintegration of the Fe-S clusters into iron and sulfide; the dotted contours in the zinc illustrate the fact that it is not known if under these conditions the zinc ion remains protein-bound or not. Detailed experimental details supporting this scheme can be found in Leal and Gomes (2008). (Leal, S. S., and C. M. Gomes. 2008. On the relative contribution of ionic interactions over iron-sulfur clusters to ferredoxin stability. *Biochim Biophys Acta* 1784 (11):1596–1600. With permission.)

isoform was always found to be more stable than its Zn^{2+}-containing counterpart: A $\Delta T_m \approx 9°C$ is determined at pH 7, a difference that becomes even more significant at extreme pH values, reaching a $\Delta T_m \approx 24°C$ at pH 2 and 10. The contribution of the Zn^{2+} site to ferredoxin stability was further resolved using selective metal chelators: During thermal unfolding the zinc scavenger TPEN significantly lowers the T_m in FdA ($\approx 10°C$) whereas it has no effect in FdB. This has clearly shown that the Zn^{2+} site contributes to ferredoxin stability but that FdB has devised a structural strategy that accounts for an enhanced stability without using a metal "cross-linker." In fact, the analysis of the FdB amino acid sequence and structural model shows that stabilization afforded by the zinc coordination has been replaced by nonpolar contacts formed between the residues that occupy the same spatial region where the zinc ligands are found in FdA (Figure 5.6) (Rocha et al. 2006).

MONITORING FERREDOXIN UNFOLDING AT DIFFERENT LEVELS OF METALLOPROTEIN ORGANIZATION

In order to analyze globally the changes in protein conformation and the disintegration of Fe-S clusters occurring during ferredoxin unfolding, a large variety of complementary biophysical methods were used to monitor ferredoxin unfolding at different levels of the organization of this metalloprotein (Todorovic et al. 2007). For this purpose, the ferredoxin thermal unfolding was monitored at pH 2.5, as under these conditions the protein fold and its iron-sulfur centers remain intact, although destabilized, making them prone to unfolding studies with T_m transitions well below 100°C (Leal and Gomes 2007; Todorovic et al. 2007).

Interestingly, calorimetric analysis showed that the thermal transition of AaFd occurs in two steps, one corresponding to an exothermic transition ($T_m \approx 55-60°C$) and the other to an endothermic one ($T_m \approx 65-70°C$). Exergonic reactions observed by

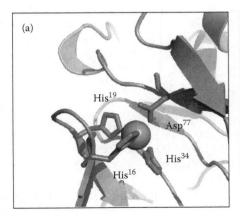
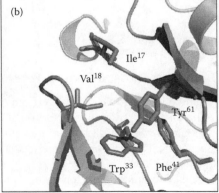

FIGURE 5.6 Cartoon representation of the zinc-binding regions of the two *S. metallicus* ferredoxin models (Rocha et al. 2006), highlighting the zinc-containing form and its His/Asp ligands (a) and the form without zinc but in the place of which a set of hydrophobic interactions effectively replace the clipping effect of the metal ion (b).

differential scanning calorimetry (DSC) are usually associated with protein aggregates, but in this case no precipitation was observed. Rather, this was interpreted as resulting from the decomposition of the Fe-S center, as exothermic transitions had been observed during DSC studies on bovine adrenodoxin, a [2Fe2S]-containing protein. Therefore the first DSC transition was assigned to the decomposition of the Fe-S clusters, and the second transition to the protein-unfolding reaction, which involves heat energy uptake (Leal and Gomes 2007). This was further investigated using other methodologies, many of which are depicted in Figure 5.1: In addition to direct quantitation of released Fe and Zn, the structural variations occurring during ferredoxin unfolding were analyzed by: (1) far-UV CD and FTIR to assess changes of the secondary structure; (2) Trp-emission and near-ultraviolet (near-UV) CD to detect tertiary contacts; (3) visible CD and absorption changes at 410 nm to monitor coarse alterations of the [3Fe4S] and [4Fe4S] cluster moieties, and (4) paramagnetic ^{1}H-NMR and RR spectroscopy to provide fine details of the disassembly of the metal centers (Todorovic et al. 2007). This approach showed that thermally induced unfolding of AaFd is initiated with the loss of α-helical content at relatively low temperatures (with onset at ~45°C), followed by the disruption of both iron-sulfur clusters (T_m ~ 55–60°C). The observed disruption of α-helices occurring at lower temperatures was hypothesized to involve essentially helices α1 and α3, which are far way from the $(\beta\alpha\beta)_2$ core that harbors the Fe-S clusters cysteine ligands. RR and NMR data indicated that the last steps in the disassembly of the clusters during thermal unfolding are the disruption of the terminal Cys^{89} ligand from the [4Fe4S] center and of the bridging ligands of the [3Fe4S] cluster. The degradation of the metal centers triggers major structural changes of the protein matrix, including loss of tertiary contacts (T_m ~ 58°C) and a change, rather than a significant net loss, of secondary structure (T_m ~ 60°C). The determined midpoint for the zinc release during the thermal transition (T_m^{Zn} ≈ 69°C) shows that disintegration of the His/Asp zinc site occurs at a temperature higher than that at which the Fe-S centers break up, suggesting that the N-terminal extension of the protein to which the Zn site is bound is probably the last region of the protein to unfold. Taken together these data allowed for outlining a mechanistic scheme that depicts the ferredoxin thermal unfolding pathway (Figure 5.7).

The Molten Globule State of Ferredoxin

Protein conformations can range from the tightly packed, homogenous native state to the unstructured, unfolded form that comprises a very large ensemble of conformations. In between these extremes, more or less structured protein states can be found, which are presumably intermediates of the folding process and encompass the so-called molten globules.

Although this designation groups a heterogeneous set of conformations, a set of common characteristic properties is more or less consensual: (1) secondary structure content comparable to that of the native state, (2) absence of most of the tertiary structure, (3) a large exposed hydrophobic surface area as a result of a loosely packed hydrophobic core, and (4) compactness comparable to that of the native state, with a slightly (<20%) increased radius.

Iron-Sulfur Clusters, Protein Folds, and Ferredoxin Stability

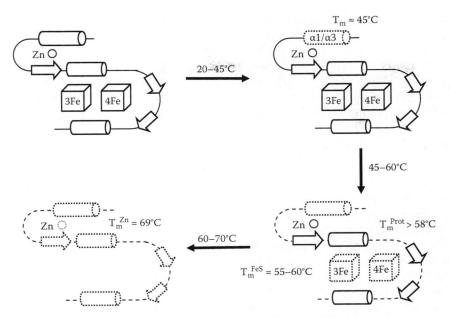

FIGURE 5.7 Scheme depicting the changes occurring during thermal unfolding of *Aa*Fd at pH 2.5. The dotted cylinders and arrows denote structural destabilization of secondary-structure elements, and the dashed line denotes protein destabilization. The dashed cubes represent the disintegration of the Fe-S clusters and the so do the dotted contours in the zinc. Detailed experimental details supporting this scheme can be found in Leal and Gomes (2007) and Todorovic et al. (2007). (Leal, S. S., and C. M. Gomes. 2007. Studies of the molten globule state of ferredoxin: structural characterization and implications on protein folding and iron-sulfur center assembly. *Proteins* 68 (3):606–16. Todorovic, S. et al. 2007. A spectroscopic study of the temperature induced modifications on ferredoxin folding and iron-sulfur moieties. *Biochemistry* 46 (37):10733–38. With permission.)

Interestingly, we have observed that the apo-ferredoxin state obtained upon cooling thermally unfolded *Aa*Fd has characteristics of a molten globule: compactness identical to the native form, similar secondary structure evidenced by far-UV CD, no tertiary contacts detected by near-UV CD, and an exposure of the hydrophobic surface evidenced by ANS binding (Leal and Gomes 2007). This was a very exciting finding, as although the role of molten globule states in the protein-folding process is far from being consensual, some models postulate that proteins fold through a common type of intermediate related to molten globules. In contrast to the native form, this apo ferredoxin state undergoes *reversible* thermal and chemical unfolding. Its conformational stability was investigated by GdmCl denaturation, and this state is ~1.5 kcal. mol^{-1} destabilized in respect to the holo ferredoxin. The conformational dynamics of the molten globule state were analyzed using the single tryptophan located nearby the Fe-S pocket as a probe: A combination of fluorescence quenching, red edge emission shift analysis, and resonance energy transfer to bound ANS evidenced a restricted mobility and confinement within a hydrophobic environment. In this respect, it can be hypothesized that the newly synthesized apo-ferredoxin, which will interact with the Fe-S assembly machinery proteins, may have features in common with the molten globule state described. In fact, structural flexibility and a nonnative conformation could be prerequisite characteristics of the target apo iron-sulfur proteins that will receive the Fe-S clusters. In this perspective, apo-collapsed nonnative states such as the one described for *AaFd* may turn out to be good working models for newly synthesized peptides of iron-sulfur proteins, providing an experimental framework that will allow investigating the mechanism of the cofactor assembly.

FUTURE PERSPECTIVE

Iron is essential for many biological processes, and in particular Fe-S clusters are fundamental to the biological activity of many different proteins and numerous cellular processes. Although our knowledge of the biological machineries involved in the cellular biogenesis of Fe-S clusters has been increasing, we still lack a mechanistic understanding of the processes through which metals are inserted into their target protein. For example, do recipient apoproteins require a particular structural feature to be able to accommodate the Fe-S cluster? Do molecular chaperones play a role or are required to maintain a specific conformation on the recipient protein? Although these processes can be successfully reproduced *in vitro* through the combining of required components, little attention has been paid to the conformational and structural features of the apo- and scaffolding proteins used in these assays. One possibility for moving forward in this field would be to analyze Fe-S biosynthesis and incorporation more from the protein perspective, combining methodologies that provide information on the protein structure (CD, FTIR) with those that monitor the formation of the Fe-S cluster (EPR, Mössbauer, visible absorption). Also, with respect to metal insertion the biological incorporation of zinc, which remains poorly understood, needs to be analyzed. In what concerns more fundamental aspects of the interplay between Fe-S clusters and the polypeptide chain, it is clear that both components contribute to the stability and folding of the conformation, but the study of the conformational properties of proteins in which Fe-S clusters undergo functional interconversions would certainly provide insights into the interactions that fine-tune this interaction.

REFERENCES

Beinert, H., R. H. Holm, and E. Munck. 1997. Iron-sulfur clusters: nature's modular, multipurpose structures. *Science* 277 (5326):653–59.

Bell, S. H., D. P. Dickson, C. E. Johnson, R. Cammack, D. O. Hall, and K. K. Rao. 1982. Mossbauer spectroscopic evidence for the conversion of [4 Fe-4 S] clusters in *Bacillus stearothermophilus* ferredoxin into [3 Fe-3 S] clusters. *FEBS Lett* 142 (1):143–46.

Botelho, H. M., S. S. Leal, A. Veith, V. Prosinecki, C. Bauer, R. Frohlich, A. Kletzin, and C. M. Gomes. 2010. Role of a novel disulfide bridge within the all-beta fold of soluble Rieske proteins. *J Biol Inorg Chem* 15 (2):271–81.

Cavagnero, S., Z. H. Zhou, M. W. Adams, and S. I. Chan. 1998. Unfolding mechanism of rubredoxin from *Pyrococcus furiosus*. *Biochemistry* 37 (10):3377–85.

Feng, Y., N. Zhong, N. Rouhier, T. Hase, M. Kusunoki, J. P. Jacquot, C. Jin, and B. Xia. 2006. Structural insight into poplar glutaredoxin C1 with a bridging iron-sulfur cluster at the active site. *Biochemistry* 45 (26):7998–8008.

Frazao, C., D. Aragao, R. Coelho, S. S. Leal, C. M. Gomes, M. Teixeira, and M. A. Carrondo. 2008. Crystallographic analysis of the intact metal centres [3Fe-4S](1+/0) and [4Fe-4S] (2+/1+) in a Zn(2+)-containing ferredoxin. *FEBS Lett* 582 (5):763–67.

Fujii, T., Y. Hata, M. Oozeki, H. Moriyama, T. Wakagi, N. Tanaka, and T. Oshima. 1997. The crystal structure of zinc-containing ferredoxin from the thermoacidophilic archaeon *Sulfolobus* sp. strain 7. *Biochemistry* 36 (6):1505–13.

Gomes, C. M., A. Faria, J. C. Carita, J. Mendes, M. Regalla, P. Chicau, H. Huber, K. O. Stetter, and M. Teixeira. 1998. Di-cluster, seven-iron ferredoxins from hyperthermophilic *Sulfolobales*. *J Biol Inorg Chem* 3 (1):499–507.

Griffin, S., C. L. Higgins, T. Soulimane, and P. Wittung-Stafshede. 2003. High thermal and chemical stability of *Thermus thermophilus* seven-iron ferredoxin. Linear clusters form at high pH on polypeptide unfolding. *Eur J Biochem* 270 (23):4736–43.

Henriques, B. J., L. M. Saraiva, and C. M. Gomes. 2005. Probing the mechanism of rubredoxin thermal unfolding in the absence of salt bridges by temperature jump experiments. *Biochem Biophys Res Commun* 333 (3):839–44.

———. 2006. Combined spectroscopic and calorimetric characterisation of rubredoxin reversible thermal transition. *J Biol Inorg Chem* 11 (1):73–81.

Higgins, C. L., J. Meyer, and P. Wittung-Stafshede. 2002. Exceptional stability of a [2Fe-2S] ferredoxin from hyperthermophilic bacterium *Aquifex aeolicus*. *Biochim Biophys Acta* 1599 (1–2):82–89.

Higgins, C. L., and P. Wittung-Stafshede. 2004. Formation of linear three-iron clusters in *Aquifex aeolicus* two-iron ferredoxins: effect of protein-unfolding speed. *Arch Biochem Biophys* 427 (2):154–63.

Iwasaki, T., and T. Oshima. 1997. A stable intermediate product of the archaeal zinc-containing 7Fe ferredoxin from *Sulfolobus* sp. strain 7 by artificial oxidative conversion. *FEBS Lett* 417 (2):223–26.

Iwasaki, T., E. Watanabe, D. Ohmori, T. Imai, A. Urushiyama, M. Akiyama, Y. Hayashi-Iwasaki, N. J. Cosper, and R. A. Scott. 2000. Spectroscopic investigation of selective cluster conversion of archaeal zinc-containing ferredoxin from *Sulfolobus* sp. strain 7. *J Biol Chem* 275 (33):25391–401.

Johnson, D. C., D. R. Dean, A. D. Smith, and M. K. Johnson. 2005. Structure, function, and formation of biological iron-sulfur clusters. *Annu Rev Biochem* 74:247–81.

Jones, K., C. M. Gomes, H. Huber, M. Teixeira, and P. Wittung-Stafshede. 2002. Formation of a linear [3Fe-4S] cluster in a seven-iron ferredoxin triggered by polypeptide unfolding. *J Biol Inorg Chem* 7 (4–5):357–62.

Kennedy, M. C., T. A. Kent, M. Emptage, H. Merkle, H. Beinert, and E. Munck. 1984. Evidence for the formation of a linear [3Fe-4S] cluster in partially unfolded aconitase. *J Biol Chem* 259 (23):14463–71.

Kletzin, A., A. S. Ferreira, T. Hechler, T. M. Bandeiras, M. Teixeira, and C. M. Gomes. 2005. A Rieske ferredoxin typifying a subtype within Rieske proteins: spectroscopic, biochemical and stability studies. *FEBS Lett* 579 (5):1020–26.

Leal, S. S., and C. M. Gomes. 2005. Linear three-iron centres are unlikely cluster degradation intermediates during unfolding of iron-sulfur proteins. *Biol Chem* 386 (12):1295–1300.

———. 2007. Studies of the molten globule state of ferredoxin: structural characterization and implications on protein folding and iron-sulfur center assembly. *Proteins* 68 (3):606–16.

———. 2008. On the relative contribution of ionic interactions over iron-sulfur clusters to ferredoxin stability. *Biochim Biophys Acta* 1784 (11):1596–1600.

Leal, S. S., M. Teixeira, and C. M. Gomes. 2004. Studies on the degradation pathway of iron-sulfur centers during unfolding of a hyperstable ferredoxin: cluster dissociation, iron release and protein stability. *J Biol Inorg Chem* 9 (8):987–96.

Lill, R., and U. Muhlenhoff. 2006. Iron-sulfur protein biogenesis in eukaryotes: components and mechanisms. *Annu Rev Cell Dev Biol* 22:457–86.

Meyer, J. 2008. Iron-sulfur protein folds, iron-sulfur chemistry, and evolution. *J Biol Inorg Chem* 13 (2):157–70.

Mihara, H., and N. Esaki. 2002. Bacterial cysteine desulfurases: their function and mechanisms. *Appl Microbiol Biotechnol* 60 (1–2):12–23.

Mitou, G., C. Higgins, P. Wittung-Stafshede, R. C. Conover, A. D. Smith, M. K. Johnson, J. Gaillard, A. Stubna, E. Munck, and J. Meyer. 2003. An Isc-type extremely thermostable [2Fe-2S] ferredoxin from *Aquifex aeolicus*. Biochemical, spectroscopic, and unfolding studies. *Biochemistry* 42 (5):1354–64.

Moczygemba, C., J. Guidry, K. L. Jones, C. M. Gomes, M. Teixeira, and P. Wittung-Stafshede. 2001. High stability of a ferredoxin from the hyperthermophilic archaeon *A. ambivalens*: involvement of electrostatic interactions and cofactors. *Protein Sci* 10 (8):1539–48.

Moura, J. J., I. Moura, T. A. Kent, J. D. Lipscomb, B. H. Huynh, J. LeGall, A. V. Xavier, and E. Munck. 1982. Interconversions of [3Fe-3S] and [4Fe-4S] clusters. Mossbauer and electron paramagnetic resonance studies of *Desulfovibrio* gigas ferredoxin II. *J Biol Chem* 257 (11):6259–67.

Pereira, M. M., K. L. Jones, M. G. Campos, A. M. Melo, L. M. Saraiva, R. O. Louro, P. Wittung-Stafshede, and M. Teixeira. 2002. A ferredoxin from the thermohalophilic bacterium *Rhodothermus marinus*. *Biochim Biophys Acta* 1601 (1):1–8.

Rocha, R., S. S. Leal, V. H. Teixeira, M. Regalla, H. Huber, A. M. Baptista, C. M. Soares, and C. M. Gomes. 2006. Natural domain design: enhanced thermal stability of a zinc-lacking ferredoxin isoform shows that a hydrophobic core efficiently replaces the structural metal site. *Biochemistry* 45 (34):10376–84.

Sanchez-Ruiz, J. M., J. L. Lopez-Lacomba, M. Cortijo, and P. L. Mateo. 1988. Differential scanning calorimetry of the irreversible thermal denaturation of thermolysin. *Biochemistry* 27 (5):1648–52.

Smith, A. D., J. N. Agar, K. A. Johnson, J. Frazzon, I. J. Amster, D. R. Dean, and M. K. Johnson. 2001. Sulfur transfer from IscS to IscU: the first step in iron-sulfur cluster biosynthesis. *J Am Chem Soc* 123 (44):11103–4.

Teixeira, M., R. Batista, A. P. Campos, C. Gomes, J. Mendes, I. Pacheco, S. Anemuller, and W. R. Hagen. 1995. A seven-iron ferredoxin from the thermoacidophilic archaeon *Desulfurolobus ambivalens*. *Eur J Biochem* 227 (1–2):322–27.

Todorovic, S., S. S. Leal, C. A. Salgueiro, I. Zebger, P. Hildebrandt, D. H. Murgida, and C. M. Gomes. 2007. A spectroscopic study of the temperature induced modifications on ferredoxin folding and iron-sulfur moieties. *Biochemistry* 46 (37):10733–38.

Urbina, H. D., J. J. Silberg, K. G. Hoff, and L. E. Vickery. 2001. Transfer of sulfur from IscS to IscU during Fe/S cluster assembly. *J Biol Chem* 276 (48):44521–26.

Watanabe, S., A. Kita, K. Kobayashi, and K. Miki. 2008. Crystal structure of the [2Fe-2S] oxidative-stress sensor SoxR bound to DNA. *Proc Natl Acad Sci U S A* 105 (11):4121–26.

Wittung-Stafshede, P., C. M. Gomes, and M. Teixeira. 2000. Stability and folding of the ferredoxin from the hyperthermophilic archaeon *Acidianus ambivalens*. *J Inorg Biochem* 78 (1):35–41.

Yabe, T., E. Yamashita, A. Kikuchi, K. Morimoto, A. Nakagawa, T. Tsukihara, and M. Nakai. 2008. Structural analysis of *Arabidopsis* CnfU protein: an iron-sulfur cluster biosynthetic scaffold in chloroplasts. *J Mol Biol* 381 (1):160–73.

6 Folding and Stability of Myoglobins and Hemoglobins

David S. Culbertson and John S. Olson

CONTENTS

Introduction to Myoglobin and Hemoglobin 97
    Globins in Biology 97
    Basic Globin Structures...98
    Factors Governing Mb and Hb Stabilities..99
Myoglobin and Monomeric Hemoglobin Unfolding..99
    Structure of Apo-Mb 99
    Previous Studies of Apo-Mb Unfolding... 100
    General Equilibrium Mechanisms for the Unfolding
    of Monomeric Holo-Globins... 102
    Holo-Mb Unfolding Studies... 107
Mammalian Hemoglobins.. 112
    Erythropoiesis 112
    Proposed Mechanism for Hb Assembly *In Vivo*.. 113
    Mechanism for Apo-Hb Dimer Unfolding.. 113
    Complete Mechanism for the Unfolding of Holo-HbA 116
Conclusions and Perspectives 117
References... 118

INTRODUCTION TO MYOGLOBIN AND HEMOGLOBIN

GLOBINS IN BIOLOGY

The hemoglobin family of proteins is ubiquitous in nature and examples are found in vertebrates, invertebrates, plants, fungi, bacteria, and archaea (Vinogradov et al. 2005). Mbs and Hbs have diverse functions ranging from gas sensing to NO scavenging to O_2 management. Vertebrate hemoglobins function primarily as O_2 transporters packaged in red blood cells and deliver O_2 from the alveolar gas spaces to striated skeletal and cardiac muscles and actively respiring neuronal tissues. The oxygenated forms of these proteins, HbO_2 and MbO_2, scavenge and detoxify NO by dioxygenating it to nitrate (Flogel et al. 2001; Eich et al. 1996; Doherty et al. 1998; Olson et al. 2004). There are also a wide variety of invertebrate animal hemoglobins, which range from small mini-globin monomers (Pesce et al. 2002) to huge molecular weight aggregates that are found in extracellular spaces of polychaete and annelide worms (Strand et al. 2004).

BASIC GLOBIN STRUCTURES

The basic globin fold consists of several α-helices that are arranged in a pseudo spherical shape, with a flexible hydrophobic packet for heme binding. Globins are divided into four groups: (1) protoglobins with more than eight α helices, (2) truncated two-on-two helical Hbs (2/2), (3) single-domain three-on-three helical globins (3/3—often with two other short helices), and (4) multiple-domain 3/3 helical globins that contain another domain involved in redox reactions or other activities, including protein kinases and DNA binding (Vinogradov et al. 2005). Mammalian Mbs and hemoglobin α, β, δ, and γ subunits are classical single-domain 3/3 globins. They have similar primary sequences and folds containing seven to eight α helices, labeled A to H, which form a hydrophobic pocket for heme binding (Figure 6.1). Helices C and D are short, variable in length, and located between the longer B and E helices, forming one side of the heme pocket. Sperm whale (sw) Mb has all eight α helices and serves as the model for the structure and function of all single domain globins (Figure 6.1). When heme is bound, globins become highly resistant to denaturation induced by chemical and thermal perturbations (Landfried et al. 2007; Hargrove and Olson 1996).

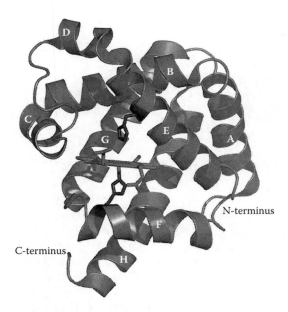

FIGURE 6.1 Basic helical structure of single-domain globins with 3/3 helix fold as found in mammalian Mb and Hb subunits. The structure shown is for native sw met-Mb (1jp6). The proximal and distal histidines coordinate heme (dark sticks), and helices A to H are labeled as well as both termini.

Folding and Stability of Myoglobins and Hemoglobins

FACTORS GOVERNING MB AND HB STABILITIES

Resistance to Oxidation, Heme Loss, and Denaturation. Oxidation of Mbs and Hbs to the ferric form disables their capacity to bind O_2 until reconversion to the ferrous forms by intracellular reductases (Xu, Quandt, and Hultquist 1992). The rate of heme loss from ferrous Mb is at least 50 to 100 times slower than from the ferric form, and globin unfolding does not appear to occur readily at pH values ≥7.0 until the heme group is lost (Hargrove, Wilkinson, and Olson 1996.). Thus under physiological conditions, auto-oxidation almost certainly occurs prior to denaturation.

Some globins have evolved to be able to retain functionality under demanding conditions and need to be resistant to oxidation, heme loss and denaturation. Sperm whales (*Physeter catodon*), as well other marine mammals that undergo extended periods of apnea when diving, posess higher skeletal muscle Mb concentration to increase O_2 delivery capacity (Kooyman and Ponganis 1998). The apo-forms of these Mbs display overall folding constants, which are 20 to 100 times greater than their terrestrial mammalian orthologues, making them much easier to study in terms of identifying stable intermediates (Scott, Paster, and Olson 2000).

Oligomerization and Hb Stability. The formation of oligomers confers enhanced stability to globins, which is strongly dependent on protein concentration, with disassembly being favored at low concentrations. In erythrocytes, tetrameric Hb is highly concentrated and present at ~5 mM on a tetramer basis. At these levels, tetrameric human hemoglobin A is highly resistant to dissociation into dimers and monomers, the fractions of which are extremely small. Organisms that use extracellular globins for O_2 transport have even larger molecular weights. The hemoglobin from *Lumbricus terrestris* is strongly resistant to denaturation as it is assembled into a giant Hb complex composed of 144 subunits coupled to a central linker protein scaffold with a total molecular weight of ~3.6 MDa (Strand et al. 2004). Oligomerization enhances the stability of proteins both by the strength of the new subunit interfaces and by the multiplication factor for the individual stabilities. For example, the folding constant for an individual subunit is taken to the fourth power when calculating the overall stability of a homo-tetramer as in the equation:

$$K_{folding}(\text{tetramer}) = K_{interfaces}(K_{folding}(\text{monover})) \tag{6.1}$$

where K interfaces can be complex functions depending on the number and types of interfaces.
This effect accounts for the markedly enhanced resistance of human α and β subunits to denaturation when they are assembly into hemoglobin tetramers (Bunn 1987; Antonini and Brunori 1971; Hargrove et al. 1997).

MYOGLOBIN AND MONOMERIC HEMOGLOBIN UNFOLDING

STRUCTURE OF APO-MB

Although ultra-high-resolution structures of holo-Mb from several species have been determined by x-ray crystallography, the crystal structure of "native" apo-Mb has not been determined. Apo-Mb can be produced *in vitro* by extraction of the heme cofactor into organic solvents at low pH (Ascoli, Fanelli, and Antonini 1981). When samples of apo-Mb are neutralized, the resultant protein has lost about a third of its helicity as measured by its CD spectrum, but the remaining tertiary and secondary structures appear to be similar to those found in the holo-form. High-resolution nuclear magnetic resonance (NMR) experiments indicate loss of structure in the F helix, the EF and FG loops, the N-terminal portions of the G helix, and the C-terminus of the H helix, whereas the A, B, C, D, E, and most of the G and H helices appear intact (Lecomte et al. 1999; Eliezer et al. 1998).

PREVIOUS STUDIES OF APO-MB UNFOLDING

Equilibrium Unfolding. The unfolding of apo-Mb has been extensively studied *in vitro*. The existence of a folding intermediate was first proposed in 1976 (Balestrieri et al. 1976), building on previous fluorescence experiments (Kirby and Steiner 1970), and has led to the analysis of apo-Mb unfolding using a three-state unfolding mechanism (Barrick and Baldwin 1993; Hargrove, Krzywda et al. 1994; Ramsay, Ionescu, and Eftink 1995). An example of a GuHCl-induced titration of sw apo-Mb from our own recent data is shown in Figure 6.2, which represents many similar curves published previously (Smith 2003; Ramsay, Ionescu, and Eftink 1995; Culbertson and Olson 2010; Scott, Paster, and Olson 2000). The acid- and urea-induced intermediates have been structurally characterized by circular

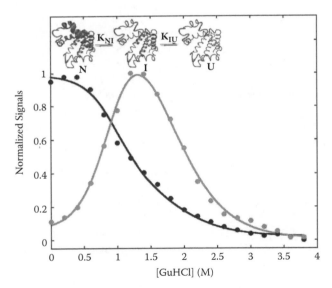

FIGURE 6.2 GuHCl-induced equilibrium unfolding of wild-type SW apo-Mb monitored by CD at 222 nm (black dots) and fluorescence at 345 nm (gray squares). The structures of the N, I, and U states are symbolically represented for their supposed secondary-structure content. The lines (black, CD data; gray, fluorescence data) represent simultaneous fits to a three-state model, yielding K_{NI} and K_{IU} values in buffer of 0.021 (±0.002) and 0.019 (±0.002), respectively. (Culbertson, D. S. and Olson, J. S. 2010. Role of heme in the unfolding and assembly of myoglobin, *Biochemistry*, in press.)

dichroism (CD), site-directed mutagenesis, and hydrogen exchange NMR experiments. The results suggest that the A, G, H, and part of the B helices remain folded and docked to each other, forming a hydrophobic core, whereas the other regions of the protein, including most of the heme pocket, are highly disordered and flexible (Hughson, Wright, and Baldwin 1990; Eliezer et al. 1998, 2000; Nishimura, Wright, and Dyson 2003; Griko et al. 1988; Hughson and Baldwin 1989; Hughson, Barrick, and Baldwin 1991).

The intermediate (I) has been shown to possess higher fluorescence than either the original native or end point unfolded states (Figure 6.2). The emission peak for this I state is at a wavelength in between those for the folded and unfolded states when the intermediate is induced by GuHCl (Ramsay, Ionescu, and Eftink 1995). One theory is that partial unfolding to the I state causes an increase in fluorescence of Trp7 due to an apolar collapse of the EF loop onto the G and H helices, which moves Lys79 and Met131 away from the tryptophan and buries the indole ring in a more hydrophobic environment (Kirby and Steiner 1970; Irace et al. 1981; Hargrove, Krzywda et al. 1994; Tcherkasskaya et al. 2000; Tcherkasskaya, Ptitsyn, and Knutson 2000; Ballew, Sabelko, and Gruebele 1996). Another theory is that hyperfluorescence may be caused by enhanced flexibility of the fluorophores in the intermediate state (Ervin et al. 2002).

The three-state N-to-I-to-U equilibrium-unfolding path is easily observed under acid and GuHCl denaturation conditions (Griko et al. 1988; Nishii, Kataoka, and Goto 1995; Barrick and Baldwin 1993; Barrick, Hughson, and Baldwin 1994) but is less readily observed when unfolding is induced by urea due to higher fluorescence emission of the unfolded state (Baryshnikova et al. 2005; Barrick and Baldwin 1993). At low pH, sw apo-Mb may populate two distinct intermediates (i.e., I_a and I_b) that differ by additional secondary structure in helix B (Loh, Kay, and Baldwin 1995; Jamin and Baldwin 1998); however, only one equilibrium state intermediate has been reported using GuHCl and urea as denaturants at neutral pH.

Analysis of GuHCl-Induced Equilibrium Unfolding of Apo-Mb. The three-state equilibrium unfolding mechanism involves consecutive N-to-I and I-to-U transitions as shown in Figure 6.2. These two transitions are well defined by the fluorescence data, which alone are sufficient to obtain stability parameters for both transitions. However, the intermediate for sw apo-Mb is also relatively well defined by far-UV CD data, and both detection signals can be fit simultaneously to expressions derived from the three-state mechanism shown in Equations 6.2, 6.3, and 6.4 (Barrick and Baldwin 1993; Hargrove, Krzywda et al. 1994; Scott, Paster, and Olson 2000; Smith 2003). In these equations, K_{NI} and K_{IU} are the equilibrium unfolding constants for the N-to-I and I-to-U transitions, and the linear dependencies of the corresponding free energies for the N-to-I and I-to-U transitions on GuHCl concentration ([X]) are represented by M_{NI} and M_{IU} in the following equations:

$$K_{NI} = K_{NI}^{0} exp^{\frac{M_{NI}[X]}{RT}}$$

$$K_{IU} = K_{IU}^{0} exp^{\frac{M_{IU}[X]}{RT}} \qquad (6.2)$$

The fractional populations of the N, I, and U states are defined as Y_N, Y_I, and Y_U and given by:

$$Y_N = \frac{[N]}{[N]+[I]+[U]} = \frac{1}{1+K_{NI}+K_{NI}K_{IU}}$$

$$Y_I = \frac{[I]}{[N]+[I]+[U]} = \frac{K_{NI}}{1+K_{NI}+K_{NI}K_{IU}}$$

$$Y_U = \frac{[U]}{[N]+[I]+[U]} = \frac{K_{NI}K_{IU}}{1+K_{NI}+K_{NI}K_{IU}} \quad (6.3)$$

The measured signal at a given [GuHCl] is the weighted sum of the individual signals, S_N, S_I, and S_U, for each state:

$$S = S_N Y_N + S_I Y_I + S_U Y_U$$

$$= \frac{S_N + S_I K_{NI}^0 \exp\left(\frac{{}^M NI[X]}{RT}\right) + S_U K_{NI}^0 K_{IU}^0 \exp\left(\frac{({}^M NI + {}^M IU)[X]}{RT}\right)}{1 + K_{NI}^0 \exp\left(\frac{{}^M NI[X]}{RT}\right) + K_{NI}^0 K_{IU}^0 \exp\left(\frac{({}^M NI + {}^M IU)[X]}{RT}\right)} \quad (6.4)$$

where [X] equals the concentration of denaturant, which in Figure 6.2 is GuHCl.

A complete derivation of this equation is given in the Appendix, derivation A. For wild-type sw apo-Mb, the K_{NI}^0 and K_{IU}^0 stability constants in buffer are 0.021 ± 0.002 and 0.019 ± 0.002, respectively, in 10 mM potassium phosphate, pH 7, 20°c. Fitting apo-Mb data to obtain both stability constants requires m_{NI} and m_{IU} to be either set to known values or varied as part of the fit. In Figure 6.2, m_{NI} and m_{IU} were set to 9.85 and 5.68 kJ mol^{-1} M^{-1}, respectively, based on a global fit of several apo-Mb mutants (Smith 2003). The results in Figure 6.2 show that the unfolding of apo-Mb is readily interpreted in terms of a three-state model with a well-defined molten globule intermediate containing roughly 50% of the helicity of the native apo-Mb state.

GENERAL EQUILIBRIUM MECHANISMS FOR THE UNFOLDING OF MONOMERIC HOLO-GLOBINS

Monomeric holo-globins and related type b heme monomers (e.g., cytochrome b_{562}) appear to unfold at equilibrium by simple two-state mechanisms at much lower pH values or higher denaturant concentrations than the first apoprotein transitions. In all cases, the holo-protein is stabilized markedly by the presence of heme (Hargrove and Olson 1996; Robinson et al. 1997). In general, denaturation of holo-Hbs occurs after the heme is oxidized, and many studies have focused on the state of hemin once unfolding of the protein has occurred. In the following sections, we have listed three distinct mechanisms for unfolding of a monomeric holo-Hb (mHb), which include no hemin (H) binding to the unfolded apo-Hb state (U), binding of hemin to the U state, and self-aggregation of hemin to from dimers (H_2). These simple mechanisms differ significantly with respect to the dependence of the unfolding curves on the absolute initial holoprotein concentration.

Simple Model with Monodisperse Free Hemin. The simplest mechanism for unfolding of a monomeric holo-globin assumes that the released hemin remains monodisperse and that, after hemin dissociation from the holo-Hb, unfolding of the resultant native apo-Hb (N) is a simple two-state process, N to U (Figure 6.3A). In this case, the equilibrium constants are defined as:

$$K_{NH} = \frac{[N][H]}{[mHb]} = K_{NH}^{0} exp\left(\frac{M_{NH}[X]}{RT}\right) \quad (6.5)$$

$$K_{NU} = \frac{[U]}{[N]} = K_{NU}^{0} exp\left(\frac{M_{NU}[X]}{RT}\right)$$

and the total protein concentration at any titration point is given by:

$$P_0 = [mHb] + [N] + [U] = [mHb]\left(1 + \frac{K_{NH}}{[H]} + \frac{K_{NH}K_{NU}}{[H]}\right) \quad (6.6)$$

K_{NH} is the equilibrium constant for heme dissociation from the native state, K_{NU} is the unfolding constant for the N to U transition, and mHb, N, U, and H represent monomeric holo-Hb, the folded native apo-state, the unfolded apo-state, and free hemin, respectively. The concentration of denaturant [X] exponentially increases the K_{NH} and K_{NU} constants as defined by m_{NH} and m_{NU}.

Derivations of the fractions of each protein species, mHb, N, U, and free H are given in the Appendix. The expressions are complex because hemin binding is a bimolecular process that is favored at high initial holo-Hb concentrations. As a result, the free concentration of hemin, H, is a function of denaturant and total protein concentration, P_0, and is defined by the following expression:

$$[H] = \frac{-(K_{NH} + K_{NH}K_{NU}) + \sqrt{(K_{NH} + K_{NH}K_{NU})^2 + 4P_0(K_{NH} + K_{NH}K_{NU})}}{2} \quad (6.7)$$

The fraction of native holo-globin is defined by:

$$Y_{mHb} = \frac{[mHb]}{[mHb] + [N] + [U]} = \frac{1}{1 + \frac{K_{NH}}{[H]} + \frac{K_{NH}K_{NU}}{[H]}} \quad (6.8)$$

where [H] is given by Equation 6.7. As shown in Figure 6.3, this simple mechanism predicts that the holo-Hb unfolding curves should be strongly dependent on the total

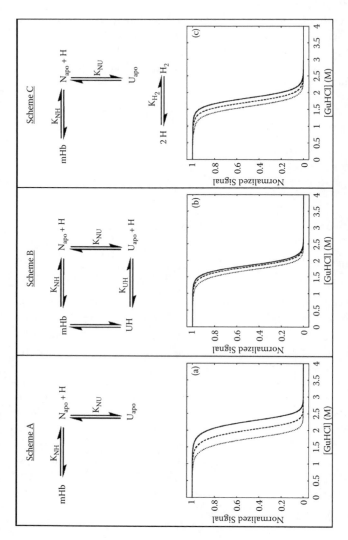

FIGURE 6.3 Effect of protein concentration on the unfolding curves of monomeric globins for three mechanisms with different fates of dissociated hemin. The upper panels describe the reaction schemes, and the lower panels show simulated unfolding curves at $P_0 = 1$ μM (dotted line), 20 μM (dashed line), and 400 μM (solid line). A. Simple mechanism where hemin remains free and monomeric after dissociation and protein unfolding (see Appendix, derivation B). The following constants were used: $K_{NU} = 1.0 \times 10^{-3}$, $m_{NU} = 13$ kJ mol^{-1} M^{-1}, $K_{NH} = 10^{-12}$ M, $m_{NH} = 15$ kJ mol^{-1} M^{-1}, and T= 20°C. B. Mechanism where dissociated hemin can bind to the unfolded state U (see Appendix, derivation C). The values from A were used and additionally: $K_{UH} = 1.0 \times 10^{-8}$ M and $m_{UH} = 7$ kJ mol^{-1} M^{-1}. C. Mechanism where dissociated hemin self-associates to form a dimer (see Appendix, derivation D). The values from A were used and additionally: $K_{H2} = 4.5 \times 10^6$ M (de Villiers et al. 2007), $m_{H2} = 4$ kJ mol^{-1} M^{-1}.

protein concentration. A marked right-shift of the GuHCl-induced unfolding curves will occur with increasing protein concentration if hemin remains mono-disperse after dissociation from the globin.

Model with Hemin Binding to the Unfolded State. An expanded mechanism would allow hemin to stay bound upon unfolding or reassociate with the unfolded state to produce a fourth UH state (Figure 6.3B). Binding of hemin to the unfolded U state was originally proposed by Hargrove et al. (1996) because they observed little protein concentration dependence for GuHCl-induced unfolding of holo-Mb, and this phenomenon was also observed for the thermal unfolding of cytochrome b_{562} (Robinson et al. 1998). Hemin is relatively insoluble, binds nonspecifically to a variety of proteins, and is found in globin precipitates in red cells (i.e., Heinz bodies) under pathological conditions (Bunn and Forget 1986).

Hemin binding to unfolded apo-Hb adds another protein species and equilibrium equation to the original mechanism, and this binding and the total protein concentration are defined by:

$$K_{UH} = \frac{[U][H]}{[UH]} = K_{UH}^{0} \exp\left(\frac{{}^{M}U_{H[X]}}{RT}\right) \tag{6.9}$$

$$P_0 = [mHb] + [N] + [U] + [UH] \tag{6.10}$$

The complete derivation for this four-state mechanism is given in the Appendix, derivation C, and again a quadratic equation in [H] is generated with an even more complex root:

$$[H] = \frac{-(K_{NH} + K_{NH}K_{NU})}{2\left(1 + \frac{K_{NH}K_{NU}}{K_{UH}}\right)} +$$

$$\frac{\sqrt{(K_{NH} + K_{NH}K_{NU})^2 + 4P_0\left(1 + \frac{K_{NH}K_{NU}}{K_{UH}}\right)(K_{NH} + K_{NH}K_{NU})}}{2\left(1 + \frac{K_{NH}K_{NU}}{K_{UH}}\right)} \tag{6.11}$$

The fraction of folded holo-mHb is defined as before but contains the additional [UH] term in the denominator:

$$Y_{mHb} = \frac{[mHb]}{[mHb] + [N] + [U] + [UH]} = \frac{1}{1 + \frac{K_{NH}}{[H]} + \frac{K_{NH}K_{NU}}{[H]} + \frac{K_{NH}K_{NU}}{K_{UH}}} \tag{6.12}$$

Allowing heme binding to the unfolded state decreases the stabilizing effect of increasing protein concentration because as P_0 increases, the fractions of mHb and UH both increase. As a result, the unfolding curves show much less dependence on globin concentration (Figure 6.3B). At very high protein concentrations, where P_0 ~100 times the K_d for hemin binding to the U state, there is little or no free hemin during unfolding, and the unfolding process becomes effectively unimolecular, mHb to UH. Under these conditions, there is little protein concentration dependence, and the process appears to be a simple two-state process (see dashed and solid lines in Figure 6.3B).

Model with Hemin Dimerization. Another potential complication is that free hemin can form dimers and higher order aggregates with itself at neutral pH, even at micromolar concentrations (Adams 1977; Brown, Dean, and Jones 1970; de Villiers et al. 2007; Asher et al. 2009). This third mechanism allows self-association of free hemin to form dimers after dissociation from the holo-protein (Figure 6.3c). A complete analysis is given in the Appendix, derivation D. In this case, the additional step is described by a hemin dimerization constant, K_{H2}, which is defined by:

$$K_{H_2} = \frac{[H_2]}{[H]^2} = K_{H_2}^0 \exp\left(\frac{M_{H_2}[X]}{RT}\right) \tag{6.13}$$

In this mechanism, the total free hemin also includes dimers (i.e., $H_{free} = [H] + 2[H_2] = [N] + [U]$). As a result of this bimolecular hemin association step, the final equation for [H] is a cubic equation, and Cardano's method was used to find the appropriate root of the equation:

$$a = 2K_{H_2}$$

$$b = 1 + 2K_{H_2}(K_{NH} + K_{NH}K_{NU})$$

$$c = K_{NH} + K_{NH}K_{NU}$$

$$d = -P_0(K_{NH} + K_{NH} \cdot K_{NU})$$

which is:

$$[H] = -\frac{b}{3a}$$
$$-\frac{1}{3a}\left(\frac{\sqrt[3]{2b^3 - 9abc + 27a^2d + \sqrt{(2b^3 - 9abc + 27a^2d)^2 - 4(b^2 - 3ac)^3}}}{2}\right)$$
$$-\frac{1}{3a}\left(\frac{\sqrt[3]{2b^3 - 9abc + 27a^2d - \sqrt{(2b^3 - 9abc + 27a^2d)^2 - 4(b^2 - 3ac)^3}}}{2}\right)$$

$$\tag{6.14}$$

This value of [H] can be used to compute the fraction of native holo-mHb as defined in the original three-state mechanism.

$$Y_{mHb} = \frac{[mHb]}{[mHb]+[N]+[U]} = \frac{1}{1+\frac{K_{NH}}{[H]}+\frac{K_{NH}K_{NU}}{[H]}} \quad (6.15)$$

In general, the computed unfolding curves for this mechanism show less dependence on protein concentration compared to the mechanism where dissociated hemin remains free and monomeric in solution. The favorable effect of increasing the initial amount of holo-Hb on resistance to hemin loss is compensated for by the increased tendency of free hemin to aggregate at higher concentrations. However, the exact dependence and shapes of the curves will be strongly dependent on how the denaturant or condition affects the hemin dimerization reaction (i.e., positive versus negative m_{H2} values). Thus any thorough analysis requires examining dependence of K_{H2} on pH, temperature, or denaturant.

The key conclusion from the theoretical curves shown in Figure 6.3 is that holo-Hb unfolding needs to be measured and analyzed carefully as function of total protein concentration. This variation is the best tool for discriminating between the three possible mechanisms shown in Figure 6.3. The 1 to 400 µM holo-globin concentrations were chosen to cover the range used previously in in vitro studies with Hbs and Mbs. In cases where unfolding curves for holo-Mb were measured as function of P_0, the observed dependence was small and similar to that shown in Figures 6.3B and 6.3C, indicating the dissociated hemin binds to the unfold state, may also dimerize, but is probably never free and monodisperse in solution (Hargrove and Olson 1996). Thus the more complex schemes in Figures 6.3B and 6.3C need to be the starting points for any analysis of monomeric holo-globin unfolding, despite the apparent simplicity of the individual unfolding curves at a fixed protein concentration.

Holo-Mb Unfolding Studies

The mechanism of holo-Mb unfolding has been addressed in only a few studies, and most analyses have been semi-empirical and assume an apparent two-state equilibrium and/or kinetic mechanism. The resistance of holo-Mb to unfolding appears to be determined primarily by heme affinity (Hargrove and Olson 1996), and therefore redox state and bound ligands are major factors influencing stability. For example, the reduced, unliganded form of Mb (deoxy-Mb [Fe^{II}]) is significantly more resistant to chemical denaturation than the oxidized form (met-Mb [Fe^{III}]) (Hargrove and Olson 1996). As described in the previous section, a quantitative analysis of the observed unfolding curves requires a detailed mechanism that accounts for the fate of the dissociated heme or hemin; i.e., does it self-associate, bind to unfolded and intermediate states, or participate in all these processes?

Heme Binding to Apo-Mb. Heme affinity is too high to measure directly in simple titration experiments. Estimates of the equilibrium association constant can be obtained from the ratio of the rates of heme association to apo-Mb and dissociation from holo-Mb. The value for sw Mb and hemin is ~10^{14} M^{-1} (Hargrove, Barrick, and Olson 1996). Three major factors appear to contribute to this high heme affinity: (1) nonspecific hydrophobic interactions between the porphyrin and the hydrophobic pocket (10^5–10^7 M^{-1}), (2) covalent bond formation between the ferric iron and the proximal His93 (10^3–10^4 M^{-1}), and (3) specific interactions with conserved amino acids in the heme pocket (10^3–10^4 M^{-1}) (Hargrove, Wilkinson, and Olson 1996; Hargrove, Barrick, and Olson 1996; Hargrove, Singleton et al. 1994).

The first kinetic studies of heme binding to apo-Mb and apo-Hb were carried out in the 1960s (Antonini and Gibson 1960; Gibson and Antonini 1960). In most cases, the reaction of free heme with apoglobin shows a biphasic time course, with a rapid bimolecular phase (K'_H ~1 × 10^8 M^{-1} s^{-1}) and a slow first-order phase, which itself shows heterogeneity but little or no dependence on apo-protein concentration. The slow phases have been attributed to the presence of heme dimers or higher aggregates that have to dissociate before binding to globin can occur (de Villiers et al. 2007). To avoid these problems, most workers have used CO-heme or dicyano-heme, which appear to be monodisperse up to concentrations of 10 μM due to the presence of axial ligands (Gibson and Antonini 1960; Rose and Olson 1983; Light et al. 1987; Hargrove, Barrick, and Olson 1996; Chiba et al. 1994; Kawamura-Konishi, Kihara, and Suzuki 1988).

General Mechanism for Holo-Mb Unfolding. Heme binds strongly and rapidly to the folded N state of apo-Mb; the key question is whether heme interacts significantly with intermediates prior to complete folding, i.e., with the I and/or U states of apo-Mb. A few studies suggest heme may interact with the kinetic intermediate during folding (Chiba et al. 1994; Eliezer et al. 1993) and nonspecifically with the unfolded state as described in the preceding section (Hargrove and Olson 1996).

A general mechanism for holo-Mb unfolding is shown in Figure 6.4 and based on the three-state mechanism of apo-Mb unfolding shown in Figure 6.2 and the idea that

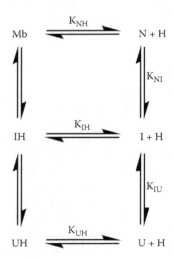

FIGURE 6.4 Scheme for equilibrium unfolding of sw holo-Mb, based on the three-state equilibrium unfolding of sw apo-Mb and heme binding to all three states.

H could have significant affinity for the I and U states (Hargrove and Olson 1996). To obtain heme dissociation constants for all three states (K_{NH}, K_{IH}, and K_{UH}) using this model, the K_{NI} and K_{IU} stability constants for the N-to-I and I-to-U states of apo-Mb need to be determined independently. These five parameters are required to define the equilibrium unfolding curves for holo-Mb. Although a simple two-state model has been used to analyze holo-Mb unfolding, such a one-step mechanism does not adequately describe the underlying processes, which are governed by both apo-Mb stabilities and heme affinities. For example, Hargrove and Olson (Hargrove and Olson 1996) showed that the unfolding of His97Asp holo-Mb cannot be analyzed by a two-state mechanism because, for this mutant, hemin dissociation occurs before unfolding of the apo-Mb intermediate (I) and two phases are observed for the CD changes.

The same mathematical approach for determining the fraction of monomeric holo-globin described in the previous section can be used for deriving expressions from the complete holo-Mb mechanism (cf. Appendix). The free concentration of heme or hemin [H] is obtained from the following root of the complex quadratic equation described in the Appendix, derivation E:

$$[H] = \frac{-(K_{NH} + K_{NH}K_{NI} + K_{NH}K_{NI}K_{IU})}{2\left(1 + \frac{K_{NH}K_{NI}}{K_{IH}} + \frac{K_{NH}K_{NI}K_{IU}}{K_{UH}}\right)} +$$

$$\frac{\sqrt{(K_{NH} + K_{NH}K_{NI} + K_{NH}K_{NI}K_{IU})^2 + 4P_0\left(1 + \frac{K_{NH}K_{NI}}{K_{IH}} + \frac{K_{NH}K_{NI}K_{IU}}{K_{UH}}\right)(K_{NH} + K_{NH}K_{NI} + K_{NH}K_{NI}K_{IU})}}{2\left(1 + \frac{K_{NH}K_{NI}}{K_{IH}} + \frac{K_{NH}K_{NI}K_{IU}}{K_{UH}}\right)}$$

(6.16)

The expression for the fraction of holo-Mb is also more complex because there are six protein forms.

$$Y_{Mb} = \frac{1}{1 + \frac{K_{NH}}{[H]} + \frac{K_{NH}K_{NI}}{[H]} + \frac{K_{NH}K_{NI}K_{IU}}{[H]} + \frac{K_{NH}K_{NI}}{K_{IH}} + \frac{K_{NH}K_{NI}K_{IU}}{K_{UH}}} \quad (6.17)$$

Expressions for the fractions of each of these species as a function of denaturant [GuHCl] in the Appendix, derivation E.

GuHCl-induced unfolding curves for sw met-Mb (Fe^{III}) are shown in Figure 6.5. In these experiments, hemin absorbance, CD, and fluorescence spectra were measured for individual equilibrated mixtures of the protein and denaturant. All three detection methods show steep unfolding curves with roughly the same midpoint [GuHCl], suggesting an apparently simple and highly cooperative two-state apparent unfolding process, despite the complexity of the scheme in Figure 6.4.

These data alone cannot define all the parameters for the general holo-Mb unfolding scheme and must be combined with apo-Mb unfolding data to obtain K_{NI} and K_{IU} independently (i.e., the fit in Figure 6.2). With these apo-Mb unfolding parameters and m-values fixed from the data in Figure 6.2 (K_{NI}^0 and K_{IU}^0 = 0.021 and 0.019,

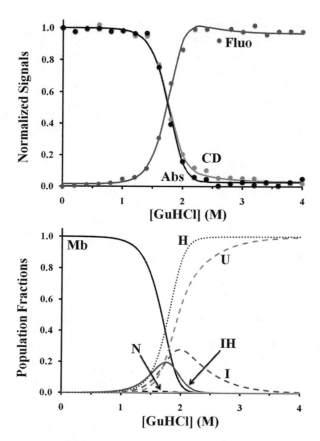

FIGURE 6.5 Upper panel: Holo-Mb unfolding curves fitted to the six-state scheme in Figure 6.4 with $K_{NH}^0 = 8.8 \times 10^{-14}$ M, $K_{UH}^0 = 1.0 \times 10^{-6}$ M, and $K_{IH}^0 = 1.4 \times 10^{-11}$ M, and m_{NH}, m_{IH}, and m_{UH} are set to 18.3, 16.6, and 8.0 kJ mol^{-1} M^{-1}, respectively. The apo-Mb unfolding parameters were taken from the fits in Figure 6.3, and $P_0 = 10$ μM. Lower panel: Theoretical curves for holo-Mb (Mb), native apo-Mb (N), free hemin (H), holo-intermediate (IH), apo-intermediate (I), unfolded holo-state (UH), and unfolded apo-state (U).

respectively), it is possible to simultaneously fit the observed holo-Mb unfolding curves and obtain values for the K_{NH} and K_{IH} (K_{UH} is fixed to 10^{-6} M) dissociation constants with high precision (lines in Figure 6.5). The correspondence between the observed data and the computed curves is remarkably good, considering that the key apo-Mb unfolding parameters were fixed (Figure 6.5a). The fractions of each intermediate during unfolding are shown in the lower panel. Clearly the only major protein species at 10 μM P_o are Mb and U, with only a small amount of I and little of IH.

The dissociation constant K_{NH}^0 for holo-Mb was calculated to be to be 8.8×10^{-14} M at pH 7, which agrees with the affinities reported previously for wild-type sw met-Mb (Hargrove and Olson 1996; Hargrove, Barrick, and Olson 1996). In addition, the fit was also able to estimate K_{IH}^0 to be 1.4×10^{-11} M despite the lack of an apparent inflection point in the curves. Thus our data suggest that the affinity of the

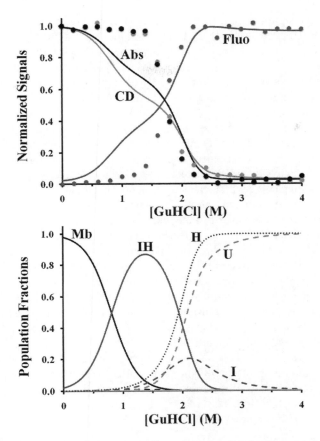

FIGURE 6.6 Upper panel: Holo-Mb unfolding curves fitted to the six-state scheme in Figure 6.4 with K_{IH}^0 set equal to K_{NH}^0, which implies strong hemin binding to the apo-Mb intermediate. The curves were computed for $P_0 = 10$ μM, using the apo-Mb unfolding parameters in Figure 6.2, and K_{NH}^0 and $K_{IH}^0 = 1.6 \times 10^{-12}$ M, $K_{UH}^0 = 1.0 \times 10^{-6}$ M, and m_{NH}, m_{IH}, and m_{UH} are set to 18.3, 16.6, and 8.0 kJmol^{-1} M^{-1}, respectively. Lower panel: Theoretical curves for holo-Mb (Mb), native apo-Mb (N), free hemin (H), holo-intermediate (IH), apo-intermediate (I), unfolded holo-state (UH), and unfolded apo-state (U).

apo-Mb intermediate for heme is ~160-fold weaker than that of the fully folded apo-form. If the N and I states are forced to have equal affinities for hemin, the fit to the observed data is very poor, because equal hemin affinity for the N and I states forces holo-Mb unfolding to resemble a three-state process with the buildup of a large amount of the IH intermediate upon unfolding (Figure 6.6). Thus when analyzed using the complete scheme shown in Figure 6.4, the observed data demonstrate that the affinity of hemin for the I state is much less than that for the N state and that hemin binds well only to fully folded wild-type apo-Mb.

K_{UH}^0 cannot be determined exactly at one protein concentration. However, the estimate of ~10^{-6} M, used in the fits, is realistic based on affinity constants for non-specific hemin binding to unfolded apo-cytochrome b_{562} (Robinson et al. 1997) and to native bovine serum albumin (Marden et al. 1989; Hargrove, Barrick, and Olson 1996). The fitted, relative values for K_{NH}^0, K_{IH}^0, and K_{UH}^0 demonstrate that it is the affinity of hemin for the N state that dictates the overall stability of the holoprotein as previously suggested by a less rigorous analysis (Hargrove and Olson 1996). However, we have recently shown that the IH state is populated in mutant Mbs with unstable apo-N states or poor hemin affinity, and that this species is a hemichrome (Culbertson and Olson 2010).

The dependence of the unfolding curve on holo-protein concentration can be a useful tool to determine the fate of hemin after dissociation. Reduced dependence indicates hemin binding to the unfolded state and self-association to form hemin dimers. Previous equilibrium studies of aquomet-Mb unfolding showed very little right-shifting of the unfolding curve with increasing protein concentration (Hargrove and Olson 1996). This lack of protein concentration dependence is readily explained by nonspecific hemin binding to the intermediate and unfolded states (i.e., Figure 6.3, scheme B). Hemin dimerization can also account for part of the lack for protein concentration dependence. The exact cause of the lack of protein concentration dependence and well-defined heme dissociation equilibrium constants can be determined by analyses of unfolding curves over a wider range of protein concentration and more careful examinations of the dependence of hemin dimerization on the nature of the denaturant.

The globins involved in O_2 transport and storage are highly concentrated *in vivo*. Myoglobin is present in skeletal muscles at ~0.6mM (Bekedam et al. 2009) and in heart myocytes at ~1.5mM (Nemeth and Laury 1984). These concentrations are 100- to 1000-fold greater than those used in most biophysical studies (low μM range). Thus *in vivo*, unfolding of native holo-Mb is effectively a first-order process. Mb will unfold into UH, with little net loss of hemin even if K_{UH} is as large as 10^{-6} M. Thus the extent of unfolding of holo-Mb will be governed by both strong preferential affinity of the folded state for hemin, K_{NH} versus K_{UH}, and the overall stability of the apoprotein.

MAMMALIAN HEMOGLOBINS

ERYTHROPOIESIS

Even though Hb has been extensively studied for over a century, the detailed mechanisms involved in Hb assembly during erythropoiesis remain unclear. Hb is the main component in mature erythrocytes and reaches a concentration of 20–22 mM on a per heme basis. Among developmental variants, hemoglobin A (HbA, A for adult) is the most prevalent in human erythrocytes (97%) and is composed of two α and two β subunits. The β subunits found in HbA are replaced by δ and γ subunits in hemoglobin A_2 (HbA_2) and fetal hemoglobin (HbF), respectively.

Expression of Hb subunits and biosynthesis of heme are strictly regulated and any deviation from the 2:2:4 ratio can lead to severe cytotoxic effects (Chen 2007). The synthesis of heme is tightly regulated by several control points in its anabolic and catabolic pathways and, importantly, these levels regulate gene transcription and translation of Hb chains (Chen 2007; Taketani 2005; Han et al. 2001). The α and β chains are translated independently on ribosomes in the pre-erythroid cell cytoplasm, whereas the final iron insertion step in heme synthesis occurs in the inner membrane of mitochondria. It is still uncertain whether apo-Hb chains fold co-translationally or post-translationally, and whether heme binds before or after subunit assembly into dimers. *In vitro* studies have demonstrated that without heme present, globin synthesis does not proceed, and the resultant apo-chains are unstable and rapidly aggregate to form insoluble precipitates (Chen 2007). However, the question of when heme binds remains elusive.

Proposed Mechanism for Hb Assembly *In Vivo*

At the minimum, formation of functional Hb involves folding of monomers, association of monomers and dimers, and heme binding. A simplified model for the assembly of HbA is shown in Figure 6.7. This scheme assumes that heme binding occurs only after apo-subunits assemble to an $\alpha_1\beta_1$ dimer. This idea is based on the observation that heme removal from tetrameric hemoglobin results in the formation of $\alpha_1\beta_1$ apo-dimers (Winterhalter and Huehns 1964; Yip, Waks, and Beychok 1972), which have 48% of the helicity of the holoprotein (Hrkal and Vodrazka 1967) and are still capable of reacting rapidly with free heme to reconstitute intact and functional holo-tetramers (Winterhalter and Huehns 1964; Winterhalter and Colosimo 1971). When heme is removed from isolated α and β subunits, almost all secondary structure is lost, the apo-proteins precipitate at temperatures greater than 15°C, and reconstitution with added heme does not occur in the absence of partner subunits (Waks, Yip, and Beychok 1973). The scheme in Figure 6.7 has served as our working hypothesis for examining apo- and holo-Hb unfolding curves and designing mutants that will express and assemble more readily during heterologous expression in bacteria (Graves et al. 2008).

Until recently, heme-proteins were believed to fold without the help of molecular chaperones. The discovery of the alpha hemoglobin stabilizing protein, AHSP, by Weiss and coworkers (Gell et al. 2002; Kihm et al. 2002) has refocused attention on the assembly of Hb. AHSP is an erythroid protein that binds to α chains and inhibits heme loss and precipitation. The affinity of AHSP for α chains is much less that of β chains, and thus addition of β subunits to α:AHSP complexes leads to dissociation and uptake of α chains by the β subunits to produce intact hemoglobin. These properties suggest AHSP may be a chaperone or escort protein for α subunits during hemoglobin assembly (Yu et al. 2007; Bank 2007; Weiss et al. 2005; Mollan et al. 2010); however, its exact role during erythropoiesis remains unclear.

Mechanism for Apo-Hb Dimer Unfolding

As described earlier, removal of heme from native HbA generates an $\alpha_1\beta_1$ apodimer. The unfolding of this apo-Hb $\alpha_1\beta_1$ dimer into denatured monomeric subunits can occur by several mechanisms. The three most likely pathways for unfolding are shown in Figure 6.8. For simplicity, the α and β subunits are assumed to have similar properties. In mechanism A, the dimer is assumed to unfold and dissociate into unfolded monomers in a single concerted process (Figure 6.8a). In mechanism B, the dimer dissociates into folded monomers, which in turn unfold via a monomeric intermediate into completely unfolded monomers (Figure 6.8b). In mechanism C,

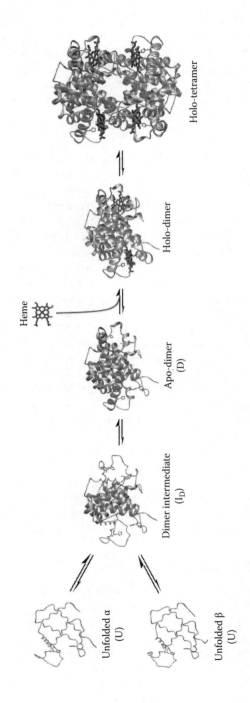

FIGURE 6.7 Proposed hemoglobin assembly scheme. Unfolded subunits partially fold and assemble, followed by further folding, heme binding, and association of $\alpha_1\beta_1$ dimers.

Folding and Stability of Myoglobins and Hemoglobins

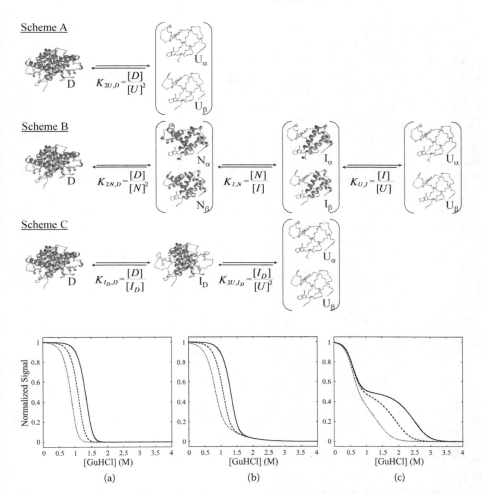

FIGURE 6.8 Possible unfolding mechanisms for apo-Hb dimers and simulations of GuHCl-induced equilibrium unfolding based on these mechanisms. D, I_D, N, I, and U represent the dimer, dimer intermediate, native, intermediate, and unfolded states. (Top) A. Model in which the apo-Hb dimer dissociates and unfolds in one step. $K_{2U,D} = 6.25 \times 10^{10}$ M; $m_{2U,D} = 30$ kJ mol^{-1} M^{-1}. B. Model in which the dimer unfolds to two folded states followed by partial unfolding to an intermediate I and then complete unfolding. $K_{2N,D} = 1.0 \times 10^8$ M, $m_{2N,D} = 2$ kJ mol^{-1} M^{-1}, $K_{I,N} = 5$, $m_{I,N} = 9$ kJ mol^{-1} M^{-1}, $K_{U,I} = 5$, $m_{U,I} = 5$ kJ mol^{-1} M^{-1}. C. Model involving a partially folded dimer intermediate and then dissociation into unfolded monomers. $K_{ID,D} = 100$, $m_{ID,D} = 18$ kJ mol^{-1} M^{-1}, $K_{2U,ID} = 6.25 \times 10^8$ M, and $m_{2U,ID} = 12$ kJ mol^{-1} M^{-1}. The K and m values in all three models were set to estimated values that were mathematically equivalent and gave midpoint [GuHCl]s similar to those observed in experiments by Graves et al. (2008) at 5°C, pH 7 for P_0 ~3–10 μM. The α and β subunits were created with PyMol and PDB entry 1ird and, for these mechanisms, are assumed to be equivalent in stability. (Bottom) Simulations for the unfolding of the apo-Hb dimer according to schemes A, B, and C are shown in panels (a), (b), and (c) respectively. Protein concentration was varied from 1 (dotted) to 20 (dashed) to 400 (solid line) μM. Equations for all three models are given in the Appendix, derivations F, G, and H.

the dimer first unfolds to an intermediate with ~50% loss of helicity but with the subunits still forming an $\alpha_1\beta_1$ interface. Then, this dimeric I_D state dissociates into completely unfolded monomers (Figure 6.8c). Because dissociation into monomers occurs in each case, all three mechanisms predict a strong dependence on total protein concentration, with increased resistance to unfolding at high P_0. Derivations of expressions for the unfolding curves predicted by each mechanism are given in the Appendix, derivatives F, G, and H. These equations were used to simulate GuHCl-induced unfolding curves as a function of total protein concentration, P_0, and results at 1, 20, and 400 µM subunits are shown in Figure 6.8, panels (a), (b), and (c).

The unfolding curves predicted by mechanism A are steep, as expected for a simple two-state mechanism (i.e., D or 2U), and show a marked right-shift to greater stability at higher protein concentrations. Although more complex with four states, mechanism B also predicts steep unfolding curves if reasonable values of the folding parameters are used. In this mechanism, the major resistance to unfolding is dimer dissociation into the much less stable monomers, which unfold completely at the denaturant concentrations required to disrupt the interface of the original apodimer. Again, a strong dependence on protein concentration is predicted. In contrast, mechanism C predicts broad unfolding curves, which become biphasic at high protein concentrations. In this case, the initial step involves unimolecular conformational changes, which are independent of protein concentration because the dimer remains intact. The second step in mechanism C is bimolecular involving dissociation into unstable monomers, which is markedly inhibited at high protein concentrations.

Preliminary data collected by Philip E. Graves (Graves 2008) and the authors indicate that the apo-Hb dimer unfolding does show broad unfolding curves that begin to show two distinct phases of roughly equal amplitudes at high apo-hemoglobin concentrations. These data suggest that mechanism C applies to apo-HbA. The simplest structural interpretation is that, like apo-Mb, the initial additions of denaturant cause the unfolding of the heme pocket and its associated B, C, D, and E helices, whereas the A, G, and H helices remain intact in both the α and β subunits, preserving most of the $\alpha_1\beta_1$ interface, which involves helices B, G, and H in the intact Hb tetramer. Further addition of denaturant simultaneously disrupts the dimer interface and unfolds the A, G, and H helices leading to completely unfolded subunits. However, these results and interpretations are very preliminary, and much more work is needed to validate these ideas.

COMPLETE MECHANISM FOR THE UNFOLDING OF HOLO-HBA

The simple mechanism for Hb assembly shown in Figure 6.7 was based in part on the broad apo-Hb unfolding curves, which have been observed in our initial experiments and are simulated by mechanism C in Figure 6.8. In this mechanism, heme binding occurs only after the folded apodimer is formed. However, studies of holo-Mb unfolding suggest that heme can bind nonspecifically to folding intermediates including the completely unfolded state, albeit much more weakly than to the N folded states. If heme binding to all species involved in tetrameric hemoglobin formation is allowed, the much more complex eight-state mechanism shown in Figure 6.9 has to be considered.

Folding and Stability of Myoglobins and Hemoglobins

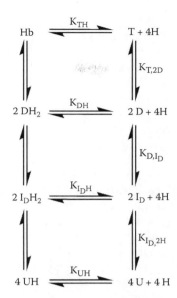

FIGURE 6.9 Proposed general mechanism for HbA unfolding and assembly. The scheme is based on both apo and holo forms of unfolded monomers (U), dimeric intermediates (I_D), fully folded dimers (D), and tetramers (T). The apo-Hb equilibrium constants that need to be defined are $K_{T,2D}$, $K_{D,ID}$, and $K_{ID,2U}$ and the heme dissociation constants for each holo state K_{TH}, K_{DH}, K_{IDH}, and K_{UH}.

This model is complex, requiring a definition of the apo-Hb folding constants describing conversion of completely unfolded apo-subunits to apo-tetramers plus heme dissociates equilibrium constraints for each of the four globin species shown in Figure 6.9. Deriving the expressions for the folded species is much more difficult than for Mb, which involves only monomeric species, and requires finding the roots of depressed quartic equations due to tetramer formation and heme dissociation. The real situation may be even more complex if differences between the heme affinities and stabilities of the α and β subunits are large and if heme dimerization occurs. Thus even the scheme in Figure 6.9 is a simplification. However, this scheme and the additional complications will have to be addressed quantitatively in order to understand fully *in vivo* assembly of Hb in pre-erythroid cells.

CONCLUSIONS AND PERSPECTIVES

In this chapter, we have examined mechanisms for combined analyses of both apo- and holo-globin equilibrium unfolding. Monomeric holo-globin unfolding can be modeled using apoprotein unfolding curves to define key unimolecular equilibrium parameters and then determining the equilibrium dissociation constants for hemin binding to the apo-globin states. In the case of Mb, its apo-form unfolds in a three-state process. Heme binds avidly and specifically to the fully folded N state of apo-Mb, weaker to the intermediate, and even weaker and non specifically to the unfolded state. Because the affinity of heme for the N state is so high ($K_{NH} \leq 10^{-13}$ M), the unfolding of holo-Mb is governed primarily by the free energy required to dissociate the heme group and is less affected by the stabilities of the apo-Mb N and I states. The lack of a strong dependence of holo-Mb unfolding on total protein concentration indicates that heme binds to the intermediate and unfolded states, and that dissociated heme may self-associate. Most of models derived for the monomeric globins apply to the unfolding of other proteins with small molecule or metal cofactors. The results in Figure 6.3 indicate that complete descriptions will require analyses of unfolding as function of protein concentration to discriminate between specific mechanisms.

Although there is no general agreement on the mechanism of apo- and holo-Hb folding and assembly, significant progress is being made. There is initial evidence that supports a model in which unfolded α and β chains first form an apo-globin dimer intermediate with structured A, G, and H helices and an almost intact $\alpha_1\beta_1$ interface. The heme pocket regions in this intermediate then fold and bind heme to form a holo-dimer, which, in turn, self-assembles into holo-tetramers. This model needs to be tested via parallel experiments, using GuHCl, urea, and acid as denaturants under rigorously controlled conditions, and protein concentration needs to be varied systematically. Again, the models for human hemoglobin apply to other oligomeric proteins with cofactors. The importance of working with these complex models is fourfold. These equations provide the framework for understanding: (1) the basic principles of heme protein assembly, (2) clinically important defects in hemoglobin assembly during erythropoiesis in humans (i.e., hemoglobinopathies, thalassemia syndromes, anemias), (3) the conditions that induce hemoglobin denaturation *in vivo* (i.e., agents causing oxidative stress, acidosis, fevers), and (4) the factors that increase yields of holo-globins during heterologous expression in *Escherichia coli*.

REFERENCES

Adams, P. A. 1977. The kinetics of the recombination reaction between apomyoglobin and alkaline haematin. *Biochem J* 163 (1):153–58.

Antonini, E., and M. Brunori. 1971. *Hemoglobin and myoglobin in their reactions with ligands*. Amsterdam: North-Holland.

Antonini, E., and Q. H. Gibson. 1960. Some observations on the kinetics of the reactions with gases of natural and reconstituted haemoglobins. *Biochem J* 76:534–38.

Ascoli, F., M. R. Fanelli, and E. Antonini. 1981. Preparation and properties of apohemoglobin and reconstituted hemoglobins. *Methods Enzymol* 76:72–87.

Asher, C., de Villiers, K.A., and Egan, T.J. 2009. Inorg *Chem* 48, 7994–8003.

Balestrieri, C., G. Colonna, A. Giovane, G. Irace, and L. Servillo. 1976. Equilibrium evidence of non-single step transition during guanidine unfolding of apomyoglobins. *FEBS Lett* 66 (1):60–64.

Ballew, R. M., J. Sabelko, and M. Gruebele. 1996. Direct observation of fast protein folding: the initial collapse of apomyoglobin. *Proc Natl Acad Sci U S A* 93 (12):5759–64.

Bank, A. 2007. AHSP: a novel hemoglobin helper. *J Clin Invest* 117 (7):1746–49.

Barrick, D., and R. L. Baldwin. 1993. Three-state analysis of sperm whale apomyoglobin folding. *Biochemistry* 32 (14):3790–96.

Barrick, D., F. M. Hughson, and R. L. Baldwin. 1994. Molecular mechanisms of acid denaturation. The role of histidine residues in the partial unfolding of apomyoglobin. *J Mol Biol* 237 (5):588–601.
Baryshnikova, E. N., B. S. Melnik, A. V. Finkelstein, G. V. Semisotnov, and V. E. Bychkova. 2005. Three-state protein folding: experimental determination of free-energy profile. *Protein Sci* 14 (10):2658–67.
Bekedam, M.A. et al. 2009. *J Appl Physiol* 107(4), 1138–1143.
Brown, S. B., T. C. Dean, and P. Jones. 1970. Aggregation of ferrihaems. Dimerization and protolytic equilibria of protoferrihaem and deuteroferrihaem in aqueous solution. *Biochem J* 117 (4):733–39.
Bunn, H. F. 1987. Subunit assembly of hemoglobin: an important determinant of hematologic phenotype. *Blood* 69 (1):1–6.
Bunn, H. F. and, and B. G. Forget. 1986. *Hemoglobin: molecular, genetic and clinical aspects*, ed. W. B. S. Company, 2nd ed. Philadelphia: Elsevier Health Sciences.
Chen, J. J. 2007. Regulation of protein synthesis by the heme-regulated eIF2alpha kinase: relevance to anemias. *Blood* 109 (7):2693–99.
Chiba, K., A. Ikai, Y. Kawamura-Konishi, and H. Kihara. 1994. Kinetic study on myoglobin refolding monitored by five optical probe stopped-flow methods. *Proteins* 19 (2):110–19.
Culbertson, D. S. and Olson, J. S. 2010. Role of heme in the unfolding and assembly of myoglobin, *Biochemistry*, in press.
de Villiers, K. A., C. H. Kaschula, T. J. Egan, and H. M. Marques. 2007. Speciation and structure of ferriprotoporphyrin IX in aqueous solution: spectroscopic and diffusion measurements demonstrate dimerization, but not mu-oxo dimer formation. *J Biol Inorg Chem* 12 (1):101–17.
Doherty, D. H., M. P. Doyle, S. R. Curry, R. J. Vali, T. J. Fattor, J. S. Olson, and D. D. Lemon. 1998. Rate of reaction with nitric oxide determines the hypertensive effect of cell-free hemoglobin. *Nat Biotechnol* 16 (7):672–76.
Eich, R. F., T. Li, D. D. Lemon, D. H. Doherty, S. R. Curry, J. F. Aitken, A. J. Mathews, K. A. Johnson, R. D. Smith, G. N. Phillips Jr., and J. S. Olson. 1996. Mechanism of NO-induced oxidation of myoglobin and hemoglobin. *Biochemistry* 35 (22):6976–83.
Eliezer, D., K. Chiba, H. Tsuruta, S. Doniach, K. O. Hodgson, and H. Kihara. 1993. Evidence of an associative intermediate on the myoglobin refolding pathway. *Biophys J* 65 (2):912–17.
Eliezer, D., J. Chung, H. J. Dyson, and P. E. Wright. 2000. Native and non-native secondary structure and dynamics in the pH 4 intermediate of apomyoglobin. *Biochemistry* 39 (11):2894–2901.
Eliezer, D., J. Yao, H. J. Dyson, and P. E. Wright. 1998. Structural and dynamic characterization of partially folded states of apomyoglobin and implications for protein folding. *Nat Struct Biol* 5 (2):148–55.
Ervin, J., E. Larios, S. Osvath, K. Schulten, and M. Gruebele. 2002. What causes hyperfluorescence: folding intermediates or conformationally flexible native states? *Biophys J* 83 (1):473–83.
Flogel, U., M. W. Merx, A. Godecke, U. K. Decking, and J. Schrader. 2001. Myoglobin: a scavenger of bioactive NO. *Proc Natl Acad Sci U S A* 98 (2):735–40.
Gell, D., Y. Kong, S. A. Eaton, M. J. Weiss, and J. P. Mackay. 2002. Biophysical characterization of the alpha-globin binding protein alpha-hemoglobin stabilizing protein. *J Biol Chem* 277 (43):40602–9.
Gibson, Q. H., and E. Antonini. 1960. Kinetic studies on the reaction between native globin and haem derivatives. *Biochem J* 77:328–41.
Graves, P. E. 2008. Enhancing stability and expression of recombinant human hemoglobin in *E. coli*: progress in the development of a recombinant HBOC source. MS diss., Rice University.

Graves, P. E., D. P. Henderson, M. J. Horstman, B. J. Solomon, and J. S. Olson. 2008. Enhancing stability and expression of recombinant human hemoglobin in *E. coli*: progress in the development of a recombinant HBOC source. *Biochim Biophys Acta* 1784 (10):1471–79.

Griko, Y. V., P. L. Privalov, S. Y. Venyaminov, and V. P. Kutyshenko. 1988. Thermodynamic study of the apomyoglobin structure. *J Mol Biol* 202 (1):127–38.

Han, A. P., C. Yu, L. Lu, Y. Fujiwara, C. Browne, G. Chin, M. Fleming, P. Leboulch, S. H. Orkin, and J. J. Chen. 2001. Heme-regulated eIF2alpha kinase (HRI) is required for translational regulation and survival of erythroid precursors in iron deficiency. *Embo J* 20 (23):6909–18.

Hargrove, M. S., D. Barrick, and J. S. Olson. 1996. The association rate constant for heme binding to globin is independent of protein structure. *Biochemistry* 35 (35):11293–99.

Hargrove, M. S., S. Krzywda, A. J. Wilkinson, Y. Dou, M. Ikeda-Saito, and J. S. Olson. 1994. Stability of myoglobin: a model for the folding of heme proteins. *Biochemistry* 33 (39):11767–75.

Hargrove, M. S., and J. S. Olson. 1996. The stability of holomyoglobin is determined by heme affinity. *Biochemistry* 35 (35):11310–18.

Hargrove, M. S., E. W. Singleton, M. L. Quillin, L. A. Ortiz, G. N. Phillips Jr., J. S. Olson, and A. J. Mathews. 1994. His64(E7)→Tyr apomyoglobin as a reagent for measuring rates of hemin dissociation. *J Biol Chem* 269 (6):4207–14.

Hargrove, M. S., T. Whitaker, J. S. Olson, R. J. Vali, and A. J. Mathews. 1997. Quaternary structure regulates hemin dissociation from human hemoglobin. *J Biol Chem* 272 (28):17385–89.

Hargrove, M. S., A. J. Wilkinson, and J. S. Olson. 1996. Structural factors governing hemin dissociation from metmyoglobin. *Biochemistry* 35 (35):11300–309.

Hrkal, Z., and Z. Vodrazka. 1967. A study of the conformation of human globin in solution by optical methods. *Biochim Biophys Acta* 133 (3):527–34.

Hughson, F. M., and R. L. Baldwin. 1989. Use of site-directed mutagenesis to destabilize native apomyoglobin relative to folding intermediates. *Biochemistry* 28 (10):4415–22.

Hughson, F. M., D. Barrick, and R. L. Baldwin. 1991. Probing the stability of a partly folded apomyoglobin intermediate by site-directed mutagenesis. *Biochemistry* 30 (17):4113–18.

Hughson, F. M., P. E. Wright, and R. L. Baldwin. 1990. Structural characterization of a partly folded apomyoglobin intermediate. *Science* 249 (4976):1544–48.

Irace, G., C. Balestrieri, G. Parlato, L. Servillo, and G. Colonna. 1981. Tryptophanyl fluorescence heterogeneity of apomyoglobins. Correlation with the presence of two distinct structural domains. *Biochemistry* 20 (4):792–99.

Jamin, M., and R. L. Baldwin. 1998. Two forms of the pH 4 folding intermediate of apomyoglobin. *J Mol Biol* 276 (2):491–504.

Kawamura-Konishi, Y., H. Kihara, and H. Suzuki. 1988. Reconstitution of myoglobin from apoprotein and heme, monitored by stopped-flow absorption, fluorescence and circular dichroism. *Eur J Biochem* 170 (3):589–95.

Kihm, A. J., Y. Kong, W. Hong, J. E. Russell, S. Rouda, K. Adachi, M. C. Simon, G. A. Blobel, and M. J. Weiss. 2002. An abundant erythroid protein that stabilizes free alpha-haemoglobin. *Nature* 417 (6890):758–63.

Kirby, E. P., and R. F. Steiner. 1970. The tryptophan microenvironments in apomyoglobin. *J Biol Chem* 245 (23):6300–6306.

Kooyman, G. L., and P. J. Ponganis. 1998. The physiological basis of diving to depth: birds and mammals. *Annu Rev Physiol* 60:19–32.

Landfried, D. A., D. A. Vuletich, M. P. Pond, and J. T. Lecomte. 2007. Structural and thermodynamic consequences of b heme binding for monomeric apoglobins and other apoproteins. *Gene* 398 (1–2):12–28.

Lecomte, J. T., S. F. Sukits, S. Bhattacharya, and C. J. Falzone. 1999. Conformational properties of native sperm whale apomyoglobin in solution. *Protein Sci* 8 (7):1484–91.

Light, W. R., R. J. Rohlfs, G. Palmer, and J. S. Olson. 1987. Functional effects of heme orientational disorder in sperm whale myoglobin. *J Biol Chem* 262 (1):46–52.

Loh, S. N., M. S. Kay, and R. L. Baldwin. 1995. Structure and stability of a second molten globule intermediate in the apomyoglobin folding pathway. *Proc Natl Acad Sci U S A* 92 (12):5446–50.

Marden, M. C., E. S. Hazard, L. Leclerc, and Q. H. Gibson. 1989. Flash photolysis of the serum albumin-heme-CO complex. *Biochemistry* 28 (10):4422–26.

Mollan, T. L., X. Yu, M. J. Weiss, and J.S. Olson. 2010. The role of alpha-hemoglobin stabilizing protein in redox chemistry, denaturation, and hemoglobin assembly. *Antioxid Redox Sign* 12 (2):219–31.

Nemeth, P. M., and O. H. Lowry. 1984. Myoglobin levels in individual human skeletal muscle fibers of different types. *J Histochem Cytochem* 32 (11):1211–16.

Nishii, I., M. Kataoka, and Y. Goto. 1995. Thermodynamic stability of the molten globule states of apomyoglobin. *J Mol Biol* 250 (2):223–38.

Nishimura, C., P. E. Wright, and H. J. Dyson. 2003. Role of the B helix in early folding events in apomyoglobin: evidence from site-directed mutagenesis for native-like long range interactions. *J Mol Biol* 334 (2):293–307.

Olson, J. S., E. W. Foley, C. Rogge, A. L. Tsai, M. P. Doyle, and D. D. Lemon. 2004. No scavenging and the hypertensive effect of hemoglobin-based blood substitutes. *Free Radic Biol Med* 36 (6):685–97.

Pesce, A., M. Nardini, S. Dewilde, E. Geuens, K. Yamauchi, P. Ascenzi, A. F. Riggs, L. Moens, and M. Bolognesi. 2002. The 109 residue nerve tissue minihemoglobin from *Cerebratulus lacteus* highlights striking structural plasticity of the alpha-helical globin fold. *Structure* 10 (5):725–35.

Ramsay, G., R. Ionescu, and M. R. Eftink. 1995. Modified spectrophotometer for multi-dimensional circular dichroism/fluorescence data acquisition in titration experiments: application to the pH and guanidine-HCl induced unfolding of apomyoglobin. *Biophys J* 69 (2):701–7.

Robinson, C. R., Y. Liu, R. O'Brien, S. G. Sligar, and J. M. Sturtevant. 1998. A differential scanning calorimetric study of the thermal unfolding of apo- and holo-cytochrome b562. *Protein Sci* 7 (4):961–65.

Robinson, C. R., Y. Liu, J. A. Thomson, J. M. Sturtevant, and S. G. Sligar. 1997. Energetics of heme binding to native and denatured states of cytochrome b562. *Biochemistry* 36 (51):16141–46.

Rose, M. Y., and J. S. Olson. 1983. The kinetic mechanism of heme binding to human apohemoglobin. *J Biol Chem* 258 (7):4298–4303.

Scott, E. E., E. V. Paster, and J. S. Olson. 2000. The stabilities of mammalian apomyoglobins vary over a 600-fold range and can be enhanced by comparative mutagenesis. *J Biol Chem* 275 (35):27129–36.

Smith, L. 2003. The effects of amino acid substitution on apomyoglobin stability, folding intermediates, and holoprotein expression. PhD diss., Rice University.

Strand, K., J. E. Knapp, B. Bhyravbhatla, and W. E. Royer Jr. 2004. Crystal structure of the hemoglobin dodecamer from *Lumbricus erythrocruorin:* allosteric core of giant annelid respiratory complexes. *J Mol Biol* 344 (1):119–34.

Taketani, S. 2005. Aquisition, mobilization and utilization of cellular iron and heme: endless findings and growing evidence of tight regulation. *Tohoku J Exp Med* 205 (4):297–318.

Tcherkasskaya, O., V. E. Bychkova, V. N. Uversky, and A. M. Gronenborn. 2000. Multisite fluorescence in proteins with multiple tryptophan residues. Apomyoglobin natural variants and site-directed mutants. *J Biol Chem* 275 (46):36285–94.

Tcherkasskaya, O., O. B. Ptitsyn, and J. R. Knutson. 2000. Nanosecond dynamics of tryptophans in different conformational states of apomyoglobin proteins. *Biochemistry* 39 (7):1879–89.

Vinogradov, S. N., D. Hoogewijs, X. Bailly, R. Arredondo-Peter, M. Guertin, J. Gough, S. Dewilde, L. Moens, and J. R. Vanfleteren. 2005. Three globin lineages belonging to two structural classes in genomes from the three kingdoms of life. *Proc Natl Acad Sci U S A* 102 (32):11385–89.

Waks, M., Y. K. Yip, and S. Beychok. 1973. Influence of prosthetic groups on protein folding and subunit assembly. Recombination of separated human alpha-and beta-globin chains with heme and alloplex interactions of globin chains with heme-containing subunits. *J Biol Chem* 248 (18):6462–70.

Weiss, M. J., S. Zhou, L. Feng, D. A. Gell, J. P. Mackay, Y. Shi, and A. J. Gow. 2005. Role of alpha-hemoglobin-stabilizing protein in normal erythropoiesis and beta-thalassemia. *Ann N Y Acad Sci* 1054:103–17.

Winterhalter, K. H., and A. Colosimo. 1971. Chromatographic isolation and characterization of isolated chains from hemoglobin after regeneration of sulfhydryl groups. *Biochemistry* 10 (4):621–24.

Winterhalter, K. H., and E. R. Huehns. 1964. Preparations, properties, and specific recombination of alpha-beta-globin subunits. *J Biol Chem* 239:3699–3705.

Xu, F., K. S. Quandt, and D. E. Hultquist. 1992. Characterization of NADPH-dependent methemoglobin reductase as a heme-binding protein present in erythrocytes and liver. *Proc Natl Acad Sci U S A* 89 (6):2130–34.

Yip, Y. K., M. Waks, and S. Beychok. 1972. Influence of prosthetic groups on protein folding and subunit assembly. I. Conformational differences between separated human alpha- and beta- globins. *J Biol Chem* 247 (22):7237–44.

Yu, X., Y. Kong, L. C. Dore, O. Abdulmalik, A. M. Katein, S. Zhou, J. K. Choi, D. Gell, J. P. Mackay, A. J. Gow, and M. J. Weiss. 2007. An erythroid chaperone that facilitates folding of alpha-globin subunits for hemoglobin synthesis. *J Clin Invest* 117 (7):1856–65.

Section II

Mechanisms of Metal Transporters and Assembly

7 Frataxin
An Unusual Metal-Binding Protein in Search of a Function

Annalisa Pastore

CONTENTS

Cellular Localization and Maturation of Frataxin 126
Sequence and Structure Conservation of Frataxin 126
Same Fold, Different Stabilities 130
A Protein in Search of a Function: Same Fold, Similar Function(s)? 131
Frataxin Is an Unusual Iron-Binding Protein 132
Frataxin as a Ferritin-Like Protein Implicated in Oxidative Damage 134
Frataxin as an Iron Chaperone 136
A Different Point of View: Frataxin as an Inhibitor 137
References 139

Friedreich's ataxia (FRDA) is a neurodegenerative disease caused by deficiency of frataxin, a protein finally identified in 1996 after years of attempts to find the FRDA gene. In the vast majority of patients (96–98%), defective expression of frataxin is due to a homozygous GAA triplet repeat expansion within the first intron of the FXN gene, located on chromosome 9q13 (Campuzano et al. 1996). Anomalous expansion of GAA repeats determines the formation of a triple helix non-B DNA structure, resulting in inhibition of frataxin mRNA transcription (Sakamoto et al. 2001). A small percentage of FRDA patients (approximately 2–4%) are compound heterozygotes with missense mutations in the gene (Cossée et al. 1999). They carry intronic GAA expansions on one allele and a point mutation, which is mainly located in the conserved C-terminal region of frataxin, within exons of the other allele.

Frataxin is an essential protein, as assessed by independent evidence. Absence or deficiency of frataxin has been shown to be lethal (Cossée et al. 2000; Puccio et al. 2001). It is ubiquitously expressed, reaching the highest concentration in the heart, spinal cord, and dorsal root ganglia (Koutnikova et al. 1997). The expression levels correlate with the pattern of neuronal degeneration, cardiomyopathy, and increased risk of diabetes.

The functional role of frataxin has been debated for a long time, but remains controversial. The aim of this review is assessing critically the current hypotheses in the hope of shedding light into its link to pathology.

CELLULAR LOCALIZATION AND MATURATION OF FRATAXIN

Mitochondrial localization of frataxin was predicted shortly after identification of the protein on the basis of evolutionary considerations (Gibson et al. 1996). Frataxin homologues (the CyaY proteins) were identified in purple bacteria, the closest relatives of the mitochondrial genome. The presence of frataxin in these organisms, together with a number of similarities in the pathological symptoms, suggested that chronic peroxidative mitochondrial damage could be the cause of neuronal death. A mitochondrial localization was rapidly confirmed experimentally in yeast and human cells (Koutnikova et al. 1997; Babcock et al. 1997; Campuzano et al. 1997; Foury and Cazzalini 1997; Wilson and Roof 1997).

We now know that eukaryotic frataxin is nuclearly encoded, translated in the cytoplasm, and then imported into mitochondria (Koutnikova et al. 1997). In this final compartment, a two-step proteolytic processing removes the transit sequence to produce the mature protein.

The human frataxin gene encodes a 210 amino acid protein, representing the precursor form of frataxin (Campuzano et al. 1996). The nature of the mature form of the human protein has, however, been doubtful for a long time. The first clue on frataxin maturation emerged when an interaction between the mouse frataxin precursor and mouse mitochondrial processing peptidase (MPP) was observed in a two-hybrid system (Koutnikova, Campuzano, and Koenig 1998). Processing of human frataxin *in vitro* by recombinant rat MPP was successively shown to occur through first generating a 19 kDa intermediate form cleaved between Gly41 and Leu42 further matured by cleavage between Ala55 and Ser56 (Branda et al. 1999; Cavadini et al. 2000). Much more recently, however, it was found, by analyzing the *in vivo* processing of frataxin in human cells, that the relevant cleavage responsible for the generation of the major form of mature frataxin *in vivo* occurs between Lys80 and Ser81 (Condò et al. 2007). The resulting protein, 130 amino acids long, was shown to be fully functional and able to rescue aconitase defects in frataxin-deficient cells and the lethal phenotype of fibroblasts completely deleted for frataxin (Condò et al. 2007; Schmucker et al. 2008). A possible reason for the initial confusion is that the migration profile of frataxin depends on the experimental conditions. Different truncated forms, such as 56–210 and 78–210, can be generated when the normal maturation process of frataxin is impaired.

SEQUENCE AND STRUCTURE CONSERVATION OF FRATAXIN

Frataxins are small proteins (100–200 amino acids) that do not share detectable sequence homology with proteins of known function. Bacterial frataxins contain only a conserved sequence approximately 100–130 residues long, whereas eukaryotic frataxins have an additional N-terminal tail, which contains the mitochondrial import signal (Gibson et al. 1996). They are overall rather acidic proteins with isoelectric points around 4.5 (Adinolfi et al. 2002).

Frataxin

Frataxins share a block of approximately 100 amino acids that is highly conserved from bacteria to humans, sharing a sequence identity of ~25% but a similarity of 40–70% (Figure 7.1). Among the residues completely conserved are a number of acidic residues and three tryptophan residues. This conserved region must be functionally important. A homologue of frataxin was recently identified in *Trichomonas vaginalis* (Dolezal et al. 2007), a unicellular organism that inhabits oxygen-poor environments and has an energy metabolism that relies on glycolytic breakdown in the cytosol. This discovery provides additional evidence not only of a common evolutionary history of frataxins, but also of their essentiality of this pathway for eukaryotes.

Extensive structural and biochemical studies have been carried out for the human (hfra), yeast (Yfh1), and bacterial orthologues (CyaY). Hfra was studied in at least three different forms: a truncated construct (hfra[92–210]), which spans the evolutionary conserved domain (Gibson et al. 1996); two extended forms (hfra[75–210], and hfra[78–210]), close to what we now know spans the major cellular form of mature frataxin (Condò et al. 2007); and a further extended construct (hfra[56–210]), which corresponds to MPP processing (Cavadini et al. 2000).

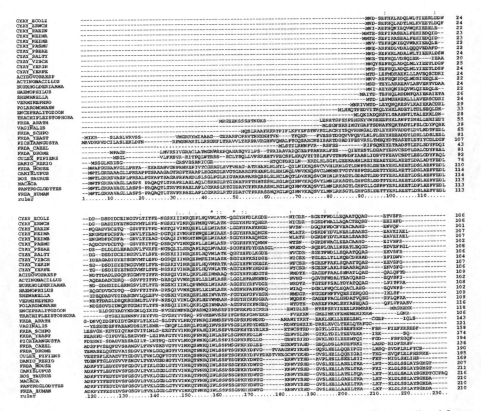

FIGURE 7.1 Multiple alignment of the frataxin family. Bacterial sequences are listed first. The alignment was achieved by the clustalx software. Conserved and semiconserved residues are marked on the top of the alignment.

The structure of the evolutionary conserved domain was first established by nuclear magnetic resonance (NMR) studies of the human orthologue (Musco et al. 2000). The crystal structures of hfra(92–210) and *Escherichia coli* CyaY were published immediately after (Dhe-Paganon et al. 2000; Cho et al. 2000). Now, several years after the first publications, other structures are available: the NMR structures of CyaY and Yfh1, together with two complexes of CyaY with metals and the structures of two assemblies of Yfh1 (He et al. 2004; Nair et al. 2004; Pastore, Franzese et al. 2007; Karlberg et al. 2008; Prischi et al. 2009). They all share a very similar fold, which directly reflects the high degree of sequence conservation (Figure 7.2). The various structures superpose with root mean square deviations (RMSDs) of 1.7–2.4 Å.

The fold of the conserved domain consists in a globular slightly elongated domain in which two helices pack against a 5- to 7-residue β-sheet (Figure 7.3). The two helices are N- and C-terminal to the β-sheet. Very few noticeable structural differences

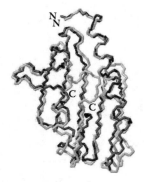

FIGURE 7.2 Structural superposition of the evolutionary conserved domains of hfra (1ekg, in gray) and CyaY (1ew4, in black). The two x-ray structures superpose with 1.7 Å RMSD.

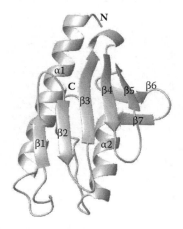

FIGURE 7.3 X-ray structure of the evolutionary conserved domain of hfra (1ekg). The secondary structure elements are marked together with the positions of the N- and C-termini.

Frataxin

can be observed between eukaryotic and prokaryotic orthologues. In hfra, there is a four-residue insertion in the loop between α1 and β1, and residues 180–181 are structured and form an extra β-strand. The corresponding region (84–85) of CyaY has no regular structure. The C-termini have different lengths, with that of Yfh1 being shorter. The N-terminal extension (75–92) was shown to be disordered (Musco et al. 2000; Prischi et al. 2009).

Despite its apparent simplicity, no other protein could for a long time be found in the protein structure database (PDB) with the same fold. Only recently, solution of the crystal structure of complex I of the *Thermus thermophilus* mitochondrial respiratory chain has shown that the protein Nqo15 has a domain, N3, with a fold similar to that observed in frataxin despite lack of detectable homology (Sazanov and Hinchliffe 2006). The crystal structures superimpose with RMSD values of 2.5–3.3 Å.

Indications of which residues are essential for folding (structural residues) or for functional purposes (functional residues) could be obtained by mapping the conserved residues into the structure (Musco et al. 2000; Dhe-Paganon et al. 2000). Structurally important residues are expected to be buried in the hydrophobic core, thus contributing to the fold stability. Conservation of exposed residues is instead an indication of groups that are involved in protein functions. It was immediately recognized that frataxin conserved and semiconserved residues cluster onto the same surface, i.e., that containing α1 and the β-sheet (Musco et al. 2000). A ridge of semiconserved acidic residues is distributed along the first helix, forming a highly negatively charged surface (Figure 7.4). Adjacent to this patch, on the first strand of the β-sheet, is a well-exposed aspartate (D31 in *E. coli*), which is one of the most conserved residues. Also on the β-sheet is the exposed and completely conserved tryptophan (W61 in *E. coli*, W155 in hfra).

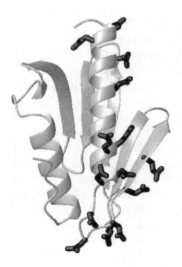

FIGURE 7.4 Representation of the semiconserved acidic ridge on α1 and β1 as mapped on the x-ray structure of CyaY (1ew4). The side chains of aspartates and glutamates are shown in black.

The importance of the conserved residues in α1 and β1 was clearly demonstrated by showing that, although substitution of specific acidic residues by alanine in the acidic ridge of Yfh1 leads to loss of iron binding *in vitro* but has no effect on Fe-S cluster synthesis *in vivo* (Aloria et al. 2004; Gakh et al. 2006), more drastic changes of the electrostatic properties of the molecule induce a phenotype. When these are introduced, the effect is clear (Foury, Pastore, and Trincal 2007): Substitution of two or four of the acidic residues into lysine or alanine, respectively, impairs Fe-S cluster assembly, weakens interaction with Fe-S cluster proteins, and increases oxidative damage. This demonstrated conclusively that the acidic ridge is essential for the Yfh1 function and is likely to be involved in iron-mediated protein–protein interactions.

Another interesting observation is that missense mutations found in heterozygous FRDA individuals all affect conserved residues (Cossée et al. 1999). Mutation of the conserved exposed tryptophan to an arginine, for instance, leads to a particularly severe FRDA phenotype (Labuda, Poirier, and Pandolfo 1999).

SAME FOLD, DIFFERENT STABILITIES

Analysis of the thermodynamic stability of a protein is important for understanding the fold determinants. The evolutionary conserved domain, despite being highly conserved, has a large variability in stability in different species. The factors that influence the thermal stability of the domain in CyaY, Yfh1, and hfra have been investigated in detail in an attempt to correlate these properties to disease (Adinolfi et al. 2004). Whereas CyaY and hfra (both hfra[92–210] and hfra[75–210]) are thermally stable well-behaved proteins, Yfh1 is, under the same conditions, only marginally stable having a high temperature melting point around 35°C and undergoing a low melting temperature around 0°C (Pastore, Martin et al. 2007). An important factor that dictates this behavior was found to be the length of the proteins: The C-terminus has, as mentioned earlier, a structural role by inserting in between the two helices and packing against the hydrophobic core. It was shown that production of hfra(92–210) and CyaY mutants with a truncated C-terminus leads to a significant drop of the high temperature melting point, whereas lengthening of the short C-terminus of Yfh1 leads to an increased melting temperature (Adinolfi et al. 2004). A role of the C-terminus in the fold stability was suggested to be directly biologically relevant because an alternatively spliced transcript of the frataxin gene, the isoform A1, which is expressed at low levels in humans, encodes a 196-residue protein. This length is comparable to that of Yfh1 (Pianese et al. 2002). Its sequence is identical to that of the main isoform only up to residue 160 and has no homology from there on. It is therefore possible that hfra could exist in two alternative folds.

The nonconserved N-terminus present only in eukaryotes does not contribute to stability (Musco et al. 2000). How can full-length Yfh1 be so intrinsically unstable? The protein is partially unfolded at temperatures as low as 35°C, which is close to the optimal growth temperature of yeast. It was by analyzing the behavior of Yfh1 that we realized that frataxins are strongly stabilized by cations, with a stronger effect, detectable already at low ion:protein ratios, for different divalent cations (Adinolfi et al. 2004). This could be put into direct relationship with the capacity of the protein to bind nonselectively iron and other cations.

A strong stabilizing effect was also observed for phosphate (Adinolfi et al. 2004). This was interpreted as a way of mimicking the effect of an interaction between frataxins and mitochondrial DNA, suggesting a role of the protein in DNA protection similar to that proposed for Dps, a protein that protects DNA in starved bacterial cells (Ren et al. 2003). However, this is likely to be only a secondary and nonspecific function since frataxins are proteins poorly expressed.

Another reason for studying the frataxin stability is to explain the role of the missense mutations observed in FRDA heterozygous patients (Cossée et al. 1999). A detailed comparison of the conformational properties of the wild-type and mutant proteins showed that, under physiological conditions, several of the clinically occurring mutants retain a native fold but are differentially destabilized as reflected both by their reduced thermodynamic stability and by a higher tendency toward proteolytic digestion (Musco et al. 2000; Correia et al. 2006, 2008). Increased chain flexibility was observed in specific regions of the proteins. These results suggested the hypotheses that mutations in FRDA could cause a combination of reduced efficiency of protein folding and accelerated frataxin degradation. Therefore although different from other neurodegenerative diseases due to toxic aggregation, FRDA could, at least partially, be caused by protein misfolding due to destabilization of the frataxin fold.

A PROTEIN IN SEARCH OF A FUNCTION: SAME FOLD, SIMILAR FUNCTION(S)?

The cellular function of frataxin has been controversial from the very beginning. There are several reasons for believing that frataxin must have one or more main functions common to all the members of the frataxin family:

1. Sequence conservation is so high among the family that there is little doubt that the proteins share the same fold. Bacterial and eukaryotic proteins share 30% identity and as high as 40–60% similarity.
2. Analysis of the sequence conservation strongly indicates that the functional region involves conserved and exposed residues on $\alpha 1$ and $\beta 1$ and some of the β-sheet residues. The only way to explain this pattern of conservation is by assuming that the conserved residues are functionally relevant. The conserved exposed region cries for a common binding partner.
3. Yeast knockout studies have shown a clear, though partial, rescuing of the phenotype when adding bacterial frataxin (CyaY). An only partial rescuing can be easily accounted for by considering the distance between bacteria and yeast (partial rescuing of the phenotype was also observed between yeast and human frataxin), but the presence of an effect strongly supports the concept of a common function.

These considerations provide convincing evidence that frataxins must share the same overall function(s) even though within possible specific species adaptation. This concept is not new and holds for many proteins. It is well known, for instance, that single-function proteins in prokaryotes find their eukaryotic orthologues as multifunctional and multidomain proteins (for a review, see for instance Brocchesi and Karlin 2005). Another and possibly even more appropriate example to understand what we mean with specific species adaptation is that of ferritins. In a recent review, Elyzabeth Theil says:

> Ferritins are protein nanocages, found in animals, plants, bacteria, and archaea, that convert iron and oxygen to ferric oxy biominerals in the protein central cavity.... The number of genes, temporal regulation, tissue distribution in multi-cellular organisms, and gene product size (maxi-ferritins have 24 subunits and mini-ferritins, or Dps proteins, have 12 subunits and are restricted to bacteria and archaea) suggest an overwhelming diversity and variability. However, common themes of regulation and function are described which indicate not only that the three-dimensional protein structure and the functions of the ferritins are conserved, but also that broad features of genetic regulation are conserved relative to organismal and/or community needs. (Theil 2007)

We expect a similar scenario to apply to frataxin.

The functions suggested for frataxin are all related to iron and cover different aspects of the metabolism of this difficult cation, from heme biogenesis (Yoon and Cowan 2004) to Fe-S cluster formation (Huynen et al. 2001). Within this frame, frataxin has been implicated as an iron scavenger (Cavadini et al. 2000), an iron chaperone (Yoon and Cowan 2003), a controller of oxidative stress, and, more recently, as an inhibitor/regulator (Wong et al. 1999; Tan et al. 2001; Condò 2007; Cossée et al. 2000; Simon et al. 2004; Adinolfi et al. 2009).

FRATAXIN IS AN UNUSUAL IRON-BINDING PROTEIN

An involvement in iron metabolism was suggested almost from the beginning, but how and what this involvement would be has been a matter of extensive debate and is still unresolved. Frataxin-defective organisms, from unicellular yeast to humans, exhibit metabolic disturbances mainly caused by intramitochondrial iron accumulation (Babcock et al. 1997; Bradley et al. 2000), loss of Fe-S cluster-dependent enzymes (Rotig et al. 1997), reduced oxidative phosphorylation (Ristow et al. 2000), and altered antioxidant defenses (Chantrel-Groussard et al. 2001; Jiralerspong et al. 2001).

No direct interactions with iron were initially reported, mostly because of the low affinities of binding involved (Musco et al. 2000). It later became clear that frataxins are iron-binding proteins but with very unusual and unprecedented properties.

Monomeric CyaY and Yhf1 were both shown to bind two Fe^{2+} ions (Bou-Abdallah et al. 2004; Cook et al. 2006) with comparable dissociation constants (Kd) (Bou-Abdallah et al. 2004; Cook et al. 2006) (Table 7.1). This suggested the presence of a di-iron binding center. CyaY can also bind up to 6 and 25 Fe^{3+} atoms/monomer under aerobic and anaerobic conditions, respectively (Bou-Abdallah et al. 2004). More complex is the

TABLE 7.1
Summary of the Frataxin Interactions and Their Affinities

Species	Partner	Technique	Kd (µM)	N	Ref.
CyaY	Fe^{2+}	Calorimetry	3.8 ± 0.1	2.2 ± 0.2[a]	Bou-Abdallah et al. 2004
CyaY	Fe^{3+}	UV-vis, pH-stat, fluorescence	n.d.	6[b]	Bou-Abdallah et al. 2004
Yfh1	Fe^{2+}	Calorimetry	3.0 ± 0.4 2.0 ± 0.2	2[c]	Cook et al. 2006
Hfra(78–210)	Fe^{2+}	Fluorescence	55	6–7	Yoon and Cowan 2003
Hfra(78–210)	Fe^{3+}	Fluorescence	11.7	6.4 ± 0.2	Yoon and Cowan 2003
Hfra(78–210)	Fe^{2+}	Calorimetry	10–55	6–7	Yoon and Cowan 2003
Hfra(78–210)	Fe^{3+}	Calorimetry	10.2	6.5 ± 0.2	Yoon and Cowan 2003
Hfra(46–210)	Fe^{2+}	Calorimetry	0.3–0.4[d]	1.0	Yoon et al. 2007
Hfra(78–210) (His-tagged)	Fe^{2+}	Calorimetry	4600 ± 200	6.8 ± 0.3	Huang et al. 2008
Hfra(46–210)	Isu		5.4[e]	—	—
Hfra(78–210)	Isu(D39A)	Calorimetry	0.15	0.94	Yoon and Cowan 2003
Holo-Hfra (78–210)	Isu	Fluorescence	0.67	—	Yoon and Cowan 2003
Holo-Hfra (78–210)	Isu(D39A) (1:1)	Fluorescence	0.45	—	Yoon and Cowan 2003
Hfra(78–210) (His-tagged)	Isu + Fe^{2+}	Calorimetry	0.030 ± 0.006	0.96 ± 0.01	Huang et al. 2008
Hfra(78–210) (His-tagged)	Isu + Fe^{3+} (1:1)	Calorimetry	0.021	—	Huang et al. 2008
Hfra(78–210) (His-tagged)	Isu + Fe^{2+} (1:7)	Calorimetry	0.012	—	Huang et al. 2008
Hfra(47–210) (His-tagged)[f]	Human ferrochelatase	Calorimetry	0.017	0.49 ± 0.05	Yoon and Cowan 2004
Yfh1	Yeast ferrochelatase	Biacore	0.040	—	Lesuisse et al. 2003

Note: When not indicated, no experimental errors were reported in the original papers.

[a] Ultrafiltration experiments indicated that there are additional weaker Fe^{2+} binding sites.
[b] Additional ions bind at high ion:protein ratios.
[c] The value calculated from curve fitting is 1.5. However, saturation was estimated to be reached around 2:1 Fe^{2+}:protein ratios.
[d] According to whether the protein was His tagged or not.
[e] The interaction occurs only in the presence of a large excess of ferrous ion.
[f] This sample naturally N-terminally degraded to hfra(81–210) without His-tag.

situation for hfra. Different constructs were reported to bind both Fe^{2+} and Fe^{3+} cations with stoichiometries of 2 to 7 cations per monomer (Yoon and Cowan 2003; Yoon, Dizin, and Cowan 2007; Huang, Dizin, and Cowan 2008). The discrepancies between the dissociation constants observed for the different constructs are however very large, shading some doubts onto the biological significance of these results.

The regions involved in iron binding were mapped by NMR spectroscopy (Nair et al. 2004; He et al. 2004; Pastore et al. 2007a). This technique is in principle not ideal for studies involving iron because of the paramagnetism of this element, which can induce major broadening effects so strong as to completely bleach the signal (Bertini and Luchinat 1996). The spectra were nevertheless of acceptable quality due to trapping of the cation to specific sites. The effects observed were invariably mapped to equivalent regions of the different orthologues, which contain the semi-conserved acidic surface on α1 and β1 (Figure 7.4). These studies clearly showed a primary binding site that saturates at low cation:protein molar ratios. Additional binding sites could be observed upon further iron addition.

Iron coordination of hfra and Yfh1 was independently studied by x-ray absorption spectroscopy (XAS) and extended x-ray absorption fine structure (EXAFS) analysis (Cook et al. 2006). The iron bound to the monomeric proteins was confirmed to be in a ferrous state if the samples are prepared anaerobically or in the presence of the reducing agent dithionite. This would disagree with a putative role of frataxins as ferroxidases (Park et al. 2002, 2003; O'Neill et al. 2005). Fe^{2+} is in a high-spin state and coordinated in a highly centrosymmetric metal ligand six-coordination geometry (Cook et al. 2006). No evidence was found for metal–metal interactions, thus not supporting the possibility of a di-iron group previously suggested as the bound species (Nair et al. 2004).

Most of the known examples of iron-binding proteins, setting aside heme-containing ones, which need to have pockets to host the cofactor, coordinate the ion through at least one cysteine. The thiol group is coadjuvated by histidines and acidic groups. Frataxins have no conserved cysteine and/or histidine and coordinate iron solely by negatively charged groups. There are no examples in the PDB of such iron coordination, especially considering that these groups are all exposed on the protein surface.

Another unusual property is a lack of binding selectivity. CyaY, Yfh1, and to a minor extent, hfra are able to bind almost any divalent and trivalent cation, ranging from Ca^{2+} and Co^{2+} to Al^{3+} and various lanthanides (Nair et al. 2004; Pastore, Franzese et al. 2007). All these cations induce chemical shift perturbation effects at the same resonances, indicating competition for the same binding site. Zn^{2+} has a strong tendency to promote protein aggregation. This behavior has however been observed for other systems (Pastore and de Chiara, unpublished observations) and seems to be an intrinsic feature of this element.

FRATAXIN AS A FERRITIN-LIKE PROTEIN IMPLICATED IN OXIDATIVE DAMAGE

Even when iron binding was conclusively demonstrated *in vitro*, this still provided no clue into the frataxin function *in vivo*. The first hypothesis about the cellular function of frataxin was suggested by the Isaya's group (Adamec et al. 2000; Gahk et al. 2002): It was suggested that frataxin acts as an iron scavenger able to form large spherical assemblies that can trap iron in a bioavailable form. They could indeed show, using a vast number of complementary techniques, that, when exposed to iron excesses, Yfh1 oligomerizes and forms large aggregates of

48 subunits. These assemblies can bind up to 50 iron atoms per protein monomer. The bound iron is predominately in a high-spin Fe(III) state (Nichol et al. 2003).

Such large aggregates are able to sequester iron cations in a nontoxic and yet available form that could control cellular oxidative stress by reducing production of reactive oxygen species. These properties are strongly reminiscent of those of ferritin, an iron scavenger known to help clearance of iron excess in the cell (Harrison and Arosio 1996). In support of this hypothesis is the observation that frataxin deficiency leads to oxidative damage in humans (Emond et al. 2000; Schultz et al. 2000), mice (Ristow et al. 2000; Thierbach et al. 2005), yeast (Karthikeyan et al. 2000), and *Caenorhabditis elegans* (Vazquez-Manrique et al. 2006). Mitochondrial iron overload, resulting from a frataxin deficiency, leads to oxidative damage in mitochondrial and nuclear DNA as well as to Fe-S clusters in mitochondrial aconitase and other respiratory enzymes (Babcock et al. 1997; Cavadini et al. 2000; Foury 1999; Karthikeyan, Lewis, and Resnick 2002). Accordingly, it was shown that frataxin protects DNA against iron-induced oxidative damage (Gakh et al. 2006; O'Neill et al. 2005).

A ferritin-like model able to retain the bound metal in a bioavailable form could also be supported by the ferroxidase activity reported for both Yfh1 and hfra (Park et al. 2002, 2003; O'Neill et al. 2005). A ferroxidase center was mapped in Yfh1 to the residues Asp79 and Asp82 present in α1. Yeast mutants targeting these residues have reduced ferroxidase activity and assemble at a slower rate than the wild-type protein (Gakh et al. 2006). It was sugested that Asp86, Glu89, Glu90, Glu93, Asp101, and Glu103, which all map in α1 or β1, participate in the iron mineralization chemistry observed under iron-overloaded conditions and in the metal-sequestering properties of Yfh1 oligomers (Gakh et al. 2006).

There are, however, several considerations against this being the major function of frataxin, which could explain the pattern of sequence conservation discussed earlier. Oligomer formation is efficiently competed by other divalent metal ions or by the presence of physiological salt concentrations (Adinolfi et al. 2002). A tendency to aggregate *in vitro* was also found for CyaY (Adinolfi et al. 2002) but not for hfra. The only conditions under which hfra self-assembles forming 59-mer aggregates able to bind approximately 10 iron/protein monomer is when it is overexpressed in *E. coli* (Cavadini et al. 2002). Other divalent cations, such as magnesium or calcium, compete efficiently with self-assembly at concentrations comparable to those found in mitochondria (Adinolfi et al. 2002). An important role in *in vitro* aggregation was attributed to the N-terminal extension present only in eukaryotes (O'Neill et al. 2005). It was also shown that self-assembly is unessential when the protein functions as an iron chaperone capacity (i.e., during heme and Fe-S cluster assembly) (Aloria et al. 2004; Adinolfi et al. 2009). Equally nonconserved are the ferroxidase activity properties, which could not be observed for CyaY (Bou-Abdallah et al. 2004). Finally, the most problematic aspect of a ferritin-like hypothesis is probably that, while explaining iron accumulation in shortage of frataxin, it does not justify why the protein is essential since mitochondria have a specialized ferritin (Levi et al. 2001).

FRATAXIN AS AN IRON CHAPERONE

A quite different hypothesis, suggested by Cowan and coworkers, is that frataxins are iron chaperones that help to solubilize the cation and deliver it to the different processes in which it is required (Yoon and Cowan 2003). This hypothesis is mostly independent if not in conflict with iron-dependent self-assembly and relies on the observation that, as discussed previously, frataxin binds iron also at low stoichiometric ratios (Yoon and Cowan 2003; Nair et al. 2004; Bou-Abdallah et al. 2004; Cook et al. 2006). Based on the identification of proteins interacting with frataxin, it was suggested that frataxin is the chaperone in two different metabolic pathways: iron-sulfur cluster formation and heme assembly. An involvement in more than one pathway is reasonable since in prokaryotes frataxin is not part of a specific operon.

A link between frataxin and Fe-S cluster proteins was first hinted at by the evidence that both FRDA patients and yeast and knock-out mice have deficiency in Fe-S cluster proteins (Foury and Cazzalini 1997; Babcock et al. 1997; Foury 1999; Rotig et al. 1997; Puccio et al. 2001). This could however be only an indirect effect rather than a direct cause. A bioinformatics analysis of the available genomes reinforced the hint by showing a clear-cut correlation between the presence of the frataxin gene and that of one of the two chaperonins, the hscb protein, of the Fe-S cluster machinery (Huynen et al. 2001). Direct interaction was however finally observed experimentally not with hscb but with IscS/IscU, the complex central to Fe-S formation (Gerber, Muhlenhoff, and Lill 2003; Layer et al. 2006). IscS (or Nifs in eukaryotes) is the desulfurase that converts cysteine into alanine, forming a persulfide necessary for the Fe-S cluster. IscU (or Isu in eukaryotes) is a donor on which the cluster is transiently assembled (for a review see Johnson et al. 2005). GST pull-down and coprecipitation on isolated yeast mitochondria extracts were used to identify the Isu1/Nfs1 complex (Gerber, Muhlenhoff, and Lill 2003). Interestingly, the interaction was significantly stronger when physiologic concentrations of ferrous iron were maintained during complex isolation. Independent support to these results was given by clear indications of a genetic link between the frataxin and the *isc* genes (Foury and Talibi 2001; Ramazzotti, Vanmansart, and Foury 2004).

While these observations provide convincing evidence of an involvement of frataxin in de novo Fe-S cluster synthesis, there has been for a long time some confusion about which of the two proteins of the IscU/IscS complex is directly recognized by frataxin. The original report of a GST pull-down in yeast is very carefully phrased and speaks only about the complex (Gerber, Muhlenhoff, and Lill 2003): "Several lines of evidence indicate that Yfh1 specifically binds to the core ISC-assembly complex Isu1/Nfs1 ... Yfh1 was affinity purified with an Isu1-GST fusion protein, and the Isu1/Nfs1 complex was also found to associate with GST-Yfh1." After this study, different reports have shown that hfra binds Isu with high affinity, although only in the presence of iron (Yoon and Cowan 2003; Huang, Dizin, and Cowan 2008). Pull-down experiments on bacterial extracts in which the *isc* operon is overexpressed and using CyaY as the bait, on the other hand, have demonstrated a direct interaction with IscS but not with IscU (Layer et al. 2006; Adinolfi et al. 2009). An easy way to solve this debate is to bear in mind that Isu1/Nfs1 (or IscU/IscS in bacteria) form a relatively tight complex (K_d 1 μM) with a 1:1 stoichiometric ratio (Agar et al. 2000; Urbina et al. 2001). It is therefore difficult to discriminate with the techniques used whether the interaction is with Nfs1, with Isu1, or with both.

A link between frataxin and ferrochelatase, the enzyme that catalyzes the final step of heme biosynthesis and helps selective insertion of the ferrous ion into porphyrin (Taketani 2005), was first suggested by Foury and Cazzalini (Foury and Cazzalini 1997). Decrease in cytochrome c oxidase activity and severe deficiency of cytochromes was also observed in frataxin-deleted yeast strains (Foury and Cazzalini 1997; Lesuisse et al. 2003). Reintroduction of frataxin expression was shown to induce recovery of cytochrome production (Lesuisse et al. 2003). More recently, it was shown that frataxin deficiency leads to down-regulation of mitochondrial transcripts and to kinetic inhibition of the heme pathway (Schoenfeld et al. 2005).

Direct high-affinity interaction between frataxin and ferrochelatase with a K_d of 40 nM was proven by surface plasmon resonance studies carried out on recombinant Yfh1 and yeast ferrochelatase (Lesuisse et al. 2003). A stoichiometric ratio of one frataxin monomer per ferrochelatase dimer was found. Iron did not seem to be necessary for this tight interaction. NMR chemical shift perturbation, used to map the interaction onto the surface of Yfh1, showed effects localized predominantly on the α1 helix but also distributed in the β6-loop-α2 region (directly opposite from α1) or affecting residues on the surface of the β-sheet (He et al. 2004).

These studies were extended to the human proteins (Yoon and Cowan 2004). A K_d of 17 nM was estimated by calorimetry and fluorescence studies. The interaction could be detected only in the presence of iron in agreement with NMR evidence (Bencze et al. 2007).

A DIFFERENT POINT OF VIEW: FRATAXIN AS AN INHIBITOR

A new perspective into the frataxin function was recently suggested by our group. We found that, far from being an iron chaperone, frataxin acts as an inhibitor of the Fe-S cluster formation, leading to an entirely new working hypothesis (Adinolfi et al. 2009). Using the bacterial model system, we assumed that, if frataxins were iron chaperones, i.e., molecules that "neutrally" escort iron to its final destination, the presence of CyaY should either not perturb the enzymatic reaction rates (if the reaction rate limiting step does not depend on iron delivery) or, in the most favorable case, enhance reaction speed. Much to our surprise, we observed that, far from facilitating iron delivery, the presence of CyaY affects negatively the Fe-S reaction kinetics, thus producing a classical inhibitory effect (Stryer 1995). After dissecting the reconstitution pathway, we could show that CyaY binds IscS in an iron-dependent way and that this interaction is crucial for the frataxin effect on Fe-S cluster formation. CyaY influences the Fe-S cluster formation rates but does not act directly on the enzymatic reaction of the desulphurase IscS, which continues to convert cysteine to alanine. CyaY also does not compete for the binding site of IscU on IscS. The effect of CyaY on the kinetics of the processes does not depend on the specific nature of the acceptor.

Therefore IscS is not just an ancillary protein to a CyaY/IscU complex but the template necessary for the function of CyaY, which must consist in a regulation of Fe-S cluster formation. This hypothesis explains the following experimental observations:

1. Why frataxin is an essential protein: The ferritin-like hypothesis cannot explain it because in mitochondria there is a ferritin anyway. The low iron affinity and specificity is also not easily in agreement with the chaperone hypothesis.
2. Why frataxin has low iron affinities: This is required for a protein that is an iron sensor.
3. Why the long-term accumulation of FRDA symptoms: As a sensor even small frataxin concentrations would be sufficient for cell viability, but frataxin deficiency would eventually trigger the long-term effects.
4. Why frataxin is associated with oxidative stress and with the observation that time-dependent intramitochondrial iron accumulation in frataxin-deficient organisms is observed after onset of the pathology and after inactivation of the Fe-S-dependent enzymes: As a regulator that tunes the quantity of Fe-S cluster formed to the concentration of the apo acceptors, any reduction or depletion of the frataxin level will upset this equilibrium and lead to an imbalance of the Fe-S clusters produced as respect to the apo acceptors.

The concept that CyaY is not simply an iron chaperone and evidence that CyaY binds to IscS but not to IscU could, at first glance, appear to be in contrast with the existing literature. The apparent discrepancy should however be evaluated more critically. The effect of frataxin as a chaperone was proposed by Yoon and Cowan (Yoon and Cowan 2003). However, the authors compared the action of frataxin on the reaction rates of *non enzymatic* reconstitution in the presence and in the absence of Isu (the eukaryotic orthologue of IscU). It is very reasonable to find that absence of Isu abolishes almost completely Fe-S reconstitution if frataxin does not bind Fe-S itself and acts specifically on *enzymatic* cluster formation.

We do not observe a pairwise interaction with IscU, either in the absence or in the presence of iron, whereas a direct interaction between frataxin and Isu-like proteins has been reported for both the yeast and the human proteins (Ramazzotti, Vanmansart, and Foury 2004; Yoon and Cowan 2003; Huang, Dizin, and Cowan 2008). An involvement of the frataxin β-sheet, which is not part of the interface with IscS, is also well supported by independent studies (Leidgens 2008; Wang and Craig 2008) in which mutation of residues in and around the β-sheet does have deleterious effects on cell growth and on Fe-S cluster formation without affecting iron binding and/or oligomerization. This discrepancy could be explained by assuming that the binary frataxin/Isu complex from eukaryotes is stable enough to be isolated, but the interaction with IscU in prokaryotes could be observed only in the context of a ternary complex (IscS/IscU/frataxin). Accordingly, we have observed an effect on the reconstitution kinetics of the CyaY_W61R mutation, a residue that is not part of the interface.

A specific and direct pairwise interaction with IscS is, on the other hand, well in agreement with evidence that formation of a complex between Yfh1 and Isu in yeast does not require previous formation of Fe-S cluster on Isu1, thus suggesting an involvement of frataxin in the early steps of the cluster formation (Gerber, Muhlenhoff, and Lill 2003). An elegant study on the bacterial system has also provided convincing evidence that interaction of CyaY with the Fe-S machinery is mediated by IscS and not by IscU (Layer et al. 2006).

Based on our results, we proposed that the function of frataxin is that of the gatekeeper of Fe-S cluster formation which could sense the levels of iron. The concept of frataxin as an iron switch is already present in a far-sighted study of frataxin expression in erythroid cells in which the authors suggested that the frataxin role is that of a switch between two different metabolic pathways (Becker et al. 2002).

The question that remains open from our studies is about the specific mechanism by which frataxin operates. Its regulatory role could be achieved in at least two possible ways, both compatible with the experimental techniques used in our study. According to the simplest scenario, frataxins could block *tout court* cluster formation at high iron concentrations, thus making sure that no clusters are made if there is not an appropriate acceptor. This could, for instance, be achieved by blocking the site on IscU where the cluster is mounted or even just by sequestering iron. The second scenario is that frataxin intervenes in determining the molecularity of the cluster shifting the formation of 2Fe-2S to that of 4Fe-4S, again in an iron-dependent fashion. The latter hypothesis is consistent with the observed decrease of aconitase activity in yeast frataxin-depleted mitochondria (Foury, Pastore, and Trincal 2007) and in FRDA patients (Rotig et al. 1997). It is however more difficult to understand how this mechanism, which requires an ability of frataxin to stabilize one specific cluster, could be achieved at the molecular level. The techniques so far used (circular dichroism [CD] and absorbance spectroscopies) cannot conclusively discriminate between the two hypotheses. The final word on this point will come from resonance Raman and Mössbauer studies and from the high-resolution structure of frataxin binary and ternary complexes. These studies are currently being carried out in our group.

REFERENCES

Adamec, J., F. Rusnak, W. G. Owen, S. Naylor, L. M. Benson, A. M Gacy, and G. Isaya. 2000. Iron-dependent self-assembly of recombinant yeast frataxin: implications for Friedreich ataxia. *Am J Hum Genet* 67 (3):549–62.

Adinolfi, S., C. Iannuzzi, F. Prischi, C. Pastore, S. Iametti, S. R. Martin, F. Bonomi, and A. Pastore. 2009. Bacterial frataxin CyaY is the gatekeeper of iron-sulfur cluster formation catalyzed by IscS. *Nat. Struct Mol Biol* 16 (4):390–96.

Adinolfi, S., M. Nair, A. Politou, E. Bayer, S. Martin, P. Temussi, and A. Pastore. 2004. The factors governing the thermal stability of frataxin orthologues: how to increase a protein's stability. *Biochemistry* 43 (21):6511–18.

Adinolfi, S., M. Trifuoggi, A. Politou, S. Martin, and A. Pastore. 2002. A structural approach to understanding the iron-binding properties of phylogenetically different frataxins. *Hum Mol Genet* 11 (16):1865–77.

Agar, J. N., P. Yuvaniyama, R. F. Jack, V. L. Cash, A. D. Smith, D. R. Dean, and M. K. Johnson. 2000. Modular organization and identification of a mononuclear iron-binding site within the NifU protein. *J Biol Inorg Chem* 5 (2):167–77.

Aloria, K., B. Schilke, A. Andrew, and E. A. Craig. 2004. Iron-induced oligomerization of yeast frataxin homologue Yfh1 is dispensable *in vivo*. *EMBO Rep* 5 (11):1096–1101.

Babcock, M., D. de Silva, R. Oaks, A. S. Davis-Kaplan, S. Jiralerspong, L. Montermini, M. Pandolfo, and J. Kaplan. 1997. Regulation of mitochondrial iron accumulation by Yfh1p, a putative homolog of frataxin. *Science* 276 (5319):1709–12.

Becker, E. M., J. M. Greer, P. Ponka, and D. R. Richardson. 2002. Erythroid differentiation and protoporphyrin IX down-regulate frataxin expression in Friend cells: characterization of frataxin expression compared to molecules involved in iron metabolism and hemoglobinization. *Blood* 99 (10):3813–22.

Bencze, K. Z., T. Yoon, P. B. Bradley, J. A. Cowan, and T. L. Stemmler. 2007. Human frataxin: iron structure and ferrochelatase binding surface. *Chem Commun* 14:1798–1800.

Bertini, I., and C. Luchinat. 1996. NMR in paramagnetic molecules in biological systems. In *NMR of paramagnetic substances*, ed. A. B. P. Lever. Amsterdam: Coord.Chem.Rev. 150, Elsevier, 1–300.

Bou-Abdallah, F., S. Adinolfi, A. Pastore, T. M. Laue, and N. D. Chasteen. 2004. Iron binding and oxidation properties of the bacterial frataxin cyay of *Escherichia coli*. *J Mol Biol* 341:605–15.

Bradley, J. L., J. C. Blake, S. Chamberlain, P. K. Thomas, J. M. Cooper, and A. H. Schapir. 2000. Clinical, biochemical and molecular genetic correlations in Friedreich's ataxia. *Hum Mol Genet* 9 (2):275–82.

Branda, S. S., P. Cavadini, J. Adamec, F. Kalousek, F. Taroni, and G. Isaya. 1999. Yeast and human frataxin are processed to mature form in two sequential steps by the mitochondrial processing peptidase. *J Biol Chem* 274 (32):22763–69.

Brocchesi, L., and S. Karlin. 2005. Protein length in eukaryotic and prokaryotic proteomes. *Nucleic Acids Res* 33 (10):3390–3400.

Campuzano, V., L. Montermini, Y. Lutz, L. Cova, C. Hindelang, S. Jiralerspong, Y. Trottier, S. J. Kish, B. Faucheux, P. Trouillas, F. J. Authier, A. Durr, J. L. Mandel, A. Vescovi, M. Pandolfo, and M. Koenig. 1997. Frataxin is reduced in Friedreich ataxia patients and is associated with mitochondrial membranes. *Hum Mol Genet* 6 (11):1771–80.

Campuzano, V., L. Montermini, M. D. Molto, L. Pianese, M. Cosseé, F. Cavalcanti, E. Monros, F. Rodius, F. Duclos, F. Monticelli, F. Zara, J. Cañizares, H. Koutnikova, S. I. Bidichandani, C. Gellera, A. Brice, P. Trouillas, G. De Michele, A. Filla, R. De Frutos, F. Palau, P. I. Patel, S. Di Donato, J.-L. Mandel, S. Cocozza, M. Koenig, and M. Pandolfo. 1996. Friedreich's ataxia: autosomal recessive disease caused by an intronic GAA triplet repeat expansion. *Science* 271 (5254):1423–27.

Cavadini, P., J. Adamec, F. Taroni, O. Gakh, and G. Isaya. 2000. Two-step processing of human frataxin by mitochondrial processing peptidase. Precursor and intermediate forms are cleaved at different rates. *J Biol Chem* 275:41469–75.

Cavadini, P., H. A. O'Neill, O. Benada, and G. Isaya. 2002. Assembly and iron-binding properties of human frataxin, the protein deficient in Friedreich ataxia. *Hum Mol Genet* 11 (3):217–27.

Chantrel-Groussard, K., V. Geromel, H. Puccio, M. Koenig, A. Munnich, A. Rotig, and P. Rustin. 2001. Disabled early recruitment of antioxidant defenses in Friedreich's ataxia. *Hum Mol Genet* 10 (19):2061–67.

Cho, S. J., M. G. Lee, J. K. Yang, J. Y. Lee, H. K. Song, and S. W. Suh. 2000. Crystal structure of *Escherichia coli* CyaY protein reveals a previously unidentified fold for the evolutionarily conserved frataxin family. *Proc Natl Acad Sci U S A* 97 (16):8932–37.

Condò, I., N. Ventura, F. Malisan, A. Ruffini, B. Tomassini, and R. Testi. 2007. *In vivo* maturation of human frataxin. *Hum Mol Genet* 16 (13):1534–40.

Cook, J. D., K. Z. Bencze, A. D. Jankovic, A. K. Crater, C. N. Busch, P. B. Bradley, A. J. Stemmler, M. R. Spaller and T. L. Stemmler. 2006. Monomeric yeast frataxin is an iron binding protein. *Biochemistry* 45 (25):7767–77.

Correia, A., S. Adinolfi, A. Pastore, and C. Gomes. 2006. Conformational stability of human frataxin and effect of Friedreich's ataxia related mutations on protein folding. *Biochem J* 398 (3):605–11.

Correia, A. R., C. Pastore, S. Adinolf, A. Pastore, and C. M. Gomes. 2008. Implications of Friedreich's ataxia mutations in frataxin structure and dynamics. *FEBS Lett* 275 (40):3680–90.

Cossée, M., A. Dürr, M. Schmitt, N. Dahl, P. Trouillas, P. Allinson, M. Kostrzewa, A. Nivelon-Chevallier, K. H. Gustavson, A. Kohlschütter, U. Müller, J. L. Mandel, A. Brice, M. Koenig, F. Cavalcanti, A. Tammaro, G. De Michele, A. Filla, S. Cocozza, M. Labuda, L. Montermini, J. Poirier, and M. Pandolfo. 1999. Friedreich's ataxia: point mutations and clinical presentation of compound heterozygotes. *Ann Neurol* 45 (2):200–206.

Cossée, M., H. Puccio, A. Gansmuller, H. Koutnikova, A. Dierich, M. LeMeur, K. Fischbeck, P. Dolle, and M. Koenig. 2000. Inactivation of the Friedreich ataxia mouse gene leads to early embryonic lethality without iron accumulation. *Hum Mol Genet* 9 (8):1219–26.

Dhe-Paganon, S., R. Shigeta, Y. I. Chi, M. Ristow, and S. E. Shoelson. 2000. Crystal structure of human frataxin. *J Biol Chem* 275 (40):30753–56.

Dolezal, P., A. Dancis, E. Lesuisse, R. Sutak, I. Hrdý, T. M. Embley, and J. Tachezy. 2007. Frataxin, a conserved mitochondrial protein in the hydrogenosome of *Trichomonas vaginalis*. *Eukaryot Cell* 6 (8):1431–38.

Emond, M., G. Lepage, M. Vanasse, and M. Pandolfo. 2000. Increased levels of plasma malondialdehyde in Friedreich ataxia. *Neurology* 55:1752–53.

Foury, F. 1999. Low iron concentration and aconitase deficiency in a yeast frataxin homologue deficient strain. *FEBS Lett* 456:281–84.

Foury, F., and O. Cazzalini. 1997. Deletion of the yeast homologue of the human gene associated with Friedreich's ataxia elicits iron accumulation in mitochondria. *FEBS Lett* 411 (2–3):373–77.

Foury, F., A. Pastore, and M. Trincal. 2007. Acidic residues of yeast frataxin have an essential role in Fe-S cluster assembly. *EMBO Rep* 8 (2):194–99.

Foury, F., and D. Talibi. 2001. Mitochondrial control of iron homeostasis. A genome wide analysis of gene expression in a yeast frataxin-deficient strain. *J Biol Chem* 276:7762–68.

Gakh, O., J. Adamec, A. M. Gacy, R. D. Twesten, W. G. Owen, and G. Isaya. 2002. Physical evidence that yeast frataxin is an iron storage protein. *Biochemistry* 41 (21):6798–6804.

Gakh, O., S. Park, G. Liu, L. Macomber, J. A. Imlay, G. C. Ferreira, and G. Isaya. 2006. Mitochondrial iron detoxification is a primary function of frataxin that limits oxidative damage and preserves cell longevity. *Hum Mol Genet* 15 (3):467–79.

Gerber, J., U. Muhlenhoff, and R. Lill. 2003. An interaction between frataxin and Isu1/Nfs1 that is crucial for Fe/S cluster synthesis on Isu1. *EMBO Rep* 4 (9):906–11.

Gibson, T. J., E. V. Koonin, G. Musco, A. Pastore, and P. Bork. 1996. Friedreich's ataxia protein: phylogenetic evidence for mitochondrial dysfunction. *Trends Neurosci* 19 (11):465–68.

Harrison, P., and P. Arosio. 1996. The ferritins: molecular properties, iron storage functions and cellular regulation. *Biochem Biophys Acta* 1275:161–203.

He, Y., S. L. Alam, S. V. Proteasa, Y. Zhang, E. Lesuisse, A. Dancis, and T. L. Stemmler. 2004. Yeast frataxin solution structure, iron binding, and ferrochelatase interaction. *Biochemistry* 43 (51):16254–62.

Huang, J., E. Dizin, and J. A. Cowan. 2008. Mapping iron binding sites on human frataxin: implications for cluster assembly on the ISU Fe-S cluster scaffold protein. *J Biol Inorg Chem* DOI 10.1007/s00775-008-0369-4.

Huynen, M. A., B. Snel, P. Bork, and T. J. Gibson. 2001. The phylogenetic distribution of frataxin indicates a role in Fe-S cluster protein assembly. *Hum Mol Genet* 10 (21):2463–68.

Jiralerspong, S., B. Ge, T. J. Hudson, and M. Pandolfo. 2001. Manganese superoxide dismutase induction by iron is impaired in Friedreich ataxia cells. *FEBS Lett* 509 (1):101–5.

Johnson, D. J., D. R. Dean, A. D. Smith, and M. K. Johnson. 2005. Structure, function, and formation of biological Fe-S clusters. *Annu Rev Biochem* 74:247–81.

Karlberg, T., U. Schagerlof, O. Gakh, S. Park, U. Ryde, M. Lindahl, K. Leath, E. Garman, G. Isaya, and S. Al-Karadaghi. 2006. The structures of frataxin oligomers reveal the mechanism for the delivery and detoxification of iron. *Structure* 14 (10):1535–46.

Karthikeyan, G., L. K. Lewis, and M. A. Resnick. 2002. The mitochondrial protein frataxin prevents nuclear damage. *Hum Mol Genet* 11 (11):1351–62.

Koutnikova, H., V. Campuzano, F. Foury, P. Dolle, O. Cazzalini, and M. Koenig. 1997. Studies of human, mouse and yeast homologues indicate a mitochondrial function for frataxin. *Nat Genet* 16 (4):345–51.

Koutnikova, H., V. Campuzano, and M. Koenig. 1998. Maturation of wild-type and mutated frataxin by the mitochondrial processing peptidase. *Hum Mol Genet* 7 (9):1485–89.

Labuda, M., J. Poirier, and M. Pandolfo. 1999. A missense mutation (W155R) in an American patient with Friedreich ataxia. *Hum Mutat* 13:506–7.

Layer, G., S. Ollagnier de Choudens, Y. Sanakis, and M. Fontecave. 2006. Fe-S cluster biosynthesis: characterization of *Escherichia coli* cyay as an iron donor for the assembly of [2Fe-2S] clusters in the scaffold ISCU. *J Biol Chem* 281 (24):16256–63.

Leidgens, S. 2008. *The function of yeast frataxin in iron-sulfur cluster biogenesis: a systematic mutagenesis of the solvent-exposed side chains of the beta-sheet platform*. PhD thesis, Louvain la Neuf (Belgium).

Lesuisse, E., R. Santos, B. F. Matzanke, S. A. Knight, J. M. Camadro, and A. Dancis. 2003. Iron use for haeme synthesis is under control of the yeast frataxin homologue (Yfh1). *Hum Mol Genet* 12 (8):879–89.

Levi, S., B. Corsi, M. Bosisio, R. Invernizzi, A. Volz, D. Sanford, P. Arosio, and J. Drysdale. 2001. A human mitochondrial ferritin encoded by an intronless gene. *J Biol Chem* 276 (27):24437–40.

Musco, G., G. Stier, B. Kolmerer, S. Adinolfi, S. Martin, T. Frenkiel, T. Gibson, and A. Pastor. 2000. Towards a structural understanding of Friedreich's ataxia: the solution structure of frataxin. *Structure Fold Des* 8 (7):695–707.

Nair, M., S. Adinolfi, C. Pastore, G. Kelly, P. Temussi, and A. Pastore. 2004. Solution structure of the bacterial frataxin ortholog, CyaY: mapping the iron binding sites. *Structure* 12 (11):2037–48.

Nichol, H., O. Gakh, H. A. O'Neill, I. J. Pickering, G. Isaya, and G. N. George. 2003. Structure of frataxin iron cores: an x-ray absorption spectroscopic study. *Biochemistry* 42 (20):5971–76.

O'Neill, H. A., O. Gakh, S. Park, J. Cui, S. M. Mooney, M. Sampson, G. C. Ferreira, and G. Isaya 2005. Assembly of human frataxin is a mechanism for detoxifying redox-active iron. *Biochemistry* 44 (2):537–45.

Park, S., O. Gakh, S. M. Mooney, and G. Isaya. 2002. The ferroxidase activity of yeast frataxin. *J Biol Chem* 277 (41):38589–95.

Park, S., O. Gakh, H. A. O'Neill, A. Mangravita, H. Nichol, G. C. Ferreira, and G. Isaya. 2003. Yeast frataxin sequentially chaperones and stores iron by coupling protein assembly with iron oxidation. *J Biol Chem* 278 (33):31340–51.

Pastore, C., M. Franzese, F. Sica, P. Temussi, and A. Pastore. 2007. Understanding the binding properties of an unusual metal-binding protein: a study of bacterial frataxin. *FEBS J* 274 (16):4199–4210.

Pastore, A., S. R. Martin, A. Politou, K. C. Kondapalli, R. Stemmler, and P. A. Temussi. 2007. Unbiased cold denaturation: low and high temperature unfolding of yeast frataxin under physiological conditions. *J Am Chem Soc* 129 (17):5374–75.

Pianese, L., A. Tammaro, M. Turano, I. De Biase, A. Ponticelli, and S. Cocozza. 2002. Identification of a novel transcript of X25, the human gene involved in Friedreich ataxia. *Neurosci Lett* 320:137–40.

Prischi, F., Giannini, C., Adinolfi, S., Pastore, A. 2009. The N-terminus of mature human frataxin is intrinsically unfolded. *FEBS J* 276:6669–6676.

Puccio, H., D. Simon, M. Cosseé, P. Criqui-Filipe, F. Tiziano, J. Melki, C. Hindelang, R. Matyas, P. Rustin, and M. Koenig. 2001. Mouse models for Friedreich ataxia exhibit cardiomyopathy, sensory nerve defect and Fe-S enzyme deficiency followed by intramitochondrial iron deposits. *Nat Genet* 27 (2):181–86.

Ramazzotti, A., V. Vanmansart, and F. Foury. 2004. Mitochondrial functional interactions between frataxin and Isu1p, the Fe-S cluster scaffold protein, in *Saccharomyces cerevisiae*. *FEBS Lett* 557:215–20.

Ren, B., G. Tibbelin, T. Kajino, O. Asami, and R. Ladenstein. 2003. The multi-layered structure of Dps with a novel di-nuclear ferroxidase center. *J Mol Biol* 329 (3):467–77.

Ristow, M., M. F. Pfister, A. J. Yee, M. Schubert, L. Michael, C. Y. Zhang, K. Ueki, M. D. Michael II, B. B. Lowell, and C. R. Kahn. 2000. Frataxin activates mitochondrial energy conversion and oxidative phosphorylation. *Proc Natl Acad Sci U S A* 97 (22):12239–43.

Rotig, A., P. de Lonlay, D. Chretien, F. Foury, M. Koenig, D. Sidi, A. Munnich, and P. Rustin. 1997. Aconitase and mitochondrial iron-sulphur protein deficiency in Friedreich ataxia. *Nat Genet* 17:215–17.

Sakamoto, N., K. Ohshima, L. Montermini, M. Pandolfo, and R. D. Wells. 2001. Sticky DNA, a self-associated complex formed at long GAA_TTC repeats in intron 1 of the frataxin gene, inhibits transcription. *J Biol Chem* 276 (29):27171–77.

Sazanov, L. A. and P. Hinchliffe. 2006. Structure of the hydrophilic domain of respiratory complex I from *Thermus thermophilus*. *Science* 311 (5766):1430–36.

Schmucker, S., M. Argentini, N. Carelle-Calmels, A. Martelli, and H. Puccio. 2008. The *in vivo* mitochondrial two-step maturation of human frataxin. *Hum Mol Genet* 17 (22):3521–31.

Schoenfeld, R. A., E. Napoli, A. Wong, S. Zhan, L. Reutenauer, D. Morin, A. R. Buckpitt, F. Taroni, B. Lonnerdal, M. Ristow, H. Puccio, and G. A. Cortopassi. 2005. Frataxin deficiency alters heme pathway transcripts and decreases mitochondrial heme metabolites in mammalian cells. *Hum Mol Genet* 14 (24):3787–99.

Schultz, J. B., T. Dehmer, L. Schols, H. Mende, C. Hardt, M. Vorgerd, K. Burk, W. Matson, J. Dichgans, M. F. Beal, and M. B. Bogdanov. 2000. Oxidative stress in patients with Friedreich ataxia. *Neurology* 55:1719–21.

Simon, D., H. Seznec, A. Gansmuller, N. Carelle, P. Weber, D. Metzger, P. Rustin, M. Koenig, and H. Puccio. 2004. Friedreich ataxia mouse models with progressive cerebellar and sensory ataxia reveal autophagic neurodegeneration in dorsal root ganglia. *J Neurosci* 24 (8):1987–95.

Stryer, L. 1995. *Biochemistry*, 4th ed. New York: Freeman.

Taketani, S. 2005. Aquisition, mobilization and utilization of cellular iron and heme: endless findings and growing evidence of tight regulation. *Tohoku J Exp Med* 205 (4):297–318.

Tan, G., L. S. Chen, B. Lonnerdal, C. Gellera, F. A. Taroni, and G. A. Cortopassi. 2001. Frataxin expression rescues mitochondrial dysfunctions in FRDA cells. *Hum Mol Genet* 10 (19):2099–2107.

Theil, E. 2007. Coordinating responses to iron and oxygen stress with DNA and mRNA promoters: the ferritin story. *Biometals* 20 (3–4):513–21.

Thierbach, R., T. J. Schulz, F. Isken, A. Voigt, B. Mietzner, G. Drewes, J. C. von Kleist-Retzow, R. J. Wiesner, M. A. Magnuson, H. Puccio, A. F. Pfeiffer, P. Steinberg, and M. Ristow. 2005. Targeted disruption of hepatic frataxin expression causes impaired mitochondrial function, decreased life span and tumor growth in mice. *Hum Mol Genet* 14 (24):3857–64.

Urbina, H. D., J. J. Silberg, K. G. Hoff, and L. E. Vickery. 2001. Transfer of sulfur from IscS to IscU during Fe/S cluster assembly. *J Biol Chem* 276 (48):44521–26.

Yoon, T., and J. A. Cowan. 2003. Fe-S cluster biosynthesis. Characterization of frataxin as an iron donor for assembly of [2Fe-2S] clusters in ISU-type proteins. *J Am Chem Soc* 125:6078–84.

———. 2004. Frataxin-mediated iron delivery to ferrochelatase in the final step of heme biosynthesis. *J Biol Chem* 279 (25):25943–46.

Yoon, T., E. Dizin, and J. A. Cowan. 2007. N-terminal iron-mediated self-cleavage of human frataxin: regulation of iron binding and complex formation with target proteins. *J Biol Inorg Chem* 12 (4):535–42.

Vazquez-Manrique, R. P., P. Gonzalez-Cabo, S. Ros, H. Aziz, H. A. Baylis, and F. Palau. 2006. Reduction of Caenorhabditis elegans frataxin increases sensitivity to oxidative stress, reduces lifespan, and causes lethality in a mitochondrial complex II mutant. *Faseb J* 20:172–74.

Wang, T., and E. A. Craig. 2008. Binding of yeast frataxin to the scaffold for Fe-S cluster biogenesis, Isu. *J Biol Chem* 283 (18):12674–79.

Wilson, R. B., and D. M. Roof. 1997. Respiratory deficiency due to loss of mitochondrial DNA in yeast lacking the frataxin homologue. *Nat Genet* 16 (4):352–57.

Wong, A., J. Yang, P. Cavadini, C. Gellera, B. Lonnerdal, F. Taroni, and G/ Cortopassi. 1999. The Friedreich's ataxia mutation confers cellular sensitivity to oxidant stress which is rescued by chelators of iron and calcium and inhibitors of apoptosis. *Hum Mol Genet* 8:425–30.

8 Mechanism of Human Copper Transporter Wilson's Disease Protein

Amanda Barry, Zara Akram, and Svetlana Lutsenko

CONTENTS

Wilson's Disease (WD) Protein Is Essential for Homeostatic Control
of Copper in Humans 145
ATP7B Is a Cu-Transporting P-Type ATPase 146
Functional Expression of ATP7B 149
Structural Studies of ATP7B 151
  The N-Domain Contains Determinants for Nucleotide Selectivity,
  Binding, and Orientation 152
  The P-Domain Houses Catalytic Aspartate and Mg^{2+} 154
  The A-Domain Facilitates Dephosphorylation 154
  Cytosolic Domains Communicate during Catalytic Cycle 155
Protein-Mediated Copper-Transfer May be Unique Feature
of P_{1B}-ATPases Mechanism 156
Intracellular Localization and Trafficking of ATP7B 157
Wilson's Disease-Causing Mutations: Known and Predicted Consequences 158
  Many Mutations in ATP-Binding Domain Affect
  Residues in ATP Vicinity 158
  His1069Gln Affects Protein Dynamics and the Placement of ATP
  in the Binding Pocket 159
  WD-Causing Mutations in the A-Domain 160
  COMMD1 Interactions with ATP7B Are Enhanced by WD Mutations 160
References 161

WILSON'S DISEASE (WD) PROTEIN IS ESSENTIAL FOR HOMEOSTATIC CONTROL OF COPPER IN HUMANS

Copper is an important micronutrient that is used as a cofactor by the enzymes involved in the development and proper function of the central nervous system, the formation of connective tissue and blood vessels, pigmentation, respiration, the detoxification of reactive oxygen species, and many other physiological processes. Dietary copper is absorbed primarily through the small intestine and then utilized throughout the body by all cells and tissues. A large fraction (about 40%) of absorbed dietary copper is taken by the liver, which is the main organ regulating copper homeostasis in the human body. Liver uses copper for its metabolic needs and for the biosynthesis of copper-dependent ferroxidase ceruloplasmin (CP). Copper-bound CP is excreted into the blood (Terada et al. 1998), where it accounts for over 60% of total serum copper content. Elevation of hepatic copper results in the activation of the copper excretion mechanism and the export of excess copper into the bile (Terada et al. 1999). The biliary excretion of copper is the major route for regulation of copper content in the body.

The important role of the liver in human copper metabolism is mediated through the expression and function in hepatocytes of the ATP-driven copper transporter, ATP7B. ATP7B is responsible for the delivery of copper to CP and the excretion of excess copper into the bile. Prompt removal of nonutilized copper is essential, because copper can bind tightly and nonspecifically to many proteins as well as induce oxidation and radical mediated damage. In other tissues, the physiological role of ATP7B is less understood. In intestine and kidney, it may be involved in copper storage (Barnes et al. 2009; Weiss et al. 2008), while in mammary gland (Michalczyk et al. 2000, 2008) and placenta (Lesk and Hardman 1982), ATP7B is thought to mediate apical copper export in addition to copper delivery to CP. Inactivation of ATP7B in humans causes a severe disorder, Wilson's disease (WD), which affects predominantly children and young adults. Due to its role in WD etiology, ATP7B is often described in the literature as Wilson's disease protein (WNDP). We will use the two terms interchangeably.

WD is an autosomal recessive disorder with predominantly hepatic and/or neurological manifestations (Das and Ray 2006). WD is characterized by reduced biliary excretion of copper, marked accumulation of copper in the liver, and the absence (in the vast majority of cases) of an active holo-CP in the serum. In response to copper accumulation in the liver, WD patients develop a pronounced hepatic pathology (cirrhosis, chronic hepatitis, and progressive or fulminant liver failure). In WD, copper accumulation also occurs in the brain, kidneys, and cornea (Das and Ray 2006), where ATP7B is normally expressed. Copper misbalance in the brain of WD patients is associated with neurological defects (parkinsonian features, seizures) and psychiatric symptoms (personality changes, depression, and psychosis) (Machado et al. 2008; Strecker et al. 2006; Brewer 2005). A characteristic feature often found in WD patients with neurological symptoms is the appearance of Kayser-Fleischer rings, gold-brown deposits at the periphery of the cornea (Liu et al. 2002). The important role of ATP7B in human physiology and its association with human disease facilitated characterization of this transporter. In this review, we describe current understanding of the ATP7B transport mechanism and highlight the effect of WD-causing mutations on the structure, activity, and intracellular targeting of ATP7B.

ATP7B IS A CU-TRANSPORTING P-TYPE ATPASE

ATP7B (or WD protein) is a 165 kDa membrane protein, which has eight transmembrane segments (TMSs) and a large cytosolic portion composed of several functional domains (Figure 8.1). ATP7B binds ATP in the cytosol, hydrolyzes it, and uses the energy of ATP hydrolysis to transport copper from the cytosol into the lumen of

Mechanism of Human Copper Transporter Wilson's Disease Protein

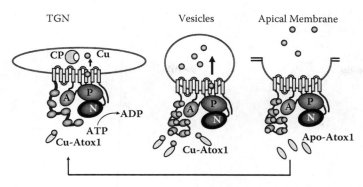

FIGURE 8.1 ATP7B is present in a cell in different locations. Under basal conditions, the predominant location of ATP7B is in the *trans*-Golgi network (TGN). Copper is delivered to ATP7B by Atox1 (Cu-Atox1). Using the energy of ATP hydrolysis ATP7B transports copper (Cu) from the cytosol into the lumen for incorporation into ceruloplasmin (CP). In elevated copper, more copper is delivered to ATP7B, ATP7B traffics to vesicles and sequesters excess vesicles in the vesicular lumen. Vesicles then fuse with the apical membrane and copper is released. Apo-Atox1 may remove regulatory copper from ATP7B facilitating the return of ATP7B to TGN under low-copper conditions.

intracellular compartments (Figure 8.1). In a cell, ATP7B is predominantly located in the *trans*-Golgi network (TGN); in this location it transports copper to copper-dependent enzymes, such as CP. Although ATP7B *in vitro* can deliver copper to tyrosinase (Guo et al. 2005), it is currently unclear whether *in vivo* the enzymes, other than CP, receive copper from ATP7B. In response to copper elevation, ATP7B traffics from the TGN to vesicles, where it sequesters copper for storage or for further export via vesicle-mediated exocytosis (Figure 8.1).

Mechanistically, ATP7B functions as a P-type ATPase. It belongs to the P_{1B} subgroup of this large class of the ATP-driven membrane-bound ion transporters, which includes CopA (another P_{1B} ATPase in archaea or bacteria), Ca^{2+}-ATPase and Na^+/K^+-ATPase (P_2-ATPases), flippases (P_4-ATPases), and others. The reaction cycle of the P-type ATPases, including ATP7B, is characterized by (1) autophosphorylation with the formation of a transient aspartyl-phosphate intermediate, (2) large and characteristic domain movements, and (3) distinct conformational states with a high (E1) and low (E2) affinity for ATP and the exported ion (Figure 8.2). It is thought that the transport cycle of ATP7B is initiated by the binding of Cu and ATP, stabilizing the E1 state (Figure 8.2). Like all P-type ATPases, ATP7B then hydrolyzes ATP with the transfer of the ATP γ-phosphate to the invariant Asp1027 in the DKTG signature motif, thus forming the E1P intermediate. This reaction requires the presence of bound copper within the transmembrane portion of the transporter. In the presence of a copper chelator, phosphorylation is not observed (Walker, Tsivkovskii, and Lutsenko 2002; Pilankatta et al. 2009).

The well-characterized P_2-type ATPases (such as Ca^{2+}-ATPase and Na^+/K^+-ATPase) have a very strict requirement for the binding of exported ion at the transmembrane sites in order to form a phosphorylated intermediate. In contrast, ATP7B can form such an intermediate in the presence of a physiologically

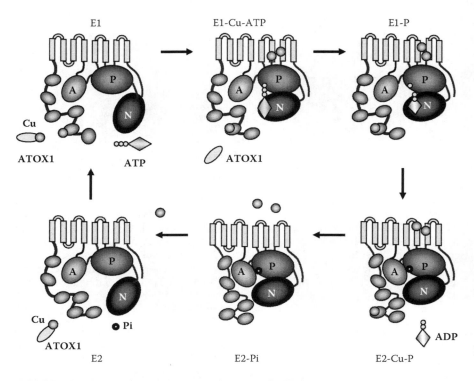

FIGURE 8.2 Cartoon illustrating the major steps of ATP7B transport cycle.

irrelevant compound, cisplatin (Safaei et al. 2008; Leonhardt et al. 2009), or when Cu-coordinating Cys residues within the membrane are mutated to Ser residues (Leonhardt et al. 2009). These observations suggest that the ATP-binding domain (ATP-BD) of ATP7B is prearranged in such a way that small changes in protein structure are sufficient to induce the transfer of γ-phosphate from ATP to the protein, and therefore the magnitude of ATP7B dynamics could be more limited (and hence require less precision) compared to other P-type pumps. In agreement with this idea, modeling studies (Efremov et al. 2004) have provided evidence for greater flexibility of the SERCA ATP-BD compared to the ATP-BD of ATP7B.

After ATP binds and a phosphorylated intermediate (E1P-Cu) is formed, ADP is released (or is competed off by high cytosolic ATP), and ATP7B undergoes conformational transition to acquire the E2P-Cu state (Figure 8.2). In this state, the transmembrane segments are thought to rearrange so that the Cu-binding site(s) become exposed to the lumen and the affinity for Cu decreases, facilitating copper release (Figure 8.2). The adherence of ATP7B to this general cycle is well established; however, the details of the process remain poorly understood. ATP7B has at least eight copper-binding sites (six sites in the cytosolic portion of the protein and likely two sites in the membrane). Presently, it is not clear how many sites should be occupied to initiate the transport cycle, and whether all bound copper is used for transport (or some copper atoms are transported while others remain bound to the regulatory sites). It is also uncertain which factors stimulate the release of copper, which is thought to bind very tightly within membrane (Gonzalez-Guerrero et al. 2008).

The fact that ATP7B can transfer copper to yeast proteins when expressed heterologously (see below) or to tyrosinase, which is not its physiological partner, indicate that specific and tight interactions of ATP7B with the acceptor protein(s) are not essential for copper release. However, in mice lacking CP, an important physiological partner of ATP7B in the liver, hepatic copper is elevated compared to control, indicating a limited capacity of ATP7B to export copper on its own (Meyer et al. 2001). Furthermore, in cell culture the R701W mutant of CP appears to induce relocalization of ATP7B from the TGN to vesicles, illustrating functional and perhaps physical interactions between the two proteins (di Patti et al. 2009). Whether the release of copper and subsequent dephosphorylation of ATP7B requires the participation of counter ions, such as a proton, is unknown. However, the stimulating effect of low pH on copper transport (Safaei et al. 2008) and dephosphorylation (our unpublished observation) suggests that a proton could be an important player in the ATP7B transport mechanism.

When Cu is released to the lumen, ATP7B adopts the E2P state (Figure 8.2). This state can also be formed (similarly to other P-type ATPases) by incubating ATP7B with inorganic phosphate, Pi (in the presence of Mg^{2+}). Addition of copper prevents phosphorylation by Pi (Tsivkovskii, Efremov, and Lutsenko 2003), presumably by shifting the conformational equilibrium from the E2 to the E1 state. Copper-dependent inhibition of Pi phosphorylation (which occurs in the E2 state), as well as copper-stimulated phosphorylation by ATP (in the E1 state), illustrate the cross talk between ligand binding sites in the membrane and in the cytosol. This tight coupling between different portions of the transporter may produce complex consequences when mutations are introduced. The opposite effects of copper on the formation of E1P and E2P intermediates can be used to discriminate between the two conformational states and to evaluate effects of different mutations on the conformational state of ATP7B. Dephosphorylation of the protein (E2) by water entering the ATP-binding site and the conversion back to the E1 state (likely due to binding of ATP present at high concentration in the cytosol) resets the cycle (Figure 8.2). It also appears that copper binding to the N-terminal domain stimulates dephosphorylation (Pilankatta et al. 2009). Whether the protein is ever present in a truly apo-form (no bound copper and no bound ATP) is an open question.

FUNCTIONAL EXPRESSION OF ATP7B

Several experimental approaches for the functional characterization of ATP7B have been described in the literature; each has its own advantages and limitations. The complementation assay in yeast allows for the rapid and efficient screening of mutations that disrupts the function of ATP7B (Forbes and Cox 1998). Yeast *Saccharomyces cerevisiae* has some attractive features for heterologous protein expression and functional studies. Yeast is genetically well characterized and, unlike bacteria, has post-translational processing mechanisms and a secretory system that in general features resembles that of high eukaryotes. An advantage of protein expression in yeast over other mammalian systems is the generation of gene cDNA constructs by the *in vivo* homolog recombination, so called Drag and Drop cloning. The homologous recombination system of *S. cerevisiae* is very efficient: as little as 20 base pairs of sequence identity is sufficient for the DNA exchange (Mezard, Pompon, and Nicolas 1992). The Drag and Drop system also allows efficient cloning and expression of heterologous genes in large-scale experiments (Mezard, Pompon, and Nicolas 1992).

The complementation assay is based on the ability of ATP7B to complement the loss of yeast copper-transporting ATPase Ccc2p in the $\Delta ccc2$ strain of *S. cerevisiae*. Without Ccc2p, yeast cells are unable to grow under iron-limiting conditions due to the disrupted delivery of copper to FET3, a copper-dependent ferroxidase required for high-affinity iron uptake. Expression of ATP7B restores copper delivery into the TGN, activating FET3 and enabling cells to grow in iron-deficient medium. By analyzing growth curves of cells expressing control ATP7B and disease-causing mutants, one can identify mutations that completely inactivate ATP7B function (no growth) and those in which ATP7B has partial activity (some growth) (Hsi et al. 2008). However, due to its indirect nature the assay does not allow for detailed studies of the ATP7B reaction mechanism or the identification of transport cycle steps that are affected by ATP7B mutations.

ATP7B has also been expressed in insect Sf9 cells using baculovirus-mediated infection (Walker, Tsivkovskii, and Lutsenko 2002). In this system, ATP7B is produced at high level (approximately 2% of total membrane protein) and is isolated in the membrane-bound form (Walker, Tsivkovskii, and Lutsenko 2002) or as a component of sealed vesicles for transport assays (Safaei et al. 2008; Leonhardt et al. 2009). This approach has permitted direct analysis of the enzymatic cycle of ATP7B (phosphorylation and dephosphorylation steps) as well as measurements of the ATP-dependent copper transport. While very useful and highly informative, this assay is laborious and time consuming. Therefore it cannot be easily applied to the routine screening and characterization of mutants.

Expression in oocytes has been demonstrated for ATP7B and a highly homologous protein, ATP7A (Lorinczi et al. 2008; Multhaup et al. 2001). The significant advantage of oocytes (compared to all other systems) is that the transporter can be targeted to a cell surface (Lorinczi et al. 2008); i.e., the luminal portion of ATP7B (or ATP7A) can be easily accessed. This property may be particularly useful for the analysis of the copper release mechanism. Another potential advantage of oocytes is the well-developed methodology for electrical measurements, including single-molecule applications. However, to date the transport function of Cu-ATPases in oocytes has not yet been demonstrated. This could be due to a very slow rate of copper transport, which would be suboptimal for electrical measurements. Alternatively, the transport of copper could be electro-neutral, if an equivalent amount of protons is transported in the opposite direction of copper. It should also be noted that oocytes have high cellular copper content in the absence of efficient copper uptake (Lorinczi et al. 2008). Thus, it is possible that most of this stored copper is utilized for cellular metabolic needs and the egg's development, and that the copper export machinery, including the copper chaperone (which facilitates ATP7B activity), might be missing or expressed poorly in oocytes. Presently, further optimization of this promising expression system is needed.

Recently, adenovirus-mediated expression of ATP7B in mammalian cells has been reported (Pilankatta et al. 2009). The obvious advantage of this protocol is the production of ATP7B in the most relevant cellular environment. Also, viral infection yields a high percentage of cells expressing ATP7B and a high level of protein expression, estimated as 10–20% of total microsomal protein (Pilankatta et al. 2009). This system allowed the measurement of copper-dependent ATPase activity using microsomal preparations from infected cells for the first time. Determination of Pi calorimetrically after ATP hydrolysis at 37°C, pH 6.0, yielded activity of 30 nmol of Pi/mg of protein/min (Pilankatta et al. 2009). It is also possible, using adenovirus-mediated expression, to analyze the partial reactions and conformational changes of ATP7B. The cost of growing mammalian cells and the overall protein yields limit (but do not hinder) potential structural applications of this excellent expression system. In addition, care should be taken when subtle kinetic properties are characterized using overexpressed ATP7B. Endogenous ATP7B has at least two distinct intracellular locations that are associated with different levels of phosphorylation of ATP7B by kinase(s). Kinase-mediated phosphorylation may affect the activity of the transporter (Valverde et al. 2008). In cells overexpressing ATP7B (where the protein is present in many locations), one analyzes a mixture of transporters that may have different levels of specific activity or kinetic behavior. The recently described Tet-regulated expression of ATP7B (Barnes et al. 2009) may represent a helpful solution when precise targeting of the transporter in cells is critical.

STRUCTURAL STUDIES OF ATP7B

It is thought that copper is transported through the membrane by entering the membrane portion of ATP7B from the cytosol (for details on copper binding, see below). Within the membrane, Cu(I) binds to the sites formed by the CPC motif in transmembrane segment TM6, the YN residues in TM7, and Met^{1359} and Ser^{1363} residues in TM8. The involvement of these residues in copper coordination has not been directly demonstrated for ATP7B. The aforementioned predictions are based on high conservation of these residues among all P-type ATPases selective for Cu(I), the role of the corresponding residues in copper coordination in the archaean orthologue CopA (Gonzalez-Guerrero et al. 2008), and the inactivating effects of the CPC > SPC mutation on copper transport in ATP7B (Forbes and Cox 1998). Copper binds to ATP7B tightly and is retained during isolation of ATP7B-containing membranes (Walker, Tsivkovskii, and Lutsenko 2002); only the addition of ATP is needed to stimulate the formation of a phosphorylated intermediate.

The cytosolic portion of ATP7B includes the domains for copper-binding/sensing (N-terminal domain, N-ATP7B, which consists of six independently folded subdomains), autocatalytic phosphorylation (P-domain), nucleotide-binding (N-domain), and energy transduction (actuator or A-domain). The A-domain regulates conformational transitions and is required for the phosphatase activity in the dephosphorylation step (Olesen et al. 2004; Kuhlbrandt 2004). Each of these domains includes a characteristic signature motif that reflects the ATP7B membership in either the P-type family of ATPases or the P_{1B} subgroup. The N-terminal domain has six GMxCxxC motifs that form six metal-binding sites (one in each subdomain). A different number of these motifs (1–6) is found in all P_{1B} ATPases involved in Cu(I) transport. The P-domain has the invariant DKTGT, TGDN, and GDGxNDx sequences that are common for all P-type ATPases and are essential for catalysis. The N-domain contains the conserved ESEHPL sequence (with an invariant ExxH motif) that is characteristic of all P_{1B}-ATPases. The A-domain has a somewhat conserved TGEA/S sequence.

THE N-DOMAIN CONTAINS DETERMINANTS FOR NUCLEOTIDE SELECTIVITY, BINDING, AND ORIENTATION

Currently, high-resolution structures have been solved for the N-domain, the A-domain, and four N-terminal subdomains of ATP7B. The N-domain is a 17 kDa protein that is composed of six antiparallel β-strands forming a β-sheet and four α-helices (Figure 8.3a). The recombinant N-domain binds ATP > ADP > AMP, and does not bind GTP (Morgan et al. 2004); i.e., it includes amino acid residues responsible for the nucleotide selectivity as well as for the recognition of the sugar moiety

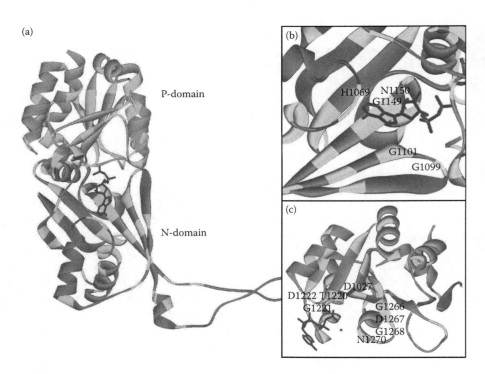

FIGURE 8.3 (a) ATP7B N-domain (Dmitriev et al. 2006) with ATP7B P-domain aligned to *Archaeoglobus fulgidus* CopA N- and P-domains bound to ADP-Mg (PDB#:3A1C) (Markus et al. 1994) using ESyPred3D. DKTGT sequence of the P-domain and SEHP sequence of the N-domain are darker in color; ADP-Mg in black; reported missense mutations in light gray (from http://www.wilsondisease.med.ualberta.ca/database.asp). (b) Zoom-in of the N-domain with residues important for ATP binding labeled. (c) ATP7B P-domain modeled to the *A. fulgidus* CopA P-domain using ESyPred3D.

Mechanism of Human Copper Transporter Wilson's Disease Protein

and phosphate tail of the adenine-containing nucleotides. Several residues of the N-domain (Gly^{1099}, Gly^{1101}, Gly^{1149}, and Asn^{1150}) show significant chemical shift upon binding of ATP indicative of nucleotide proximity (Dmitriev et al. 2006). These residues are conserved in all Cu-ATPases. In addition, mutational analysis of invariant Glu^{1064} and His^{1069} revealed their important role in ATP binding (Dmitriev et al. 2006; Morgan et al. 2004; Tsivkovskii, Efremov, and Lutsenko 2003). Mutations of these conserved residues in the N-domain in humans are associated with the WD phenotype (see Table 8.1).

Recent structural studies of the ATP7B orthologue, CopA from archaea and bacterial sources, revealed very similar structures for the N-domain with the exception of a 29-amino acid flexible loop present in human protein and not found in CopA (Dmitriev et al. 2006; Sazinsky et al. 2006; Markus et al. 1994) (Figure 8.3a). Since this loop is not important for nucleotide binding (our unpublished observation), its role is most likely in the regulation of ATP7B activity or intracellular localization. The structure of *Thermotoga maritima* CopA ATP-binding domain in a complex with a non-hydrolyzable ATP analogue illustrates that residues in the ATP-binding domain of CopA, Glu^{457}, His^{462}, and Gly^{492}

TABLE 8.1
The Main Functional Domains of ATP7B Are Targets of Numerous WD-Causing Mutations

	Residue Number	Function	Reported Disease-Causing Missense Mutations
N-terminal domain	1–631	Copper-binding regulation	$N^{41}S$, $G^{85}V$, $G^{96}D$, $D^{196}E$, $G^{333}R$, $I^{381}S$, $A^{486}S$, $L^{492}S$, $G^{515}V$, $V^{519}M$, $Y^{532}H$, $V^{536}A$, $E^{541}K$, $L^{549}P$, $G^{591}S/D$, $A^{604}P$, $R^{616}W/Q$, $G^{626}A$
A-domain	789–907	Energy transduction Phosphatase activity	$L^{795}R/F$, $A^{803}T$, $P^{840}L$, $I^{857}T$, $A^{861}T$, $G^{869}R/V$, $A^{874}V$, $V^{890}M$, $G^{891}V$ $Q^{898}R$
P-domain	1006–1035 1197–1303	Autocatalytic phosphorylation	$A^{1003}V/T$, $Q^{1004}P$, $K^{1010}R/T$, $G^{1012}R/V$, $A^{1018}V$, $V^{1024}A$, $M^{1025}R$, $D^{1027}A$, $T^{1029}A/I$, $T^{1031}S/A$, $T^{1033}A/S$, $G^{1035}V$, $Q^{1200}P$, $H^{1207}R$, $G^{1213}V$, $D^{1215}Y$, $V^{1216}M$, $T^{1220}M$, $G^{1221}E$, $D^{1222}N/Y/V$, $R^{1228}T$, $I^{1230}V$, $T^{1232}P$, $I^{1236}T$, $V^{1239}G$, $A^{1241}V$, $P^{1245}T$, $K^{1248}N$, $V^{1252}I$, $L^{1255}I$, $Q^{1256}R$, $V^{1262}F$, $G^{1266}R/K/V$, $D^{1267}A/V$, $G^{1268}R$, $N^{1270}T/S$, $D^{1271}N$, $P^{1273}S/L/Q$, $A^{1278}V$, $D^{1279}Y/G$, $G^{1281}C/D$, $G^{1287}S$, $T^{1288}R/M$, $D^{1296}N$, $V^{1297}D$, $L^{1299}F$
N-domain	1036–1196	Nucleotide binding	$V^{1036}I$, $R^{1038}K$, $R^{1041}W/P$, $L^{1043}P$, $D^{1047}V$, $P^{1052}L$, $G^{1061}E$, $A^{1063}V$, $E^{1064}K/A$, $A^{1065}P$, $E^{1068}G$, $H^{1069}Q$, $L^{1083}F$, $G^{1089}E/V$, $C^{1091}Y$, $F^{1094}L$, $Q^{1095}P$, $P^{1098}R$, $G^{1099}S$, $G^{1101}R$, $I^{1102}T$, $C^{1104}F/Y$, $V^{1106}I/D$, $V^{1109}M$, $G^{1111}D$, $H^{1115}R$, $A^{1140}V$, $Q^{1142}H$, $T^{1143}N$, $V^{1146}M$, $I^{1148}T$, $G^{1149}A$, $R^{1151}C/H$, $W^{1153}R/C$, $R^{1156}H$, $D^{1164}N$, $A^{1168}S/P$, $A^{1169}V/T$, $E^{1173}K/G$, $G^{1176}R/E$, $T^{1178}A$, $A^{1183}T/G$, $I^{1184}T$, $G^{1186}S/C$, $A^{1193}P$

(equivalent to Glu^{1064}, His^{1069}, and Gly^{1101} in ATP7B) orient and bind the adenine ring through hydrogen bonding, and Gly^{490} and Gly^{501} (equivalent to Gly^{1099} and Gly^{1149} in ATP7B) interact primarily with the ribose moiety and α-phosphate (Markus et al. 1994).

The P-Domain Houses Catalytic Aspartate and Mg^{2+}

Although no crystal structures are available for the P-domain of ATP7B, this 14 kDa domain has about 50% sequence identity to the P-domains of other P-type ATPases and most likely shares their three-dimensional fold. In CopA, an orthologue of ATP7B, the P-domain structure consists of six parallel β-sheets between six α-helices (Sazinsky et al. 2006; Markus et al. 1994) (Figure 8.3). Unlike the N-domain, which is independently folded and can be expressed as a separate protein, the P-domain has the N-domain inserted after the DKTGT-containing loop. Consequently, the P-domain has been expressed and characterized together with the N-domain. The conserved motifs of the P-domain, DKTGT (including the catalytic Asp), TGDN, and GDGxNDx, are found on three separate loops that each protrude toward ADP-Mg^{2+} bound in a pocket between the P- and N-domains (Markus et al. 1994) (Figure 8.3b). Structural analysis of β-phosphoglucomutase, which has a homologous P-domain to P-type ATPases (Lander et al. 2001), suggests that the Mg^{2+} ion is required for charge compensation and positioning of the catalytic Asp that enables phosphoryl transfer from ATP (Lahiri et al. 2003; Kuhlbrandt 2004). The binding of ATP by the isolated N-domain of ATP7B (without the P-domain containing the catalytic Asp) is independent of Mg^{2+}, however ATP hydrolysis and phosphorylation by the full-length ATP7B require Mg^{2+}.

The A-Domain Facilitates Dephosphorylation

The A-domain of ATP7B is approximately 12.6 kDa and is composed of seven anti-parallel β-strands arranged in two sheets, two additional β-strands that form a β-hairpin structure flanking the two sheets, and two α-helices at the N- and C- termini (Banci et al. 2009) (Figure 8.4). The A-domain contains the ^{858}TGE motif common to many P-type ATPases. In Ca^{2+}-ATPase, this sequence is part of a loop that comes close to the catalytic aspartate in the E2 and E2P intermediate states (Toyoshima and Inesi 2004). Recent nuclear magnetic resonance (NMR) analysis of the isolated A-domain of ATP7A showed that this loop protrudes from the structure, allowing for interactions with the phosphorylated aspartate (Banci et al. 2009). This placement is not surprising as this loop and homologous motif in SERCA has been implicated in the dephosphorylation of the protein from the E2P to the E2 state (Olesen et al. 2004). Mutation of the TGE motif in ATP7A, a close human homologue of ATP7B, is associated with stabilization of hyperphosphorylation state (Petris et al. 2002), also likely due to impaired dephosphorylation. The structure of the A-domains of ATP7A and ATP7B are nearly identical (Banci et al. 2009).

FIGURE 8.4 A-domain of human ATP7A with missense mutations in light gray (PDB#: 2KIJ).

Cytosolic Domains Communicate during Catalytic Cycle

During ATP binding and hydrolysis, the N- and P-domains work together as a clamshell first to close and allow the formation of a complex between the catalytic Asp, the γ-phosphate of ATP, and Mg^{2+} (the E1 state). After the hydrolysis, the P-type ATPases undergo a structural rearrangement that opens the ATP-binding domain, allowing ADP to dissociate and the A-domain to rotate, bringing the TGE-containing loop within reach of the phosphorylation site (Toyoshima et al. 2000; Toyoshima and Nomura 2002; Inesi et al. 2006; Hatori et al. 2007), and forming the E2 state (Figure 8.2). While these domain movements were best described for the SERCA Ca^{2+}-ATPase, recent proteolysis studies of CopA indicate that Cu-ATPases undergo similar movements. In CopA, when copper is not present the linker region between the A-domain and TM 2 is exposed, whereas in the presence of ligands stabilizing the E1-P and E2-P states, the region is resistant to cleavage (Hatori et al. 2007). The amplitude of domains movements is still a matter of study, particularly for ATP7B, for which high-resolution structure is not available. In SERCA, the first E1 structure indicated a large distance between the N- and A-domains in the presence of transported ion (Toyoshima et al. 2000; Toyoshima and Nomura 2002). However, more recent studies using fluorescence resonance energy transfer between cyan-fluorescent protein fused to the A-domain and the N-domain labeled with fluorescein isothiocyanate indicated that the mean distance between the two domains increases by about 3 Å in the presence of transported ion (Winters et al. 2008), i.e., that the average E1 structure may not be as open as previously thought.

In addition to interactions between the N-, P-, and A-domains (which appear common for all P-type ATPases), ATP7B function is regulated by the N-terminal domain. In the apo-form, the N-terminal domain interacts with the ATP-BD, decreasing affinity for ATP (Tsivkovskii, MacArthur, and Lutsenko 2001). Binding of copper induces conformational changes in the domain and weakens its interactions with ATP-BD, resulting in the increased affinity for ATP (Tsivkovskii, MacArthur, and Lutsenko 2001). Deletion of 432 residues (nucleotides 169–1458 cDNA) from the N-terminal domain of ATP7B increases the rate of catalytic phosphorylation (Huster and Lutsenko 2003), further confirming the autoinhibitory function of the N-terminus. Which regions of the N-terminus are involved in domain-domain interactions remains to be determined. It seems likely that this large domain (70 kDa, over 600 amino acid residues) has multiple points of contact with other cytosolic domains and may modulate many aspects of protein activity and/or intracellular localization. Recent studies indicate that the N-terminus contains a sequence determinant for the apical targeting of ATP7B (Braiterman et al. 2009), protein-protein interactions (Lim et al. 2006a, 2006b) and kinase-mediated phosphorylation (Lutsenko et al. 2007; Pilankatta et al. 2009), in addition to being a gateway for copper entry into the protein (see below).

PROTEIN-MEDIATED COPPER-TRANSFER MAY BE UNIQUE FEATURE OF P_{1B}-ATPASES MECHANISM

There is little doubt that, in general, the mechanism of the ATP-dependent copper transport by ATP7B follows the Post-Alberts scheme, common for P-type ATPases. However, the details of copper binding are certainly distinct. Copper is not present in the cytosol in the form of free ion; rather it is delivered to the transporter by Atox1, a small cytosolic metallochaperone. Atox1 transfers copper to the N-terminal domain of ATP7B, inducing conformational changes in the domain (Walker et al. 2004; DiDonato et al. 2000). The conformational change is likely to open access to the transmembrane portion of ATP7B. Copper is then delivered to the membrane sites either directly by metallochaperone (as was suggested for CopA [Gonzalez-Guerrero and Arguello 2008]) or by one of the N-terminal metal-binding sites, and/or through a specific pathway that remains to be identified.

The mechanisms of copper binding by ATP7B and chaperone-mediated transfer have been subjects of intensive investigations. It is now well established that the N-terminal domain of human ATP7B has six metal-binding sites, each located within an independently folded subdomain with a ferredoxin fold. The structures of individual metal-binding subdomains in two pairs (MBS3,4 and MBS5,6) and Atox1 have been solved. These studies uncovered a high similarity between the donor (Atox1) and acceptor (individual MBS) molecules, in both their structural fold and metal coordination. The copper sites in each subdomain are formed by the invariant sequence motif GMT/HCxxC, in which copper is coordinated in an almost linear fashion by sulfurs from Cys residues. In the full-length N-ATP7B, the subdomains tightly interact, as indicated by the conformational changes of the entire N-ATP7B domain upon copper binding (DiDonato et al. 2000) as well as coelution of the subdomains on gel filtration after limited proteolysis of N-ATP7B (our unpublished observation). Spectroscopic studies suggest that in N-ATP7B with four to six copper atoms bound, pairs of copper sites are facing each other, producing a characteristic spectroscopic feature indicative of the Cu-Cu proximity (Ralle, Lutsenko, and Blackburn 2004).

Whether the highly homologues subdomains show certain selectivity and/or order of copper binding is a fascinating mechanistic question. Current data indicate that the individual recombinant metal-binding subdomains (MBSs) have a comparable binding affinity for copper and that all six sites in N-ATP7B can be loaded with copper using the Cu-Atox1 complex (Walker, Tsivkovskii, and Lutsenko 2002; Wernimont, Yatsunyk, and Rosenzweig 2004; Yatsunyk and Rosenzweig 2007). Mutations of five out of six sites (in different combinations) in the recombinant N-ATP7B do not prevent Atox1-mediated transfer of copper to the remaining site, suggesting that the subdomains can act independently (Yatsunyk and Rosenzweig 2007). At the same time, the mutation of MBS2 in the full-length ATP7B (in which all cytosolic domains are present) completely disrupts the transfer of copper from Atox1 (Walker et al. 2004). Thus the interdomain interactions may contribute significantly to the selectivity and order of copper transfer. Furthermore, individual recombinant MBSs show differences in their protein dynamics and their interaction with the metallochaperone Atox1 (Banci et al. 2008, 2009; Achila et al. 2006). Molecular modeling and dynamic simulation studies (Rodriguez-Granillo, Crespo, and Wittung-Stafshede 2009) illustrate different flexibility of the metal-binding loops (MBS5 and particularly MBS4 have higher flexibility compared to MBS1, MBS2, and MBS6, whereas the MBS3 loop is the most rigid). In the full-length ATP7B, the flexibility/exposure of individual MBS can be further enhanced (or diminished), thus contributing to the observed selectivity of copper transfer.

Initial protein-protein interaction experiments using a yeast two-hybrid system (Larin et al. 1999) and subsequent NMR experiments (Banci et al. 2009; Achila et al. 2006) demonstrate that MBS5 and MBS6 do not interact strongly with Atox1. This result and the ability of recombinant MBS4 to transfer copper to the MBS5,6 pair (Achila et al. 2006) was interpreted to mean that MBS5 and MBS6 receive copper not from the chaperone but from MBS4, which shows strong interactions with Atox1. While this is a logical interpretation of the *in vitro* data, it should be noted that in cells, the ATP7B protein lacking the MBS1–4 region is functional (i.e., it receives copper, presumably from Atox1 [Guo et al. 2005]), therefore the MBS4-to-MBS5/6 transfer route is not obligatory. Furthermore, the rat orthologue of ATP7B lacks the CxxC metal-binding motif in MBS4 (although the entire subdomain is present [Tsay et al. 2004]), further illustrating the possibility that copper migration within N-ATP7B, if occurs, may bypass MBS4. Altogether, current data suggest that Atox1 docks preferentially to the MBS1-4 region and that the transfer of copper to this region is likely to serve a sensing/regulatory purpose. Whether or not copper that binds to the N-terminus is then transferred to the intramembrane sites remains to be determined.

INTRACELLULAR LOCALIZATION AND TRAFFICKING OF ATP7B

The transport function of ATP7B appears to be coupled to the transporter's ability to change its intracellular localization in response to changes in copper levels. Under basal growth conditions, ATP7B is localized within the TGN. This localization is consistent with its role in cuproenzyme biosynthesis, as ceruloplasmin was shown to incorporate copper within the Golgi network (Hellman et al. 2002). In response to elevated copper, ATP7B reversibly relocalizes to a peripheral vesicular compartment in proximity to the apical plasma membrane (Guo et al. 2005; Lutsenko et al. 2007). Mutations that inactivate the transporter usually disrupt its ability to traffic. In contrast, the TGE > AAA substitution that inhibits the dephosphorylation step (by analogy with similar mutation in ATP7A [Petris et al. 2002]) is associated with the loss of TGN retention and trafficking of ATP7B to vesicles under basal copper conditions (Cater et al. 2004). The mechanism of this constitutive trafficking is unclear; however, the mutation has a dominant phenotype and facilitates protein trafficking even in combination with other inactivating mutations such as CPC > APA (Cater et al. 2004).

WILSON'S DISEASE-CAUSING MUTATIONS: KNOWN AND PREDICTED CONSEQUENCES

Many Mutations in ATP-Binding Domain Affect Residues in ATP Vicinity

Disease mutations in ATP7B have given further insight into the function of specific protein residues. Many missense mutations have been reported in the three domains involved in ATP binding of ATP7B (see all at http://www.wilsondisease.med.ualberta.ca/database.asp; see also Table 8.1), but very few have been studied thoroughly. Due to their respective locations near conserved residues, some of these mutations are assumed to affect ATP binding or catalytic phosphorylation. For example, I1102T, the most common mutation in Indian WD patients, may affect ATP binding by perturbing the environment of ATP-binding residues G1101 and G1099. These two Gly residues are conserved in all P-type ATPases, show a considerable chemical shift upon ATP binding to ATP7B, and are in close proximity to the phosphate tail (Dmitriev et al. 2006). By mutating the hydrophobic I1102 residue to a hydrophilic polar amino acid, such as Thr, the chemical environment or, perhaps, the precise shape of the cavity surrounding the binding site for the phosphate tail of ATP would be drastically altered. In the P-domain, mutations such as T1031S/A may have a similar effect in affecting the chemical environment in vicinity to the catalytic D1027. The conserved sequence $D^{1027}KTGT^{1031}$ is present in all P-type ATPases, and mutations in the residues, such as D1027A and T1029A/I, disrupt catalytic phosphorylation (Iida et al. 1998). Furthermore, residues affecting the ^{1220}TGD motif may affect phosphorylation as do mutations in the ^{625}TGD motif of Ca^{2+}-ATPase. In this protein, mutation of the TGD motif causes inhibition of ATP binding in the presence of Mg^{2+} and perturbs the placement of the photolabeled phosphate chain in relation to the catalytic Asp (McIntosh et al. 2004). Similarly, mutations surrounding and including D1267 in the GDGxNDx motif in the hinge region of the P-domain most likely affect ATP binding and stabilization, as do mutations in D703 in the equivalent motif of Ca^{2+}-ATPase (Inesi et al. 2003). Other mutations, such as G1186S/C, may affect tertiary structure (Shah et al. 1997). Some mutations, such as C1104F, significantly alter protein folding (Morgan et al. 2004). Of the most studied mutations in the N-domain, substitutions E1064A, H1069Q, and R1151H reduce ATP affinity, but not all affect structure in a significant way (Morgan et al. 2004).

His1069Gln Affects Protein Dynamics and the Placement of ATP in the Binding Pocket

H1069Q, a common and the most studied WD-causing mutation, was first reported in 1993 using linkage disequilibrium and haplotype analysis (Atwood et al. 1998). NMR analysis of the N-domain bearing the H1069Q mutation demonstrated that this mutation does not significantly perturb the structure of the N-domain but it does prevent tight binding of ATP (Dmitriev et al. 2006). Other substitutions at this residue (Ala or Cys) do not significantly alter proteolytic resistance or folding of ATP7B (Tsivkovskii, Efremov, and Lutsenko 2003), but these mutations completely disrupt catalytic phosphorylation in the presence of ATP. More recently, structural analysis of the CopA homologue to H1069Q, H462Q, showed that the Gln partially mimics the imidazole ring of His and is able to bind nucleotide (albeit at lower affinity) but in a configuration that is less favorable for phosphoryl transfer compared to wild-type (Markus et al. 1994). These results indicate that H1069 is required for the proper orientation of ATP for ATP hydrolysis.

It is interesting that in the H1069Q and H1069A (but not H1069C), variants of the full-length ATP7B also lose their ability to become phosphorylated by inorganic phosphate (Tsivkovskii, Efremov, and Lutsenko 2003); i.e., the mutation appears to alter the close association between the N-, P-, and A-domains necessary for this reaction. While the effect on interdomain interactions is small and is not detected by proteolytic studies, it plays an important role in the cell. *In vivo* and in transfected cells, the H1069Q mutation causes the retention of ATP7B in the endoplasmic reticulum rather than proper targeting to the *trans*-Golgi membrane (Payne, Kelly, and Gitlin 1998; Huster and Lutsenko 2003). In addition, H1069Q has a shorter half-life than wild-type, indicating faster degradation (Payne, Kelly, and Gitlin 1998). Normal localization to the TGN (but not the activity) can be restored by growing cells at low temperature.

How can the aforementioned *in vivo* data (which strongly suggest the effect of H1069Q mutation on protein folding) be reconciled with the *in vitro* studies that show minimal effect of this mutation on ATP7B folding? The explanation can be found in studies of protein dynamics (Rodriguez-Granillo, Sedlak, and Wittung-Stafshede 2008). The N-domain is unstable in the absence of ATP and rapidly precipitates, whereas the binding of ATP or ADP compacts and stabilizes the protein, allowing for structure determination (Dmitriev et al. 2006). Direct comparison of protein stability in the urea-triggered unfolding experiments revealed that the apo wild-type N-domain is more stable compared to the H1069Q variant and that the stabilizing effect of ATP binding is also stronger for the wild-type protein (Rodriguez-Granillo, Sedlak, and Wittung-Stafshede 2008). Furthermore, molecular dynamic simulations revealed that H1069 contributes to maintaining the integrity of the N-domain fold by interacting with R1038, and this stabilizing interaction is disrupted in the mutant (Rodriguez-Granillo, Sedlak, and Wittung-Stafshede 2008). Thus the H1069Q substitution has numerous effects, each contributing to the phenotypic manifestation of this mutation (Rodriguez-Granillo, Sedlak, and Wittung-Stafshede 2008). The incorrect positioning of ATP appears to have an overriding consequences, since the decreasing protein dynamic (by lowering temperature) and the presence of 1 mM ATP in the cytosol are insufficient to restore the transport activity and trafficking of the H1069Q mutant in mammalian cells (Payne, Kelly, and Gitlin 1998).

WD-Causing Mutations in the A-Domain

Disease-causing mutations in the A-domain (residues 898–907 in the full-length ATP7B) have also been reported, including missense mutations, small insertions, and deletions (the largest being 2630–2656del described at the genomic DNA level) (http://www.hgmd.cf.ac.uk/). Missense mutations include L795R/F, A803T, P840L, I857T, A861T, G869R/V, A874V, V890M, G891V, and Q898R (Table 8.1). Most of these mutations are thought to affect enzyme function either by destabilizing the fold of the protein or by disturbing structural amino acid contacts (Banci et al. 2009). Ala861 is the residue in the vicinity of the TGE motif thought to be involved in the dephosphorylation of the protein from the E2P to the E2 state (Olesen et al. 2004). Banci and colleagues hypothesized that this mutation to Thr may affect the positioning of the TGE-containing loop by altering the hydrogen bond network at the A- and P-domain interface. These authors also posit, based on studies of *Enterococcus hirae* Cu-ATPase, that G869R/V may disrupt an interaction of the A-domain with the P-domain (Banci et al. 2009; Lubben et al. 1990).

Mis-targeting may selectively impair the compartment-specific functions of ATP7B. Several WD-causing mutations are associated with defects in copper-induced relocalization of ATP7B. One type of localization is the abolition of copper-induced translocation of ATP7B by the G943S mutation, which still permits cuproenzyme biosynthesis in the yeast complementation assay (Forbes and Cox 1998, 2000). Another interesting mis-targeting mutation was found in the N-terminal domain of ATP7B. The N-terminal residues 37–45 (FAFDNVGYE) are necessary for correct targeting of ATP7B to the apical membrane in elevated copper, as well as stabilizing the protein at the TGN under low copper (Braiterman et al. 2009). N41S, the WD-causing mutation found in this region, disables proper targeting and retention, causing a large fraction of the protein to move to the basolateral membrane (Braiterman et al. 2009). Loss of trafficking or partial mislocalization, rather than the loss of Cu-transport activity, may explain why, in some cases, ceruloplasmin levels are normal in patients with WD-causing mutations, as this is the case for the G943S mutation (Forbes and Cox 2000). More frequently, missense mutations affect the localization of ATP7B causing the retention of ATP7B in the endoplasmic reticulum. ER retention was reported for the two most common WD mutations H1069Q and R778L (Forbes and Cox 2000; Payne, Kelly, and Gitlin 1998).

COMMD1 Interactions with ATP7B Are Enhanced by WD Mutations

COMMD1, the protein defective in Bedlington terriers, causes a disorder characterized by hepatic copper accumulation as a consequence of decreased biliary copper excretion, the phenotype resembling WD. Similarly, the knockout of *COMMD1* gene in HEK293 cells leads to copper accumulation in these cells (Burstein et al. 1977), indicating that COMMD1 and ATP7B may work synergistically. Using pull-down experiments, COMMD1 was shown to interact with the N-terminal domain of ATP7B *in vitro* as well as *in vivo* (Tao et al. 2003), although colocalization studies did not reveal a significant overlap in cells (Burkhead et al. 2009), pointing to a transient nature of protein interactions. ATP7B trafficking is not affected in the COMMD1-deficient cells (de Bie et al. 2007). Thus the precise role of COMMD1 in copper export remains illusive. Recent studies suggest that COMMD1 forms oligomers and binds to phosphatidylinositol-4,5 biphosphate in endosomal vesicular membranes and acts as a scaffold for protein-protein interactions necessary for protein sorting (Burkhead et al. 2009). Interestingly, several mutations, particularly G85V, G591, and A604P, in the N-terminal domain of ATP7b result in enhanced interactions with COMMD1 as shown by the pull-down experiments (de Bie et al. 2007). Also, overexpression of COMMD1 stimulates degradation of wild-type ATP7B suggesting, perhaps, the role of COMMD1 in directing ATP7B to lysosomes. Since the G85V mutation results in the retention of ATP7B in the ER and marked decrease in the half-life of the protein, it was suggested that COMMD1 may regulate the ATP7B degradation through proteosomal pathway (de Bie et al. 2007). Establishing the precise role of COMMD1 in intracellular routing of ATP7B is a fascinating task for the future.

REFERENCES

Achila, D., L. Banci, I. Bertini, J. Bunce, S. Ciofi-Baffoni, and D. L. Huffman. 2006. Structure of human Wilson protein domains 5 and 6 and their interplay with domain 4 and the copper chaperone HAH1 in copper uptake. *Proc Natl Acad Sci U S A* 103 (15):5729–34.

Atwood, C. S., R. D. Moir, X. Huang, R. C. Scarpa, N. M. Bacarra, D. M. Romano, M. A. Hartshorn, R. E. Tanzi, and A. I. Bush. 1998. Dramatic aggregation of Alzheimer abeta by Cu(II) is induced by conditions representing physiological acidosis. *J Biol Chem* 273 (21):12817–26.

Banci, L., I. Bertini, F. Cantini, C. Massagni, M. Migliardi, and A. Rosato. 2009. An NMR study of the interaction of the N-terminal cytoplasmic tail of the Wilson disease protein with copper(I)-HAH1. *J Biol Chem* 284 (14):9354–60.

Banci, L., I. Bertini, F. Cantini, A. C. Rosenzweig, and L. A. Yatsunyk. 2008. Metal binding domains 3 and 4 of the Wilson disease protein: solution structure and interaction with the copper (I) chaperone HAH1. *Biochemistry* 47 (28):7423–29.

Barnes, N., M. Y. Bartee, L. Braiterman, A. Gupta, V. Ustiyan, V. Zuzel, J. H. Kaplan, A. L. Hubbard, and S. Lutsenko. 2009. Cell-specific trafficking suggests a new role for renal ATP7B in the intracellular copper storage. *Traffic* 10 (6):767–79.

Braiterman, L., L. Nyasae, Y. Guo, R. Bustos, S. Lutsenko, and A. Hubbard. 2009. Apical targeting and Golgi retention signals reside within a 9-amino acid sequence in the copper-ATPase, ATP7B. *Am J Physiol Gastrointest Liver Physiol* 296 (2):G433–44.

Brewer, G. J. 2005. Neurologically presenting Wilson's disease: epidemiology, pathophysiology and treatment. *CNS Drugs* 19 (3):185–92.

Burkhead, J. L., C. T. Morgan, U. Shinde, G. Haddock, and S. Lutsenko. 2009. COMMD1 forms oligomeric complexes targeted to the endocytic membranes via specific interactions with phosphatidylinositol 4,5-bisphosphate. *J Biol Chem* 284 (1):696–707.

Burstein, E. A., E. A. Permyakov, V. A. Yashin, S. A. Burkhanov, and A. Finazzi Agro. 1977. The fine structure of luminescence spectra of azurin. *Biochim Biophys Acta* 491 (1):155–59.

Cater, M. A., J. Forbes, S. La Fontaine, D. Cox, and J. F. Mercer. 2004. Intracellular trafficking of the human Wilson protein: the role of the six N-terminal metal-binding sites. *Biochem J* 380 (Pt 3):805–13.

Das, S. K., and K. Ray. 2006. Wilson's disease: an update. *Nat Clin Pract Neurol* 2 (9):482–93.

de Bie, P., B. van de Sluis, E. Burstein, P. V. van de Berghe, P. Muller, R. Berger, J. D. Gitlin, C. Wijmenga, and L. W. Klomp. 2007. Distinct Wilson's disease mutations in ATP7B are associated with enhanced binding to COMMD1 and reduced stability of ATP7B. *Gastroenterology* 133 (4):1316–26.

DiDonato, M., H. F. Hsu, S. Narindrasorasak, L. Que Jr., and B. Sarkar. 2000. Copper-induced conformational changes in the N-terminal domain of the Wilson disease copper-transporting ATPase. *Biochemistry* 39 (7):1890–96.

di Patti, M. C., N. Maio, G. Rizzo, G. De Francesco, T. Persichini, M. Colasanti, F. Polticelli, and G. Musci. 2009. Dominant mutants of ceruloplasmin impair the copper loading machinery in aceruloplasminemia. *J Biol Chem* 284 (7):4545–54.

Dmitriev, O., R. Tsivkovskii, F. Abildgaard, C. T. Morgan, J. L. Markley, and S. Lutsenko. 2006. Solution structure of the N-domain of Wilson disease protein: distinct nucleotide-binding environment and effects of disease mutations. *Proc Natl Acad Sci U S A* 103 (14):5302–7.

Efremov, R. G., Y. A. Kosinsky, D. E. Nolde, R. Tsivkovskii, A. S. Arseniev, and S. Lutsenko. 2004. Molecular modelling of the nucleotide-binding domain of Wilson's disease protein: location of the ATP-binding site, domain dynamics and potential effects of the major disease mutations. *Biochem J.* Aug 15; 382(Pt 1):293–305.

Forbes, J. R., and D. W. Cox. 1998. Functional characterization of missense mutations in ATP7B: Wilson disease mutation or normal variant? *Am J Hum Genet* 63 (6):1663–74.

———. 2000. Copper-dependent trafficking of Wilson disease mutant ATP7B proteins. *Hum Mol Genet* 9 (13):1927–35.

Gonzalez-Guerrero, M., and J. M. Arguello. 2008. Mechanism of Cu+-transporting ATPases: soluble Cu+ chaperones directly transfer Cu+ to transmembrane transport sites. *Proc Natl Acad Sci U S A* 105 (16):5992–97.

Gonzalez-Guerrero, M., E. Eren, S. Rawat, T. L. Stemmler, and J. M. Arguello. 2008. Structure of the two transmembrane Cu+ transport sites of the Cu+-ATPases. *J Biol Chem* 283 (44):29753–59.

Guo, Y., L. Nyasae, L. T. Braiterman, and A. L. Hubbard. 2005. NH2-terminal signals in ATP7B Cu-ATPase mediate its Cu-dependent anterograde traffic in polarized hepatic cells. *Am J Physiol Gastrointest Liver Physiol* 289 (5):G904–16.

Hatori, Y., E. Majima, T. Tsuda, and C. Toyoshima. 2007. Domain organization and movements in heavy metal ion pumps: papain digestion of CopA, a Cu+-transporting ATPase. *J Biol Chem* 282 (35):25213–21.

Hellman, N. E., S. Kono, G. M. Mancini, A. J. Hoogeboom, G. J. De Jong, and J. D. Gitlin. 2002. Mechanisms of copper incorporation into human ceruloplasmin. *J Biol Chem* 277 (48):46632–38.

Hsi, G., L. M. Cullen, G. Macintyre, M. M. Chen, D. M. Glerum, and D. W. Cox. 2008. Sequence variation in the ATP-binding domain of the Wilson disease transporter, ATP7B, affects copper transport in a yeast model system. *Hum Mutat* 29 (4):491–501.

Huster, D., and S. Lutsenko. 2003. The distinct roles of the N-terminal copper-binding sites in regulation of catalytic activity of the Wilson's disease protein. *J Biol Chem* 278 (34):32212–18.

Iida, M., K. Terada, Y. Sambongi, T. Wakabayashi, N. Miura, K. Koyama, M. Futai, and T. Sugiyama. 1998. Analysis of functional domains of Wilson disease protein (ATP7B) in *Saccharomyces cerevisiae*. *FEBS Lett* 428 (3):281–85.

Inesi, G., D. Lewis, H. Ma, A. Prasad, and C. Toyoshima. 2006. Concerted conformational effects of Ca2+ and ATP are required for activation of sequential reactions in the Ca2+ ATPase (SERCA) catalytic cycle. *Biochemistry* 45 (46):13769–78.
Inesi, G., H. Ma, S. Hua, and C. Toyoshima. 2003. Characterization of Ca2+ ATPase residues involved in substrate and cation binding. *Ann N Y Acad Sci* 986:63–71.
Kuhlbrandt, W. 2004. Biology, structure and mechanism of P-type ATPases. *Nat Rev Mol Cell Biol* 5 (4):282–95.
Lahiri, S. D., G. Zhang, D. Dunaway-Mariano, and K. N. Allen. 2003. The pentacovalent phosphorus intermediate of a phosphoryl transfer reaction. *Science* 299 (5615):2067–71.
Lander, E. S. et al. 2001. Initial sequencing and analysis of the human genome. *Nature* 409 (6822):860–921.
Larin, D., C. Mekios, K. Das, B. Ross, A. S. Yang, and T. C. Gilliam. 1999. Characterization of the interaction between the Wilson and Menkes disease proteins and the cytoplasmic copper chaperone, HAH1p. *J Biol Chem* 274 (40):28497–504.
Leonhardt, K., R. Gebhardt, J. Mossner, S. Lutsenko, and D. Huster. 2009. Functional interactions of Cu-ATPase ATP7B with cisplatin and the role of ATP7B in the resistance of cells to the drug. *J Biol Chem* 284 (12):7793–7802.
Lesk, A. M., and K. D. Hardman. 1982. Computer-generated schematic diagrams of protein structures. *Science* 216 (4545):539–40.
Lim, C. M., M. A. Cater, J. F. Mercer, and S. La Fontaine. 2006a. Copper-dependent interaction of dynactin subunit p62 with the N terminus of ATP7B but not ATP7A. *J Biol Chem* 281 (20):14006–14.
———. 2006b. Copper-dependent interaction of glutaredoxin with the N termini of the copper-ATPases (ATP7A and ATP7B) defective in Menkes and Wilson diseases. *Biochem Biophys Res Commun* 348 (2):428–36.
Liu, M., E. J. Cohen, G. J. Brewer, and P. R. Laibson. 2002. Kayser-Fleischer ring as the presenting sign of Wilson disease. *Am J Ophthalmol* 133 (6):832–34.
Lorinczi, E., R. Tsivkovskii, W. Haase, E. Bamberg, S. Lutsenko, and T. Friedrich. 2008. Delivery of the Cu-transporting ATPase ATP7B to the plasma membrane in Xenopus oocytes. *Biochim Biophys Acta* 1778 (4):896–906.
Lubben, T. H., A. A. Gatenby, G. K. Donaldson, G. H. Lorimer, and P. V. Viitanen. 1990. Identification of a groES-like chaperonin in mitochondria that facilitates protein folding. *Proc Natl Acad Sci U S A* 87 (19):7683–87.
Lutsenko, S., N. L. Barnes, M. Y. Bartee, and O. Y. Dmitriev. 2007. Function and regulation of human copper-transporting ATPases. *Physiol Rev* 87 (3):1011–46.
Machado, A. C., M. M. Deguti, L. Caixeta, M. Spitz, L. T. Lucato, and E. R. Barbosa. 2008. Mania as the first manifestation of Wilson's disease. *Bipolar Disord* 10 (3):447–50.
Markus, M. A., T. Nakayama, P. Matsudaira, and G. Wagner. 1994. 1H, 15N, 13C and 13CO resonance assignments and secondary structure of villin 14T, a domain conserved among actin-severing proteins. *J Biomol NMR* 4 (4):553–74.
McIntosh, D. B., J. D. Clausen, D. G. Woolley, D. H. MacLennan, B. Vilsen, and J. P. Andersen. 2004. Roles of conserved P domain residues and Mg2+ in ATP binding in the ground and Ca2+-activated states of sarcoplasmic reticulum Ca2+-ATPase. *J Biol Chem* 279 (31):32515–23.
Meyer, L. A., A. P. Durley, J. R. Prohaska, and Z. L. Harris. 2001. Copper transport and metabolism are normal in aceruloplasminemic mice. *J Biol Chem* 276 (39):36857–61.
Mezard, C., D. Pompon, and A. Nicolas. 1992. Recombination between similar but not identical DNA sequences during yeast transformation occurs within short stretches of identity. *Cell* 70 (4):659–70.
Michalczyk, A., E. Bastow, M. Greenough, J. Camakaris, D. Freestone, P. Taylor, M. Linder, J. Mercer, and M. L. Ackland. 2008. ATP7B expression in human breast epithelial cells is mediated by lactational hormones. *J Histochem Cytochem* 56 (4):389–99.

Michalczyk, A. A., J. Rieger, K. J. Allen, J. F. Mercer, and M. L. Ackland. 2000. Defective localization of the Wilson disease protein (ATP7B) in the mammary gland of the toxic milk mouse and the effects of copper supplementation. *Biochem J* 352 Pt 2:565–71.

Morgan, C. T., R. Tsivkovskii, Y. A. Kosinsky, R. G. Efremov, and S. Lutsenko. 2004. The distinct functional properties of the nucleotide-binding domain of ATP7B, the human copper-transporting ATPase: analysis of the Wilson disease mutations E1064A, H1069Q, R1151H, and C1104F. *J Biol Chem* 279 (35):36363–71.

Multhaup, G., D. Strausak, K. D. Bissig, and M. Solioz. 2001. Interaction of the CopZ copper chaperone with the CopA copper ATPase of *Enterococcus hirae* assessed by surface plasmon resonance. *Biochem Biophys Res Commun* 288 (1):172–77.

Olesen, C., T. L. Sorensen, R. C. Nielsen, J. V. Moller, and P. Nissen. 2004. Dephosphorylation of the calcium pump coupled to counterion occlusion. *Science* 306 (5705):2251–55.

Payne, A. S., E. J. Kelly, and J. D. Gitlin. 1998. Functional expression of the Wilson disease protein reveals mislocalization and impaired copper-dependent trafficking of the common H1069Q mutation. *Proc Natl Acad Sci U S A* 95 (18):10854–59.

Petris, M. J., I. Voskoboinik, M. Cater, K. Smith, B. E. Kim, R. M. Llanos, D. Strausak, J. Camakaris, and J. F. Mercer. 2002. Copper-regulated trafficking of the Menkes disease copper ATPase is associated with formation of a phosphorylated catalytic intermediate. *J Biol Chem* 277 (48):46736–42.

Pilankatta, R., D. Lewis, C. M. Adam, and G. Inesi. 2009. High yield heterologous expression of WT and mutant Cu+ ATPase (ATP7B, Wilson disease protein) for functional characterization of catalytic activity and serine residues undergoing copper dependent phosphorylation. *J Biol Chem* 284 (32):21307–16.

Ralle, M., S. Lutsenko, and N. J. Blackburn. 2004. Copper transfer to the N-terminal domain of the Wilson disease protein (ATP7B): X-ray absorption spectroscopy of reconstituted and chaperone-loaded metal binding domains and their interaction with exogenous ligands. *J Inorg Biochem* 98 (5):765–74.

Rodriguez-Granillo, A., A. Crespo, and P. Wittung-Stafshede. 2009. Conformational dynamics of metal-binding domains in Wilson disease protein: molecular insights into selective copper transfer. *Biochemistry* 48 (25):5849–63.

Rodriguez-Granillo, A., E. Sedlak, and P. Wittung-Stafshede. 2008. Stability and ATP binding of the nucleotide-binding domain of the Wilson disease protein: effect of the common H1069Q mutation. *J Mol Biol* 383 (5):1097–1111.

Safaei, R., S. Otani, B. J. Larson, M. L. Rasmussen, and S. B. Howell. 2008. Transport of cisplatin by the copper efflux transporter ATP7B. *Mol Pharmacol* 73 (2):461–68.

Sazinsky, M. H., A. K. Mandal, J. M. Arguello, and A. C. Rosenzweig. 2006. Structure of the ATP binding domain from the *Archaeoglobus fulgidus* Cu+-ATPase. *J Biol Chem* 281 (16):11161–66.

Shah, A. B., I. Chernov, H. T. Zhang, B. M. Ross, K. Das, S. Lutsenko, E. Parano, L. Pavone, O. Evgrafov, I. A. Ivanova-Smolenskaya, G. Anneren, K. Westermark, F. H. Urrutia, G. K. Penchaszadeh, I. Sternlieb, I. H. Scheinberg, T. C. Gilliam, and K. Petrukhin. 1997. Identification and analysis of mutations in the Wilson disease gene (ATP7B): population frequencies, genotype-phenotype correlation, and functional analyses. *Am J Hum Genet* 61 (2):317–28.

Strecker, K., J. P. Schneider, H. Barthel, W. Hermann, F. Wegner, A. Wagner, J. Schwarz, O. Sabri, and C. Zimmer. 2006. Profound midbrain atrophy in patients with Wilson's disease and neurological symptoms? *J Neurol* 253 (8):1024–29.

Tao, T. Y., F. Liu, L. Klomp, C. Wijmenga, and J. D. Gitlin. 2003. The copper toxicosis gene product Murr1 directly interacts with the Wilson disease protein. *J Biol Chem* 278 (43):41593–96.

Terada, K., T. Nakako, X. L. Yang, M. Iida, N. Aiba, Y. Minamiya, M. Nakai, T. Sakaki, N. Miura, and T. Sugiyama. 1998. Restoration of holoceruloplasmin synthesis in LEC rat after infusion of recombinant adenovirus bearing WND cDNA. *J Biol Chem* 273 (3):1815–20.

Terada, K., N. Aiba, X. L. Yang, M. Iida, M. Nakai, N. Miura, and T. Sugiyama. 1999. Biliary excretion of copper in LEC rat after introduction of copper transporting P-type ATPase, ATP7B. *FEBS Lett* 448(1):53–56.

Toyoshima, C., and G. Inesi. 2004. Structural basis of ion pumping by Ca2+-ATPase of the sarcoplasmic reticulum. *Annu Rev Biochem* 73:269–92.

Toyoshima, C., M. Nakasako, H. Nomura, and H. Ogawa. 2000. Crystal structure of the calcium pump of sarcoplasmic reticulum at 2.6 A resolution. *Nature* 405 (6787):647–55.

Toyoshima, C., and H. Nomura. 2002. Structural changes in the calcium pump accompanying the dissociation of calcium. *Nature* 418 (6898):605–11.

Tsay, M. J., N. Fatemi, S. Narindrasorasak, J. R. Forbes, and B. Sarkar. 2004. Identification of the "missing domain" of the rat copper-transporting ATPase, atp7b: insight into the structural and metal binding characteristics of its N-terminal copper-binding domain. *Biochim Biophys Acta* 1688 (1):78–85.

Tsivkovskii, R., R. G. Efremov, and S. Lutsenko. 2003. The role of the invariant His-1069 in folding and function of the Wilson's disease protein, the human copper-transporting ATPase ATP7B. *J Biol Chem* 278 (15):13302–8.

Tsivkovskii, R., B. C. MacArthur, and S. Lutsenko. 2001. The Lys1010-Lys1325 fragment of the Wilson's disease protein binds nucleotides and interacts with the N-terminal domain of this protein in a copper-dependent manner. *J Biol Chem* 276 (3):2234–42.

Valverde, R. H., I. Morin, J. Lowe, E. Mintz, M. Cuillel, and A. Vieyra. 2008. Cyclic AMP-dependent protein kinase controls energy interconversion during the catalytic cycle of the yeast copper-ATPase. *FEBS Lett* 582 (6):891–95.

Walker, J. M., D. Huster, M. Ralle, C. T. Morgan, N. J. Blackburn, and S. Lutsenko. 2004. The N-terminal metal-binding site 2 of the Wilson's disease protein plays a key role in the transfer of copper from Atox1. *J Biol Chem* 279 (15):15376–84.

Walker, J. M., R. Tsivkovskii, and S. Lutsenko. 2002. Metallochaperone Atox1 transfers copper to the NH2-terminal domain of the Wilson's disease protein and regulates its catalytic activity. *J Biol Chem* 277 (31):27953–59.

Weiss, K. H., J. Wurz, D. Gotthardt, U. Merle, W. Stremmel, and J. Fullekrug. 2008. Localization of the Wilson disease protein in murine intestine. *J Anat* 213 (3):232–40.

Wernimont, A. K., L. A. Yatsunyk, and A. C. Rosenzweig. 2004. Binding of copper (I) by the Wilson disease protein and its copper chaperone. *J Biol Chem* 279 (13):12269–76.

Winters, D. L., J. M. Autry, B. Svensson, and D. D. Thomas. 2008. Interdomain fluorescence resonance energy transfer in SERCA probed by cyan-fluorescent protein fused to the actuator domain. *Biochemistry* 47 (14):4246–56.

Yatsunyk, L. A., and A. C. Rosenzweig. 2007. Cu(I) binding and transfer by the N terminus of the Wilson disease protein. *J Biol Chem* 282 (12):862231.

Section III

Metal Ions, Protein Conformation, and Disease

9 α-Synuclein and Metals

Aaron Santner and Vladimir N. Uversky

CONTENTS

Introduction 169
(Non)-Specific Interactions of α-Synuclein with Metal Ions 171
Aluminum 174
Calcium 175
Cobalt 177
Copper 177
Iron 179
Magnesium 182
Manganese 183
Terbium 184
Zinc 184
Future Perspective 185
References 185

INTRODUCTION

α-Synuclein is a 140-amino acid protein, which is encoded by a single gene consisting of seven exons located in the chromosome 4 (Chen et al. 1995). This protein was first described by Maroteaux et al. in 1988 as a neuron-specific protein localized in the nucleus and presynaptic nerve terminals (Maroteaux, Campanelli, and Scheller 1988). It is located in close vicinity to, and loosely associated with, presynaptic vesicles and thus may play a role in synaptic release of neurotransmitters (Jakes, Spillantini, and Goedert 1994; Iwai et al. 1995). α-Synuclein has been estimated to account for as much as 1% of the total protein in soluble cytosolic brain fractions (Iwai et al. 1995). Various observations have implicated α-synuclein in the pathogenesis of several neurodegenerative disorders, including Parkinson's disease (PD), Alzheimer's disease (AD), dementia with Lewy bodies (DLB), Lewy body variant of AD (LBVAD), neurodegeneration with brain iron accumulation type 1, and several other synucleinopathies (Galvin, Lee, and Trojanowski 2001; Uversky 2007, 2008). Some of the most persuasive facts linking α-synuclein and various neuropathologies are (Trojanowski and Lee 2002): missense mutations in the α-synuclein gene cause familial PD; antibodies to α-synuclein detect Lewy bodies (LBs) and dystrophic Lewy neurites (LNs) that are hallmark lesions of PD, dementia with LBs, AD, and LBVAD; insoluble α-synuclein filaments are recovered from DLB brains and purified LBs; recombinant α-synuclein assembles into LB-like filaments; α-synuclein transgenic mice and flies develop a PD-like phenotype; cortical LBs detected with antibodies to α-synuclein correlate with dementia in PD, DLB, and LBVAD; antibodies to α-synuclein detect LBs in familial AD, sporadic AD, and elderly Down syndrome brains; α-synuclein is a building block of glial cytoplasmic inclusions (GCIs) in neurodegeneration with brain iron accumulation type 1, and multiple system atrophy (MSA); cells transfected with α-synuclein and treated with nitric oxide generators develop LB-like α-synuclein inclusions; biogenic mice overexpress mutant human amyloid precursor protein; and α-synuclein show an augmentation in α-synuclein inclusions. Thus many neurodegenerative diseases are characterized by α-synuclein pathologies resulting from deposition of α-synuclein lesions. The underlying cause of aggregation may result from structural alterations in α-synuclein, an intrinsically disordered protein (IDP) that normally possesses little or no ordered structure under the physiological conditions *in vitro* (Weinreb et al. 1996; Uversky, Li, and Fink 2001a; Bussell and Eliezer 2001; Eliezer et al. 2001; Uversky 2003, 2007, 2008).

As a link between α-synuclein aggregation and PD is one of the most frequently mentioned in literature, the majority of this chapter will be dedicated to this connection. Clinically, PD manifests as a movement disorder characterized by tremor, rigidity, and bradykinesia. These symptoms are attributed to the progressive loss of dopaminergic neurons from the substantia nigra region of the brain. Some of the surviving nigral dopaminergic neurons contain cytosolic filamentous inclusions known as LBs and LNs. The etiology of PD is unknown, but recent work suggests that there is no direct genetic basis for this disease except in extremely rare cases (Tanner et al. 1999). A positive correlation between the prevalence of PD and industrialization has been recognized (Tanner 1989), and this disorder is now considered likely to be an "environmental" disease.

The possible involvement of heavy metals in the etiology of PD follows from the results of epidemiological studies (Gorell et al. 1997, 1999; Altschuler 1999; Zayed, Campanella et al. 1990; Zayed, Ducic et al. 1990; Rybicki et al. 1993) and from the postmortem analysis of the brain tissues of PD patients (Hirsch et al. 1991; Riederer et al. 1989; Dexter et al. 1991). For example, the analysis of the PD mortality rates in Michigan (1986–1988) with respect to potential heavy-metal exposure revealed that counties with paper-, chemical-, iron-, or copper-related industrial categories had significantly higher PD death rates than counties without these industries (Rybicki et al. 1993). An epidemiological study (1987–1989) of Valleyfield in southern Quebec (Canada) established that an increased risk for PD is associated with occupational exposure to the three metals manganese, iron, and aluminum, especially when the duration of exposure is longer than 30 years (Zayed, Ducic et al. 1990). A population-based case-control study in the Henry Ford Health System (Detroit) suggested that chronic occupational exposure to manganese or copper, individually, or to dual combinations of lead, iron, and copper, is associated with PD (Gorell et al. 1997, 1999). Postmortem analysis of brain tissues from patients with PD gives further confirmation for the involvement of heavy metals in this disorder, in that a considerable increase in total iron, zinc, and aluminum content of the parkinsonian substantia nigra was observed when compared to control tissues (Hirsch et al. 1991; Dexter et al. 1989, 1991; Riederer et al. 1989). Moreover, analysis of Lewy bodies in the parkinsonian substantia nigra revealed high levels of iron and the presence of aluminum (Hirsch et al. 1991; Dexter et al. 1991).

These connections of the PD (and other synucleinopathies) pathology with α-synuclein aggregation and exposure to metal have recently attracted significant attention from researchers. More than 1200 papers have been published to analyze association of metals with PD pathogenesis, and more than 100 papers report various aspects of metal involvement in the modulation of α-synuclein behavior. Several possible mechanisms for metal-stimulated fibrillation of α-synuclein can be envisaged. The simplest would involve direct interactions between α-synuclein and the metal leading to structural changes in α-synuclein and resulting in the enhanced propensity to aggregate. These direct interactions of α-synuclein with metals can either be specific or have a nonspecific character (e.g., nonspecific electrostatic interactions). More complex models for the correlation between metals and neuronal degeneration in PD involve oxidative stress resulting from the enhanced level of redox-active metal ions (copper or iron) within the substantia nigra (Oestreicher et al. 1994; Kienzl et al. 1995; Montgomery 1995; Bush 2000). In fact, metal-induced oxidant stress can damage critical biological molecules and initiate a cascade of events including mitochondrial dysfunction, excitotoxicity, and a rise in cytosolic-free calcium, leading to cell death. It is known that increased metal accumulation promotes free-radical formation and decreased antioxidant levels, which leads to oxidative damage—a factor characteristic of the affected brain regions in PD (Oestreicher et al. 1994).

The remainder of this chapter is an overview of the current knowledge in the field of α-synuclein metalloproteomics. It begins with the analysis of nonspecific interactions between metals and α-synuclein and the consequences of these interactions on α-synuclein structure and aggregation propensity. Subsequent sections are dedicated to various aspects related to the interaction of individual metal ions with α-synuclein.

(NON)-SPECIFIC INTERACTIONS OF α-SYNUCLEIN WITH METAL IONS

There is increasing evidence that altered metal homeostasis may be involved in the progression of neurodegenerative diseases. The potential link between metal exposure, α-synuclein aggregation, and PD pathogenesis was proven by *in vitro* experiments that demonstrated that α-synuclein aggregation is facilitated in the presence of Cu^{2+} (Paik et al. 1999), and that Al^{3+} may induce structural perturbations in this protein (Paik et al. 1997). These studies will be considered in more details in the corresponding sections of this chapter.

A systematic analysis revealed that a number of mono-, di-, and trivalent metal ions such as Li^+, K^+, Na^+, Cs^+, Ca^{2+}, Co^{2+}, Cd^{2+}, Cu^{2+}, Fe^{2+}, Mg^{2+}, Mn^{2+}, Zn^{2+}, Co^{3+}, Al^{3+}, and Fe^{3+} can cause significant acceleration of the α-synuclein fibril formation *in vitro* (Uversky, Li, and Fink 2001b). Among 15 cations studied, Al^{3+} was shown to be the most effective stimulator of protein partial folding and subsequent fibrillation, along with Cu^{2+}, Fe^{2+}, Co^{3+}, and Mn^{2+} (Uversky, Li, and Fink 2001b). Earlier studies revealed that formation of a highly amyloidogenic partially folded intermediate represented a critical early stage in the fibrillation pathway of the natively unfolded α-synuclein. This partially folded α-synuclein has premolten globulelike properties and was shown to be stabilized by low pH or high temperature (Uversky, Li, and Fink 2001a). In the presence of metal cations, natively unfolded α-synuclein adopted a partially folded conformation even at neutral pH. This partially folded conformation was characterized by an increased amount of ordered secondary structure, increased protection of tyrosine residues against acrylamide quenching, and the appearance of solvent-accessible hydrophobic clusters (Uversky, Li, and Fink 2001b). Interestingly, the efficiency of a given cation to induce structural changes in α-synuclein was correlated with the cation's charge density (defined as a ratio of charge to ionic volume). Consequently, small polyvalent cations were more effective in inducing the structural changes than the large monovalent cations (Uversky, Li, and Fink 2001b). The induction of partial folding in the intrinsically disordered α-synuclein at neutral pH by cations was explained as follows. The "natively unfolded" status of α-synuclein is mainly determined by strong electrostatic repulsion between the net negative charges. To some extent this resembles the situation occurring for many proteins at low or high pH. Unfolded proteins under conditions of extreme pH can be transformed into more compact, structured conformations when the net electrostatic repulsion is reduced by counter-ion binding (Goto and Fink 1989; Goto, Calciano, and Fink 1990; Goto, Takahashi, and Fink 1990; Fink et al. 1994; Uversky et al. 1998). Thus, the metal ion–stimulated conformational change reflects the effective neutralization of Coulombic charge-charge repulsion within α-synuclein. Furthermore, polyvalent cations have the potential for cross-linking or -bridging between two or more carboxylates (Uversky, Li, and Fink 2001b).

These charge effects likely govern the ability of metal cations to accelerate α-synuclein aggregation. Importantly, both amorphous aggregates and amyloidlike fibrils were found at the end of the aggregation process. The most likely explanation is the existence of the two competing aggregation processes, fibrillation versus amorphous or "soluble" aggregate formation. Therefore, the observation of both fibrillar and amorphous aggregates for α-synuclein in the presence of metals is consistent with a minimal underlying kinetic scheme of the sort shown here (Uversky, Li, and Fink 2001b):

$$U_N \xrightleftharpoons{I} I \xrightleftharpoons{II} Nucleus \longrightarrow F$$
$$\downarrow III$$
$$A$$

The variable ratios of fibrils to amorphous aggregates detected in the presence of different cations indicated that there was an underlying branched pathway and that different cations affected the relative rates of fibril formation and nonfibrillar aggregation differently. The correlation between the amount of fibrils and the rate of fibril formation suggested that the cations with the high efficiency to accelerate α-synuclein fibrillation produced more fibrils. In other words, these cations interacted with α-synuclein in such a way as to increase the rate of the pathway leading to fibril formation (Uversky, Li, and Fink 2001b). In summary, this study revealed that the effectiveness of metal cations to induce fibrillation was correlated with the increasing ion charge density and with their ability to induce amyloidogenic partially folded species (Uversky, Li, and Fink 2001b).

Subsequent studies revealed that different cations not only accelerated the α-synuclein fibrillation to different degrees, but also resulted in the formation of fibrils with distinct morphologies (Khan et al. 2005). For example, the diameter and general structure of fibrils were not affected by the presence of Cu^{2+} and Zn^{2+}. However, fibrils grown under these conditions were noticeably shorter than fibrils produced in the absence of metals. Furthermore, these metal-enriched fibrils were efficient catalysts of the redox cycling of added Fe^{2+}. On the contrary, fibrils formed in the presence of Al^{3+} possessed a novel amyloid morphology that consisted of twisted fibrils with a periodicity of about 100 nm and were poor catalysts for the redox cycling of added Fe^{2+} (Khan et al. 2005). This study clearly illustrated that the fibril morphology depends significantly on the nature of the metal ion added.

Recently, the peculiarities of the divalent metal ions Fe^{2+}, Mn^{2+}, Co^{2+}, and Ni^{2+} binding to α-synuclein and their effects on protein aggregation were analyzed by exploiting the distinct paramagnetic properties of these metal ions (Binolfi et al. 2006). Nuclear magnetic resonance (NMR) spectroscopy revealed that these ions bind preferentially and with low affinity (millimolar) to the C-terminus of α-synuclein. The primary binding site was shown to be the ^{119}DPDNEA^{124} motif, in which Asp121 acts as the main anchoring residue. Based on these observations and on the residual dipolar coupling measurements of the protein backbone, it has been concluded that metal binding was not driven exclusively by electrostatic interactions but was mostly determined by the residual structure of the C-terminus of α-synuclein (Binolfi et al. 2006).

The stoichiometry of Cu^{2+} and Fe^{3+} binding to wild-type α-synuclein and mutant forms (A30P, A53T, E46K) associated with familial PD was studied using isothermal titration calorimetry (ITC) (Bharathi and Rao 2007). Two Cu^{2+}-binding sites were found in α-synuclein monomers (wild and mutant forms) with apparent K_B of 10^5 M and 10^4 M, respectively. However, only one Fe^{3+}-binding site with an apparent K_B of 10^5 M was found in this protein (Bharathi and Rao 2007).

The role of phosphorylation in interaction of α-synuclein with bivalent and trivalent metal ions was analyzed using a set of model peptides and phosphopeptides corresponding to the residues 119–132 of α-synuclein (^{119}DPDNEAYEMPSEEG^{132}) (Liu and Franz 2007). This study revealed that the ^{119}DPDNEA(pY)EMPSEEG^{132} phosphopeptide, where tyrosine was replaced with phosphotyrosine, possessed a marked selectivity for the trivalent metal ions Tb^{3+}, Fe^{3+}, and Al^{3+} when compared to the nonmodified peptide or the ^{119}DPDNEAYEMP(pS)EEG^{132} peptide, where serine was replaced with phosphoserine. Based on the detailed analysis, it was concluded that the phosphoester group on tyrosine provided a metal-binding anchor that was supplemented by carboxylic acid groups at positions 119, 121, and 126 to establish a multidentate ligand, whereas two glutamic acid residues at positions 130 and 131 contributed to binding additional trivalent ions. Furthermore, circular dichroism analysis showed that Fe^{3+} induced a partially folded structure in ^{119}DPDNEA(pY)EMPSEEG^{132}, whereas no change was observed for ^{119}DPDNEAYEMP(pS)EEG^{132} or for the unphosphorylated analogue (Liu and Franz 2007). Based on these observations it has been concluded that the type and location of a phosphorylated amino acid can affect a metal-binding specificity and affinity of α-synuclein as well as its overall conformation (Liu and Franz 2007).

The simultaneous presence of metals and pesticides/herbicides (another group of environmental factors linked to the PD pathogenesis) accelerated α-synuclein fibrillation synergistically (Uversky et al. 2002). In agreement with these observations, it has been recently established that herbicides preferentially bind to a partially folded intermediate conformation of α-synuclein induced by Mn^{2+}, Al^{3+}, Cd^{2+}, Cu^{2+}, and Zn^{2+} (Andre et al. 2005). Similarly, certain metals were shown to cause a significant acceleration of α-synuclein fibrillation in the presence of high concentrations of various macromolecules mostly through decreasing the fibrillation lag time. The faster fibrillation in crowded environments in the presence of heavy metals suggested a simple molecular basis for the observed elevated risk of PD due to exposure to metals (Munishkina, Fink, and Uversky 2008). Intriguingly, the addition of certain metals (Ti^{3+}, Zn^{2+}, Al^{3+}, and Pb^{2+}) was shown to overcome methionine oxidation-induced inhibition of the α-synuclein fibril formation, suggesting that a combination of oxidative stress and environmental metal pollution could play an important role in triggering the fibrillation of α-synuclein and possibly PD development as a result (Yamin et al. 2003).

Data presented in this section support the notion that many metal ions can interact with α-synuclein nonspecifically, but the binding of certain cations including Cu^{2+} and Fe^{3+} is relatively specific. Furthermore, the interaction with metals can modify α-synuclein structure, modulate its fibrillation rates, and modulate fibril/aggregate morphology. The remaining sections of this chapter are dedicated to the effects of specific metal ions on α-synuclein.

ALUMINUM

Aluminum exposure was linked to the PD pathology via epidemiological studies and the postmortem analysis of brain tissues from PD patients (Zayed, Ducic et al. 1990; Hirsch et al. 1991; Dexter et al. 1989, 1991; Riederer et al. 1989). A connection between α-synuclein and PD pathogenesis was established in 1997, when a mutation in the α-synuclein gene was identified in familial cases of early-onset PD (Polymeropoulos et al. 1997) and α-synuclein was shown to accumulate in Lewy bodies (Spillantini et al. 1997). The effect of Al^{3+} on α-synuclein structure was analyzed at this time (Paik et al. 1997). This first study revealed that the protein structure can be altered by certain environmental factors, including aluminum. This aluminum-altered conformation of α-synuclein became resistant to proteases such as trypsin, α-chymotrypsin, and calpain (Paik et al. 1997).

To gain more information on the structural consequences of Al^{+3} binding, increasing concentrations of Al^{+3} were incubated with α-synuclein and the spectral properties of the protein were analyzed (Uversky, Li, and Fink 2001b). The shapes and intensities of the far-ultraviolet circular dichroism (far-UV CD), UV absorbance, intrinsic fluorescence, and ANS fluorescence spectra changed significantly with the increase in the Al^{+3} concentration. These data indicated that the structure of natively unfolded α-synuclein was affected by interaction with the cation. Furthermore, the spectral changes induced in α-synuclein by Al^{+3} occurred simultaneously in a rather cooperative manner, were completely reversible, and were independent of protein concentration (at least in the range of 0.7–35 μM). These observations were consistent with the assumption that the Al^{+3}-induced structure formation in α-synuclein represented an intramolecular process and not self-association (Uversky, Li, and Fink 2001b). Kinetics analysis revealed that the addition of Al^{+3} induced a rapid formation of secondary structure, within the dead time of the manual mixing experiments (~5 s) (Uversky, Li, and Fink 2001b).

Structurally, the Al^{+3}-stabilized partially folded intermediate of α-synuclein possessed a strong resemblance to the amyloidogenic premolten globule form originally found at acidic pH or elevated temperatures. In agreement with this structural similarity, the rate of α-synuclein fibril formation at neutral pH was dramatically accelerated by the $AlCl_3$ addition to the α-synuclein solution. In the presence of 2.5 mM of $AlCl_3$, the lag time was approximately three times shorter, and the apparent rate of fibril formation was ~1.5-fold higher compared to protein alone. Furthermore, it was demonstrated that Al^{3+} was incorporated into the mature fibrils formed in the presence of the cation (Uversky, Li, and Fink 2001b).

In agreement with this finding, it was established that aluminum had much larger structure-forming effects at high α-synuclein concentrations. In fact, in the range of protein concentration from ~0.7–50 μM the Al^{3+}-induced changes in the spectral properties of α-synuclein were completely reversible and independent of protein content. However, at high protein concentration the Al^{3+}-induced folding of α-synuclein was shown to be accompanied by association of the partially folded intermediates (Uversky, Li, and Fink 2001b). Interestingly, these self-associated forms of partially folded α-synuclein were shown to more readily form fibrils, and possessed a larger amount of ordered secondary structure (Uversky, Li, and Fink 2001b).

The conclusion on the Al^{3+}-induced partial folding of α-synuclein was recently confirmed using selective noncovalent adduct protein probing mass spectrometry (SNAPP-MS), which utilized specific, noncovalent interactions between 18-crown-6 ether (18C6) and lysine to probe protein structure in the presence and absence of metal ions (Ly and Julian 2008). It was shown that the 18C6 SNAPP distributions for α-synuclein changed dramatically in the presence of 3 mM Al^{3+}, suggesting that Al^{3+} binding caused a significant change in the conformational dynamics of the monomeric form of this disordered protein (Ly and Julian 2008).

CALCIUM

Calcium is known to have many important and highly diverse functions in biology. For example, Ca^{2+} homeostasis and Ca^{2+} signaling events in neurons regulate multiple functions, including synaptic transmission, plasticity, and cell survival. Ca^{2+} fluxes across the plasma membrane and between intracellular compartments play critical roles in the fundamental functions of neurons, including the regulation of neurite outgrowth and synaptogenesis, synaptic transmission and plasticity, and cell survival (Mattson 2007). Therefore disturbances in Ca^{2+} homeostasis can affect the well-being of the neuron in different ways and to various degrees (Wojda, Salinska, and Kuznicki 2008). It has been hypothesized that excitotoxicity and disturbance in Ca^{2+} homeostasis are among the potential mechanisms leading to neurodegeneration (Mattson 2007). Normal substantia nigra neuron activity seems to depend on unusually high calcium entry (Schulz 2007), and α-synuclein was shown to induce Ca^{2+} influx in rat synaptoneurosomes through N-type voltage-dependent Ca^{2+} channels (Adamczyk and Strosznajder 2006). Another consequence of disturbed calcium homeostasis is the activation of calpains, a family of cysteine proteases. Calpains are elevated in the mesencephalon of patients with PD but not in other neurodegenerative disorders involving the mesencephalon (Schulz 2007). Perturbed neuronal Ca^{2+} homeostasis is also implicated in age-related cognitive impairment and AD (Bezprozvanny and Mattson 2008). Overall, neurodegenerative diseases related to aging, such as AD, PD, and Huntington's disease, are characterized by the positive feedback between Ca^{2+} homeostatic dysregulation and the aggregation of disease-related proteins such as Aβ, α-synuclein, or huntingtin (Wojda, Salinska, and Kuznicki 2008). Therefore high levels of Ca^{2+} and other metal ions may play a role in pathogenesis of these diseases. It has been proposed that a better understanding of the cellular and molecular mechanisms promoting or preventing disturbances in cellular Ca^{2+} homeostasis during aging may lead to novel approaches for therapeutic intervention in neurological disorders such as AD and PD (Mattson 2007).

It was shown using a microdialysis technique that α-synuclein binds Ca^{2+} with an IC_{50} of about 2-300 μM and that this reaction can be uninhibited by a 50-fold excess of Mg^{2+}. The Ca^{2+}-binding site was assigned to a C-terminally localized acidic 32-amino acid domain of this protein (Nielsen et al. 2001). The factional properties of α-synuclein (e.g., the ability of this protein to interact with microtubule-associated protein 1A) and the propensity of α-synuclein to aggregate were affected by Ca^{2+} binding, suggesting that Ca^{2+} ions may participate in normal α-synuclein functions in the nerve terminal and also may contribute to some pathological processes associated with α-synuclein aggregation and LB formation (Nielsen et al. 2001).

The effect of Ca^{2+} and other metals on the morphology of α-synuclein aggregates has been analyzed by atomic force microscopy (AFM) (Lowe et al. 2004). Three classes of effect were observed with different groups of metal ions. The first class included Cu^{2+}, Fe^{3+}, and Ni^{2+}, which yielded 0.8–4 nm spherical particles, similar to α-synuclein incubated without metal ions. The second class consisted of Mg^{2+}, Cd^{2+} and Zn^{2+}, which induced larger, 5–8 nm spherical oligomers. Finally, the third class included Co^{2+} and Ca^{2+}, which gave frequent annular oligomers, 70–90 nm in diameter with Ca^{2+} and 22–30 nm in diameter with Co^{2+} (Lowe et al. 2004). Each annular particle induced by Ca^{2+} appeared to be composed of a ring of six spherical particles with occasional open-ring structures composed of a chain of five spherical particles, 4 nm in height. In agreement with previous work that revealed that the acidic C terminus of α-synuclein mediates calcium binding (Nielsen et al. 2001), no annular oligomers were found when truncated α-synuclein (1–125), lacking the C-terminal 15 amino acids, was coincubated with Ca^{2+} (Lowe et al. 2004). These observations suggested that the Ca^{2+}-binding domain located within the C-terminal half of α-synuclein was involved in the formation of such annular oligomers. Importantly, soluble 30 to 50 nm-sized annular α-synuclein oligomers were isolated by mild detergent treatment from glial cytoplasmic inclusions (GCIs) purified from multiple system atrophy brain tissue (Pountney et al. 2004). As annular aggregates were specifically induced by Ca^{2+} binding, it has been hypothesized that the formation of such aggregated species inside the neurons can be influenced by the intracytoplasmic Ca^{2+} concentration (Pountney, Voelcker, and Gai 2005).

Moreover, Ca^{2+} was shown to modulate the interaction between α-synucelin and membranes (Tamamizu-Kato et al. 2006). In fact, it has been shown that in the absence of Ca^{2+} the protein was anchored to the lipid surface via the N-terminal domain. The addition of Ca^{2+} promoted effective interaction between the C-terminal domain of α-synucelin and lipid membrane. Based on these observations it has been suggested that initial lipid interaction of α-synuclein occurs via its N-terminal domain, followed by a Ca^{2+}-triggered membrane association of the acidic tail, potentially leading to α-synuclein aggregation. This model suggests that cellular Ca^{2+} dysregulation can be one of the critical factors that forments α-synuclein aggregation in PD (Tamamizu-Kato et al. 2006).

COBALT

Similar to Ca^{2+}, Co^{2+} was shown to induce the fast formation of annular aggregates of α-synuclein (Lowe et al. 2004). These Co^{2+}-induced doughnutlike oligomers had mean maximum height of 2 ± 0.8 nm and diameter ranging 22–30 nm, being a bit smaller than those formed in the presence of Ca^{2+}, which were characterized by a diameter range of 70–95 nm and a mean maximum height of 3.9 ± 1.2 nm (Lowe et al. 2004). Based on the detailed morphological analysis of metal-induced oligomers, it has been proposed that the aggregation of α-synuclein can occur as a sequential chain of events starting from monomer that assembles into ~4 nm-height spherical oligomer, to ~4-nm height annular oligomer, then to ~8 nm-height annular oligomer, from which ~8–10 nm-height linear filaments then extend. It has been also proposed that the wide range of diameters observed for annular oligomers indicated that these oligomeric species were likely not functionally defined quaternary structures, but rather were formed from competing self-association processes of either linear extension or ring closure (Lowe et al. 2004).

COPPER

It has long been believed that the concentration of free metal ions in the brain is too low to be physiologically significant and that, as a consequence, involvement of metal ions in neurological disorders may arise only from toxicological exposure to Cu, Fe, Zn, or Mn (Gaggelli et al. 2006). However, in terms of total concentrations, the brain has more than enough of the mentioned metal ions in its tissue to damage or dysregulate numerous proteins and metabolic systems (Lovell et al. 1998; Bush 2000; Tiffany-Castiglion and Qian 2001). Copper is very redox active and cannot exist in an unbound form in the cell without causing oxidative damage. Moreover, incorrectly bound Cu^{2+} may act as a catalyst for the generation of the most damaging radicals, such as the hydroxyl radical. Therefore distortions in the tight regulation of Cu^{2+} homeostasis are expected to contribute to several neurodegenerative conditions (Gaggelli et al. 2006).

The human brain is estimated to produce more than 10^{11} free radicals per day (Grune, Reinheckel, and Davies 1996), and imbalance in pro-oxidant versus anti-oxidant homeostasis results in "oxidative stress" with the generation of several potentially toxic reactive oxygen species (ROS), including both radicals (such as hydroperoxyl radical, superoxide radical, hydroxyl radical) and nonradical species (such as hydrogen peroxide) that participate in the initiation and/or propagation of radical chain reactions (Gaggelli et al. 2006). Inappropriate compartmentalization or elevation of Cu^{2+} in the cell and inappropriate binding of Cu^{2+} to cellular proteins are currently being explored as sources of pathological oxidative stress in several neurodegenerative disorders.

A first systematic analysis of the effect of Cu^{2+} on α-synuclein structure and aggregation revealed that this metal ion was able to induce an effect α-synuclein oligomerization in the presence of coupling reagents such as dicyclohexylcarbodiimide or N-(ethoxycarbonyl)-2-ethoxy-1,2-dihydroquinoline (Paik et al. 1999). This study also revealed that the full-length α-synuclein is able to bind 10.4 Cu^{2+} ions. Cu^{2+}-induced oligomerization was effectively suppressed by the truncation of the acidic C-terminus of α-synuclein by treatment with endoproteinase Asp-N, suggesting that the Cu^{2+}-induced oligomerization was dependent on the acidic C-terminal region (Paik et al. 1999).

Reactive oxygen species can initiate the covalent modification of amino acid residues in proteins, the formation of protein-protein cross-linkages, and oxidation of the protein backbone resulting in protein fragmentation. Using high-performance liquid chromatography (HPLC) and matrix-assisted laser desorption/ionization time-of-flight mass spectrometry (MALDI-TOF MS) methods, it has been shown that the major targets for modification in α-synuclein by Cu^{2+}/hydrogen peroxide oxidation are methionines, which were shown to be oxidized to methionine sulfoxides and solfones (Kowalik-Jankowska et al. 2006, 2008).

Since oxidatively modified proteins are known to possess an increased propensity to aggregate, the effect of metal-catalyzed oxidation of α-synuclein by Cu^{2+} and hydrogen peroxide on protein aggregation was examined (Paik, Shin, and Lee 2000). The oxidized protein was shown to be self-oligomerized into an SDS-resistant ladder as detected by SDS-PAGE. It has been suggested that abnormalities in Cu^{2+} and hydrogen peroxide homeostasis could play critical roles in the metal-catalyzed oxidative oligomerization of α-synuclein, which may lead to possible protein aggregation and neurodegenerations (Paik, Shin, and Lee 2000).

Cu^{2+} was shown to be one of the most potent modulators of α-synuclein structure and a very effective accelerator of α-synuclein fibrillation *in vitro* (Uversky, Li, and Fink 2001b). Using AFM-based single-molecule mechanical unfolding methodology, it has been shown that the presence of Cu^{2+} significantly enhanced the relative abundance of the "betalike" structure in the conformationally heterogeneous monomeric α-synuclein, which happened to contain three main classes of conformations (Sandal et al. 2008). Even at physiologically relevant concentrations, Cu^{2+} ions were effective in the acceleration of α-synuclein aggregation without altering the resultant fibrillar structures (Rasia et al. 2005). Using a set of spectroscopic techniques such as absorption, CD, electron paramagnetic resonance (EPR), and NMR, the primary Cu^{2+}-binding site was shown to be localized in the N-terminus of α-synuclein. This binding site involved His-50 as the anchoring residue and additional nitrogen/oxygen donor atoms in a square planar or distorted tetragonal geometry to bind Cu^{2+} (Rasia et al. 2005). The acidic C-terminus of the protein, originally thought to mediate copper binding, was shown to coordinate a second Cu^{2+} equivalent with a 300-fold reduced affinity in comparison with the N-terminal Cu^{2+}-binding site (Rasia et al. 2005). Using a set of Trp α-synuclein mutants, it has been confirmed that Cu^{2+} interacts with α-synuclein, and that Cu^{2+} binds tightly ($K_D \sim 100$ nM) near the α-synuclein's N-terminus (Lee, Gray, and Winkler 2008). Further confirmation of the N-terminal localization of a strong Cu^{2+}-binding site was obtained using EPR spectroscopy (Drew et al. 2008).

When NMR spectroscopy was used to identify independent Cu^{2+}-binding sites in α-synuclein, as many as 16 different sites capable of binding Cu^{2+} were discovered (Sung, Rospigliosi, and Eliezer 2006). Most of the sites involved negatively charged amino acid side chains, but binding was also detected at the sole histidine residue located at position 50 and to the N-terminal amino group. Importantly, the binding sites were shown to bind Cu^{2+} in the more highly structured conformation adopted by α-synuclein upon binding to detergent micelles or lipid vesicles (Sung, Rospigliosi, and Eliezer 2006).

IRON

Interconnection between the disruption of iron homeostasis, α-synuclein aggregation, and various neurodegenerative diseases is very strong. As was already pointed out, numerous epidemiological studies (Gorell et al. 1999; Altschuler 1999; Zayed, Campanella et al. 1990; Zayed, Ducic et al. 1990; Rybicki et al. 1993; Gorell et al. 1997) and the postmortem analysis of the brain tissues (Hirsch et al. 1991; Riederer et al. 1989; Dexter et al. 1991) have linked the heavy-metal exposure and metal accumulation in the brain with the PD pathogenesis. Postmortem analysis of brain tissues from patients with PD revealed that the substantia nigra of the PD brain is characterized by a shift in the Fe^{2+}/Fe^{3+} ratio in favor of Fe^{3+} and a significant increase in the Fe^{3+}-binding protein, ferritin. In parallel, significantly lower glutathione content was demonstrated (Riederer et al. 1989). It was shown that unilateral injection of $FeCl_3$ into the substantia nigra of adult rats resulted in a substantial selective decrease of striatal dopamine (95%). As a result, dopamine-related behavioral responses were significantly impaired in the iron-treated rats. Together these data supported the assumption that iron initiates dopaminergic neurodegeneration in Parkinson's disease (Youdim, Ben-Shachar, and Riederer 1991).

In fact, an important role of iron in PD pathogenesis results from its increase in substantia nigra pars compacta dopaminergic neurons and reactive microglia. Iron accumulation in dopaminergic neurons and microglia leads to a myriad of problems, such as disturbances of iron metabolism in PD including iron uptake, storage, intracellular metabolism, and release, and is thought to enhance the production of toxic reactive oxygen radicals (Berg et al. 2001). Furthermore, iron is able to induce aggregation and toxicity of α-synuclein, further suggesting that iron interaction with specific proteins, such as α-synuclein, can contribute to the process of neurodegeneration and pathogenesis of nigrostriatal injury (Berg et al. 2001). Overall, it has been established that iron levels increase with the severity of neuropathological changes in PD. The potential mechanism of this increase in the iron levels is the increased transport of this metal ion through the blood-brain barrier in late stages of PD. Iron overload may then induce progressive degeneration of nigrostriatal neurons by facilitating the formation of reactive biological intermediates, including reactive oxygen species, and the formation of cytotoxic protein aggregates (Gotz et al. 2004). Intriguingly, it has been established that intracellular iron is crucial for the 1-methyl-4-phenylpyridinium (MPP+)-induced apoptotic cell death that causes PD provoked by exposure to this neurotoxin. More specifically, MPP+-induced iron signaling was shown to be responsible for intracellular oxidant generation, α-synuclein expression, proteasomal dysfunction, and apoptosis (Kalivendi et al. 2004).

Besides PD, an interplay between iron, α-synuclein aggregation, and neurodegeneration is apparent in neurodegeneration with brain iron accumulation, type 1 (NBIA 1), or Hallervorden-Spatz syndrome. NBIA 1 is a rare neurodegenerative disorder characterized clinically by parkinsonism, cognitive impairment, pseudobulbar features, cerebellar ataxia, neuropathologically by neuronal loss, gliosis, and iron deposition in the globus pallidus, red nucleus, and substantia nigra areas of the brain. In 1998, it was shown that Lewy bodies found in NBIA 1 were positively stained with antibodies against α-synuclein, suggesting that this protein is commonly associated with the formation of LBs in other sporadic and familial neurodegenerative diseases apart from PD (Arawaka et al. 1998). This finding was later confirmed by showing that α-synuclein is accumulated in the brains of NBIA 1 patients in form of LB-like inclusions, glial inclusions, and spheroids (Galvin et al. 2000; Neumann et al. 2000). Generally, neuropathological studies show that synucleinopathies are generally associated with iron accumulation, which is consistent with a pathological link between iron and α-synuclein (Duda, Lee, and Trojanowski 2000).

Iron is a transition metal closely associated with inducing the formation of reactive oxygen species and oxidative stress. Hydrogen peroxide (H_2O_2) is a reactive oxygen species, which in the presence of redox-active metal ions (such as Fe^{2+}) can produce hydroxyl radicals *via* Fenton chemistry. Since these radicals are able to diffuse only short distances (nanometers), changes associated with the oxidative damage will be local. Therefore an additional pathological mechanism by which iron might contribute to PD is inducing aggregation of the α-synuclein, which then accumulates in LBs and LNs in PD (Wolozin and Golts 2002). For example, an analysis of human BE-M17 neuroblastoma cells overexpressing wild-type, A53T, or A30P α-synuclein revealed that iron and free-radical generators, such as dopamine or hydrogen peroxide, were able to stimulate the production of intracellular α-synuclein-containing aggregates (Ostrerova-Golts et al. 2000). The efficiency of protein aggregation was dependent on the amount and the type of α-synuclein expressed following a rank order of A53T > A30P > wild-type > untransfected. Furthermore, overexpression of α-synuclein clearly induced toxicity, and α-synuclein overexpressing BE-M17 neuroblastoma were much more vulnerable to toxicity induced by iron (Ostrerova-Golts et al. 2000).

In addition to iron-promoted α-synuclein aggregation, it has been found that α-synuclein was able produce hydrogen peroxide and hydroxyl radicals upon incubation *in vitro* in the presence of small Fe^{2+} amounts (Tabner et al. 2002). These observations suggested that one of the molecular mechanisms in the PD pathogenesis could be the direct production of hydrogen peroxide and hydroxyl radicals by α-synuclein, in a metal-dependent manner, before, during and after the abnormal protein aggregate formation (Tabner et al. 2002).

Although all metal ions studies so far were proved to promote α-synuclein aggregation and fibrillation (Uversky, Li, and Fink 2001b), the morphology of the resultant aggregates was shown to be strongly dependent on the nature of the metal ion, as revealed, for example, by the electron microscopy analysis of the effect of Cu^{2+} and Fe^{3+} on α-synuclein (wild-type, A30P, A53T, and E46K) fibril formation and morphology (Bharathi, Indi, and Rao 2007). Cu^{2+} and Fe^{3+} were shown to selectively and differentially induce the formation of distinctive fibrillar species. Cu^{2+} induced thin, long networklike fibrils with the wild type of α-synuclein, whereas the mutant forms of the protein showed amorphous aggregates with no fibrillar species. Fe^{3+} induced short and thick fibrils with both wild-type and mutant proteins (Bharathi, Indi, and Rao 2007). It has been proposed that such metal-specific fibril morphology might have relevance for better understanding the molecular mechanisms of neurodegeneration and the role of metals in this process.

Because small oligomers are widely considered as major toxic aggregate species, the formation of α-synuclein at the single particle level was recently monitored using confocal single-molecule fluorescence techniques, scanning for intensely fluorescent targets (SIFT), fluorescence intensity distribution analysis (FIDA), and AFM (Danzer et al. 2007; Kostka et al. 2008). It has been shown that both organic solvents and Fe^{3+} at low micromolar concentrations were effective inducers of α-synuclein oligomerization. The morphologies of the resulting oligomers were different, with organic solvents inducing small oligomers ("intermediate I") and iron inducing formation of larger oligomers ("intermediate II"). Importantly, Fe^{3+} caused an effect on α-synuclein aggregation only when added in the presence of intermediate concentrations of ethanol (~5%), suggesting that the effect of Fe^{3+} was dependent on the presence of the intermediate I species, therefore proposing a synergistic effect of Fe^{3+} and ethanol on α-synuclein aggregation.

Although both oligomers were on-pathway to amyloid fibrils and could effectively seed amyloid formation, Fe^{3+}-induced oligomers were SDS-resistant and could form ion-permeable pores in a planar lipid bilayer (Kostka et al. 2008). These observations provide strong support for an important role of ferric iron in the formation of toxic α-synuclein oligomer species providing a potential disease mechanism (Kostka et al. 2008).

As both protein aggregation and oxidative stress are pathological hallmarks of several neurodegenerative diseases and are intimately associated with metals (especially iron and copper), chelation-based therapy could provide a valuable therapeutic approach to such disease states (Gaeta and Hider 2005; Molina-Holgado et al. 2007; Hider et al. 2008). The important features of therapeutic iron-chelating agents were proposed to be the ability to scavenge the free redox-active iron (or copper) present in excess in the brain and to form a nontoxic iron complex, which is then excreted. A second possible mechanism of beneficial chelation would be to cap the iron at its labile binding site, preventing any mediated toxic action (Fenton activity and/or aggregation). Here, stable metal-chelator complex should favor the state in which the metal is not redox active and therefore not toxic. Furthermore, in order to prevent protein aggregation, additional interactions between the drug and the target protein are highly desirable (Gaeta and Hider 2005).

MAGNESIUM

Contrary to the metals considered so far, whose levels are typically elevated in PD brains, the content of magnesium is reduced in the brains of patients with PD. For example, analysis of the regional metal concentrations by atomic absorption and atomic emission spectroscopy in four brain regions (frontal cortex, caudate nucleus, substantia nigra, and cerebellum) revealed lower concentrations of magnesium in the caudate nucleus in parkinsonian brains (PD and parkinsonism secondary to neurofibrillary tangle disease) in comparison to control brains (Uitti et al. 1989). This finding was later confirmed by inductively coupled plasma emission spectrometry analysis of 26 regions of PD and control brains, which revealed that Mg^{2+} concentration was lower in cortex, white matter, basal ganglia, and brain stem of PD brains compared to control brains (Yasui, Kihira, and Ota 1992). Magnesium concentration in cerebrospinal fluid decreased with the duration and severity of the disease (Bocca et al. 2006). Another interesting contradiction between Mg^{2+} and other metal ions is a potential link between low Mg^{2+} intake over generations and the pathogenesis of substantia nigra degeneration in humans, which was originally proposed as one of the potential mechanisms of the pathogenesis of the Parkinsonism-dementia complex and amyotrophic lateral sclerosis of Guam and later demonstrated for rats that were exposed to low Mg^{2+} intake over two generations (Oyanagi et al. 2006).

When α-synuclein was incubated in the presence of high Mg^{2+} concentrations (>10 mM), large aggregates composed of densely packed short fibrillar elements were rapidly formed (Hoyer et al. 2002). However, the situation was different at lower Mg^{2+} concentrations. For example, to understand how changes in Fe^{2+} and Mg^{2+} levels might affect the PD pathophysiology, the binding of these ions to α-synuclein and their effect on protein aggregation have been investigated (Golts et al. 2002). This analysis revealed that Mg^{2+} affected the α-synuclein conformation differently than Fe^{2+}. Furthermore, at low concentrations Mg^{2+} inhibited α-synuclein aggregation induced either spontaneously or by incubation with iron, and no SDS-resistant dimers were formed in the presence of Mg^{2+} (Golts et al. 2002). Later it was also shown that Mg^{2+} was able to modulate the interaction between α-synuclein and several herbicides (e.g., lindane, 2,4-D-isopropylester, atrazine, chlordimeform, and paraquat) and was able to successfully inhibit the herbicide-induced aggregation and fibrillation of α-synuclein (Andre et al. 2005). In this essence, Mg^{2+} was very different from other metals, such as Al^{3+}, Cd^{2+}, Mn^{2+}, Cu^{2+}, and Zn^{2+}, all of which dramatically enhanced the herbicide interaction with α-synuclein and strongly promoted the herbicide-induced a-synuclein aggregation. Furthermore, when the effect of Mg^{2+} on herbicide-α-synuclein binding was examined in the presence of 4 μM of other metals (Al^{3+}, Cd^{2+}, Mn^{2+}, Cu^{2+}, and Zn^{2+}), a significant reduction in the apparent association constant K of herbicide-α-synuclein binding was detected with the increasing Mg^{2+} concentration. This clearly suggested that the apparent affinity of Mg^{2+} and magnesium for α-synuclein is strong enough to inhibit the conformation change induced by metals, and that the interaction of α-synuclein with different ligands (including metal ions) produced various (nonidentical) conformational changes in this protein (Andre et al. 2005). In line with these observations, a study was recently conducted to clarify the effects of Mg^{2+} administration in a rat MPP^+ model of PD (Hashimoto et al. 2008). It was shown that Mg^{2+} might protect dopaminergic neurons in the substantia nigra from degeneration, significantly inhibiting the toxicity of MPP^+ at concentration of Mg^{2+} to 1.2 mM, and completely eliminating the decrease in the number of dopaminergic neurons at an Mg^{2+} concentration of 4 mM (Hashimoto et al. 2008). These observations suggest that the interaction of Mg^{2+} with α-synuclein might play a neuroprotective role.

MANGANESE

Manganese is a trace element that is ubiquitous in the environment, as it occurs in water, soil, air, and food. Manganese intoxication was first described almost 175 years ago in brownstone millers and in workers involved in mining and processing manganese ores who inhaled toxic amounts of manganese dust (Couper 1837). Chronic manganism produces an irreversible syndrome that bears a striking resemblance to PD, including fixed gaze, bradykinesia, postural difficulties, rigidity, and tremor (Cotzias 1958). Parkinsonism due to the chronic manganese intoxication can be separated from PD by presence of dystonia and mental-status changes (Barbeau 1984). Manganism is also characterized by a particular propensity to fall backward, failure to achieve a sustained therapeutic response to levodopa, and failure to detect a reduction in fluorodopa uptake by positron emission tomography (Calne et al. 1994). Furthermore, LBs have never the been observed in manganese-induced parkinsonism, and the major effects of manganese toxicity were found in the cells of the striatum and globus pallidus, which are not dopaminergic (Bleecker 1988).

Although manganism is clearly different from PD, and a link between Mn^{2+} exposure and PD is still controversial, some facts suggested that α-synuclein can be involved in pathogenesis. This follows from the intriguing suggestion that nigral dopaminergic neurons overexpressing α-synuclein (or its mutations) may need an additional noxious factor ("second hit") to trigger nigral death (Matsuoka et al. 2001). In fact, coincubation of α-synuclein with Mn^{2+} induced partial folding of the protein and its effective fibrillation (Uversky, Li, and Fink 2001b). Mn^{2+} was shown to be one of the most effective promoters of α-synuclein aggregation *in vitro* along with Al^{3+}, Fe^{3+}, Cu^{2+}, and Co^{2+}. Furthermore, the presence of Mn^{3+} induced immediate di-tyrosine formation, suggesting that this cation is responsible for the metal-induced oxidation of the protein (Uversky, Li, and Fink 2001b). Interestingly, when SK-N-MC neuroblastoma cells stably expressing the human dopamine transporter were transfected with human α-synuclein, and cells were exposed to 30–300 μM $MnCl_2$, the viability of cells overexpressing α-synuclein after 72 h of exposure to Mn^{2+} was dramatically reduced, suggesting that Mn^{2+} may cooperate with α-synuclein in triggering neuronal cell death such as seen in manganese parkinsonism (Pifl et al. 2004).

TERBIUM

Tb^{3+} was used a luminescent probe to analyze the peculiarities of metal binding to a peptide derived from α-synuclein (Fujiwara et al. 2002; Liu and Franz 2005). α-Synuclein fragment 119–132 (^{119}DPDNEAYEMPSEEG^{132}) was chosen for this analysis because its arrangement of carboxylate groups is similar to Ca^{2+}-binding loops and because it contains two identified phosphorylation sites, Tyr125 and Ser129 (Fujiwara et al. 2002). A series of peptides were synthesized, including the nonmodified fragment ^{119}DPDNEAYEMPSEEG^{132}, the ^{119}DPDNEAYEMP(pS)EEG^{132} phosphopeptide (pS129), the ^{119}DPDNEA(pY)EMPSEEG^{132} phosphopeptide, and several truncated forms of the tyrosine phophsophopeptide. It has been shown that ^{119}DPDNEAYEMPSEEG^{132} and ^{119}DPDNEAYEMP(pS)EEG^{132} peptides possessed only a weak Tb^{3+} binding, whereas ^{119}DPDNEA(pY)EMPSEEG^{132} showed tight 1:1 binding together with 2:1 and 3:1 Tb:peptide adducts (Fujiwara et al. 2002). These data clearly showed that the phosphorylated amino acid must be appropriately positioned among additional ligating residues to establish this phosphorylation-dependent metal binding (Fujiwara et al. 2002).

ZINC

Zinc is one of potential environmental factors exposure to which might favor PD (Rybicki et al. 1993; Gorell et al. 1997, 1999; Lovell et al. 1998). The analysis of the parkinsonian substantia nigra revealed the enhanced level of zinc in comparison with the control tissues (Dexter et al. 1989, 1991; Riederer et al. 1989). Zinc was shown to be an effective promoter of α-synuclein aggregation *in vitro* (Kim et al. 2000; Uversky, Li, and Fink 2001b). Furthermore, the presence of Zn^{2+} among several other metals (Ti^{3+}, Al^{3+}, and Pb^{2+}) was shown to effectively overcome the α-synuclein fibrillation induced by the methionines oxidation (Yamin et al. 2003). It was proposed that this metal ion–induced acceleration of fibrillation may be due to the relatively strong coordination of the zinc (and other metal) ion between (at least) two methionine sulfoxides. This bridging was assumed to aid the protein in adopting a necessary conformation for fibrillation, or possibly to involve intermolecular cross-bridging, which could facilitate association and subsequent fibrillation (Yamin et al. 2003).

When the fibrillation of the oxidized form of wild-type α-synuclein and its Met-minus mutants, M5L, M116L, M127L, M116L/M127L, and M5L/M116L/M127L, were studied, the presence of Zn^{2+} was shown to induce a dramatic acceleration of fibril formation for all of the oxidized proteins studied and in a mutation-dependent manner (Hokenson et al. 2004). In fact, all of the oxidized mutants fibrillate faster than nonmodified wild-type protein. Furthermore, the efficiency of mutation-induced acceleration of fibrillation seems to be proportional to the amount of substituted methionines, since the rates of fibrillation for the oxidized α-synucleins in the presence of Zn^{2+} can be arranged in the following order (from faster to slower rates of fibrillation): M5L/M116L/M127 > M116L/M127L > M5L ≈ M116L ≈ M127L (Hokenson et al. 2004). This suggested that methionine sulfoxides of the oxidized α-synuclein are not directly involved in the coordination of Zn^{2+}. For example, a mutant with triple Met

→ Leu substitutions fibrillates faster than wild-type protein but would be expected to fibrillate slower or not at all if the methionine sulfoxides were involved in strong intermolecular metal interactions. Interaction with Zn^{2+} decreased the propensity of the Leu-substituted α-synucleins to oligomerize (Hokenson et al. 2004). It was also concluded that the effectiveness of Zn^{2+} in accelerating fibrillation of both oxidized wild-type and mutated α-synucleins indicated that the presence of Zn^{2+} caused partitioning in favor of the fibrillation pathway. Therefore Zn^{2+} overcame the factors that lead to population and stabilization of the soluble oligomers (Hokenson et al. 2004). Zinc was also one of the several metals shown to effectively promote interaction of α-synuclein with herbicides and to dramatically accelerate the herbicide-induced fibrillation of this protein (Andre et al. 2005).

FUTURE PERSPECTIVE

Significant knowledge is already accumulated in the field α-synucelin metalloproteomics. It is recognized now that metals play crucial role in α-synuclein homeostasis, modulating its structural properties and aggregation propensity. Although the majority of metal ions interact with α-synuclein nonspecifically, this protein possesses relatively high affinity to Cu^{2+} and Fe^{3+}. This field attracted significant attention primarily because of the link between the α-synuclein misfolding and aggregation and a number of neurodegenerative diseases and because of the potential link between the exposure to heavy metals and the pathogenesis of some of the synucleinopathies. Much remains to be learned in order to better understand this intriguing interconnection between metals, α-synuclein, and pathogenesis of neurodegeneration. One of the exciting directions in the field is elucidation of the effect of protein posttranslational modifications (e.g., phosphorylation) and oxidative modifications on peculiarities of α-synuclein interaction with various metals. Synergistic effects of metals and other environmental toxins (such as pesticides and herbicides) on α-synuclein structure and aggregation propensity also warrants further study. The effect of macromolecular crowding on the peculiarities of α-synuclein interaction with various metals should be investigated. Detailed structural characterization of α-synuclein complexes with various metals and other environmental agents by high-resolution techniques (e.g., NMR) will shed more light on the molecular mechanisms of synucleinopathies. Better connection between the extensive *in vitro* data on the effect of metal ions on α-synuclein structure and aggregation and the cellular processes triggered by the exposure to heavy atoms should be established. Other interesting developments are expected in nanotechnology, where α-synuclein fibrils can be used as templates for the bottom-up device fabrication. In fact, α-synuclein fibrils have already been used as a scaffold for palladium, gold, and copper nanoparticle chain synthesis, generating metal-coated fibers with reproducible average diameters between 50 and 200 nm (Colby et al. 2008).

REFERENCES

Adamczyk, A., and J. B. Strosznajder. 2006. Alpha-synuclein potentiates Ca2+ influx through voltage-dependent Ca2+ channels. *Neuroreport* 17 (18):1883–86.

Altschuler, E. 1999. Aluminum-containing antacids as a cause of idiopathic Parkinson's disease. *Med Hypotheses* 53 (1):22–23.

Andre, C., T. T. Truong, J. F. Robert, and Y. C. Guillaume. 2005. Effect of metals on herbicides-alpha-synuclein association: a possible factor in neurodegenerative disease studied by capillary electrophoresis. *Electrophoresis* 26 (17):3256–64.

Arawaka, S., Y. Saito, S. Murayama, and H. Mori. 1998. Lewy body in neurodegeneration with brain iron accumulation type 1 is immunoreactive for alpha-synuclein. *Neurology* 51 (3):887–89.

Barbeau, A. 1984. Manganese and extrapyramidal disorders (a critical review and tribute to Dr. George C. Cotzias). *Neurotoxicology* 5 (1):13–35.

Berg, D., M. Gerlach, M. B. Youdim, K. L. Double, L. Zecca, P. Riederer, and G. Becker. 2001. Brain iron pathways and their relevance to Parkinson's disease. *J Neurochem* 79 (2):225–36.

Bezprozvanny, I., and M. P. Mattson. 2008. Neuronal calcium mishandling and the pathogenesis of Alzheimer's disease. *Trends Neurosci* 31 (9):454–63.

Bharathi, S. S. Indi, and K. S. Rao. 2007. Copper- and iron-induced differential fibril formation in alpha-synuclein: TEM study. *Neurosci Lett* 424 (2):78–82.

Bharathi, and K. S. Rao. 2007. Thermodynamics imprinting reveals differential binding of metals to alpha-synuclein: relevance to Parkinson's disease. *Biochem Biophys Res Commun* 359 (1):115–20.

Binolfi, A., R. M. Rasia, C. W. Bertoncini, M. Ceolin, M. Zweckstetter, C. Griesinger, T. M. Jovin, and C. O. Fernandez. 2006. Interaction of alpha-synuclein with divalent metal ions reveals key differences: a link between structure, binding specificity and fibrillation enhancement. *J Am Chem Soc* 128 (30):9893–9901.

Bleecker, M. L. 1988. Parkinsonism: a clinical marker of exposure to neurotoxins. *Neurotoxicol Teratol* 10 (5):475–78.

Bocca, B., A. Alimonti, O. Senofonte, A. Pino, N. Violante, F. Petrucci, G. Sancesario, and G. Forte. 2006. Metal changes in CSF and peripheral compartments of parkinsonian patients. *J Neurol Sci* 248 (1–2):23–30.

Bush, A. I. 2000. Metals and neuroscience. *Curr Opin Chem Biol* 4 (2):184–91.

Bussell, R., Jr., and D. Eliezer. 2001. Residual structure and dynamics in Parkinson's disease-associated mutants of alpha-synuclein. *J Biol Chem* 276 (49):45996–6003.

Calne, D. B., N. S. Chu, C. C. Huang, C. S. Lu, and W. Olanow. 1994. Manganism and idiopathic parkinsonism: similarities and differences. *Neurology* 44 (9):1583–86.

Chen, X., H. A. de Silva, M. J. Pettenati, P. N. Rao, P. St George-Hyslop, A. D. Roses, Y. Xia, K. Horsburgh, K. Ueda, and T. Saitoh. 1995. The human NACP/alpha-synuclein gene: chromosome assignment to 4q21.3-q22 and TaqI RFLP analysis. *Genomics* 26 (2):425–27.

Colby, R., J. Hulleman, S. Padalkar, J. C. Rochet, and L. A. Stanciu. 2008. Biotemplated synthesis of metallic nanoparticle chains on an alpha-synuclein fiber scaffold. *J Nanosci Nanotechnol* 8 (2):973–78.

Cotzias, G. C. 1958. Manganese in health and disease. *Physiol Rev* 38 (3):503–32.

Couper, J. . 1837. On the effects of black oxide of manganese inhaled into the lungs. *Br Ann Med Pharmacol* 1:41–42.

Danzer, K. M., D. Haasen, A. R. Karow, S. Moussaud, M. Habeck, A. Giese, H. Kretzschmar, B. Hengerer, and M. Kostka. 2007. Different species of alpha-synuclein oligomers induce calcium influx and seeding. *J Neurosci* 27 (34):9220–32.

Dexter, D. T., A. Carayon, F. Javoy-Agid, Y. Agid, F. R. Wells, S. E. Daniel, A. J. Lees, P. Jenner, and C. D. Marsden. 1991. Alterations in the levels of iron, ferritin and other trace metals in Parkinson's disease and other neurodegenerative diseases affecting the basal ganglia. *Brain* 114 (Pt 4):1953–75.

Dexter, D. T., F. R. Wells, A. J. Lees, F. Agid, Y. Agid, P. Jenner, and C. D. Marsden. 1989. Increased nigral iron content and alterations in other metal ions occurring in brain in Parkinson's disease. *J Neurochem* 52 (6):1830–36.

Drew, S. C., S. L. Leong, C. L. Pham, D. J. Tew, C. L. Masters, L. A. Miles, R. Cappai, and K. J. Barnham. 2008. Cu2+ binding modes of recombinant alpha-synuclein--insights from EPR spectroscopy. *J Am Chem Soc* 130 (24):7766–73.

Duda, J. E., V. M. Lee, and J. Q. Trojanowski. 2000. Neuropathology of synuclein aggregates. *J Neurosci Res* 61 (2):121–27.

Eliezer, D., E. Kutluay, R. Bussell Jr., and G. Browne. 2001. Conformational properties of alpha-synuclein in its free and lipid-associated states. *J Mol Biol* 307 (4):1061–73.

Fink, A. L., L. J. Calciano, Y. Goto, T. Kurotsu, and D. R. Palleros. 1994. Classification of acid denaturation of proteins: intermediates and unfolded states. *Biochemistry* 33 (41):12504–11.

Fujiwara, H., M. Hasegawa, N. Dohmae, A. Kawashima, E. Masliah, M. S. Goldberg, J. Shen, K. Takio, and T. Iwatsubo. 2002. Alpha-synuclein is phosphorylated in synucleinopathy lesions. *Nat Cell Biol* 4 (2):160–64.

Gaeta, A., and R. C. Hider. 2005. The crucial role of metal ions in neurodegeneration: the basis for a promising therapeutic strategy. *Br J Pharmacol* 146 (8):1041–59.

Gaggelli, E., H. Kozlowski, D. Valensin, and G. Valensin. 2006. Copper homeostasis and neurodegenerative disorders (Alzheimer's, prion, and Parkinson's diseases and amyotrophic lateral sclerosis). *Chem Rev* 106 (6):1995–2044.

Galvin, J. E., B. Giasson, H. I. Hurtig, V. M. Lee, and J. Q. Trojanowski. 2000. Neurodegeneration with brain iron accumulation, type 1 is characterized by alpha-, beta-, and gamma-synuclein neuropathology. *Am J Pathol* 157 (2):361–68.

Galvin, J. E., V. M. Lee, and J. Q. Trojanowski. 2001. Synucleinopathies: clinical and pathological implications. *Arch Neurol* 58 (2):186–90.

Golts, N., H. Snyder, M. Frasier, C. Theisler, P. Choi, and B. Wolozin. 2002. Magnesium inhibits spontaneous and iron-induced aggregation of alpha-synuclein. *J Biol Chem* 277 (18):16116–23.

Gorell, J. M., C. C. Johnson, B. A. Rybicki, E. L. Peterson, G. X. Kortsha, G. G. Brown, and R. J. Richardson. 1997. Occupational exposures to metals as risk factors for Parkinson's disease. *Neurology* 48 (3):650–58.

———. 1999. Occupational exposure to manganese, copper, lead, iron, mercury and zinc and the risk of Parkinson's disease. *Neurotoxicology* 20 (2–3):239–47.

Goto, Y., L. J. Calciano, and A. L. Fink. 1990. Acid-induced folding of proteins. *Proc Natl Acad Sci U S A* 87 (2):573–77.

Goto, Y., and A. L. Fink. 1989. Conformational states of beta-lactamase: molten-globule states at acidic and alkaline pH with high salt. *Biochemistry* 28 (3):945–52.

Goto, Y., N. Takahashi, and A. L. Fink. 1990. Mechanism of acid-induced folding of proteins. *Biochemistry* 29 (14):3480–88.

Gotz, M. E., K. Double, M. Gerlach, M. B. Youdim, and P. Riederer. 2004. The relevance of iron in the pathogenesis of Parkinson's disease. *Ann N Y Acad Sci* 1012:193–208.

Grune, T., T. Reinheckel, and K. J. Davies. 1996. Degradation of oxidized proteins in K562 human hematopoietic cells by proteasome. *J Biol Chem* 271 (26):15504–9.

Hashimoto, T., K. Nishi, J. Nagasao, S. Tsuji, and K. Oyanagi. 2008. Magnesium exerts both preventive and ameliorating effects in an *in vitro* rat Parkinson disease model involving 1-methyl-4-phenylpyridinium (MPP+) toxicity in dopaminergic neurons. *Brain Res* 1197:143–51.

Hider, R. C., Y. Ma, F. Molina-Holgado, A. Gaeta, and S. Roy. 2008. Iron chelation as a potential therapy for neurodegenerative disease. *Biochem Soc Trans* 36 (Pt 6):1304–8.

Hirsch, E. C., J. P. Brandel, P. Galle, F. Javoy-Agid, and Y. Agid. 1991. Iron and aluminum increase in the substantia nigra of patients with Parkinson's disease: an X-ray microanalysis. *J Neurochem* 56 (2):446–51.

Hokenson, M. J., V. N. Uversky, J. Goers, G. Yamin, L. A. Munishkina, and A. L. Fink. 2004. Role of individual methionines in the fibrillation of methionine-oxidized alpha-synuclein. *Biochemistry* 43 (15):4621–33.

Hoyer, W., T. Antony, D. Cherny, G. Heim, T. M. Jovin, and V. Subramaniam. 2002. Dependence of alpha-synuclein aggregate morphology on solution conditions. *J Mol Biol* 322 (2):383–93.

Iwai, A., E. Masliah, M. Yoshimoto, N. Ge, L. Flanagan, H. A. de Silva, A. Kittel, and T. Saitoh. 1995. The precursor protein of non-A beta component of Alzheimer's disease amyloid is a presynaptic protein of the central nervous system. *Neuron* 14 (2):467–75.

Jakes, R., M. G. Spillantini, and M. Goedert. 1994. Identification of two distinct synucleins from human brain. *FEBS Lett* 345 (1):27–32.

Kalivendi, S. V., S. Cunningham, S. Kotamraju, J. Joseph, C. J. Hillard, and B. Kalyanaraman. 2004. Alpha-synuclein up-regulation and aggregation during MPP+-induced apoptosis in neuroblastoma cells: intermediacy of transferrin receptor iron and hydrogen peroxide. *J Biol Chem* 279 (15):15240–47.

Khan, A., A. E. Ashcroft, V. Higenell, O. V. Korchazhkina, and C. Exley. 2005. Metals accelerate the formation and direct the structure of amyloid fibrils of NAC. *J Inorg Biochem* 99 (9):1920–27.

Kienzl, E., L. Puchinger, K. Jellinger, W. Linert, H. Stachelberger, and R. F. Jameson. 1995. The role of transition metals in the pathogenesis of Parkinson's disease. *J Neurol Sci* 134 Suppl:69–78.

Kim, T. D., S. R. Paik, C. H. Yang, and J. Kim. 2000. Structural changes in alpha-synuclein affect its chaperone-like activity *in vitro*. *Protein Sci* 9 (12):2489–96.

Kostka, M., T. Hogen, K. M. Danzer, J. Levin, M. Habeck, A. Wirth, R. Wagner, C. G. Glabe, S. Finger, U. Heinzelmann, P. Garidel, W. Duan, C. A. Ross, H. Kretzschmar, and A. Giese. 2008. Single particle characterization of iron-induced pore-forming alpha-synuclein oligomers. *J Biol Chem* 283 (16):10992–11003.

Kowalik-Jankowska, T., A. Rajewska, E. Jankowska, and Z. Grzonka. 2008. Products of Cu(II)-catalyzed oxidation of alpha-synuclein fragments containing M1-D2 and H50 residues in the presence of hydrogen peroxide. *Dalton Trans* (6):832–38.

Kowalik-Jankowska, T., A. Rajewska, E. Jankowska, K. Wisniewska, and Z. Grzonka. 2006. Products of Cu(II)-catalyzed oxidation of the N-terminal fragments of alpha-synuclein in the presence of hydrogen peroxide. *J Inorg Biochem* 100 (10):1623–31.

Lee, J. C., H. B. Gray, and J. R. Winkler. 2008. Copper(II) binding to alpha-synuclein, the Parkinson's protein. *J Am Chem Soc* 130 (22):6898–99.

Liu, L. L., and K. J. Franz. 2005. Phosphorylation of an alpha-synuclein peptide fragment enhances metal binding. *J Am Chem Soc* 127 (27):9662–63.

———. 2007. Phosphorylation-dependent metal binding by alpha-synuclein peptide fragments. *J Biol Inorg Chem* 12 (2):234–47.

Lovell, M. A., J. D. Robertson, W. J. Teesdale, J. L. Campbell, and W. R. Markesbery. 1998. Copper, iron and zinc in Alzheimer's disease senile plaques. *J Neurol Sci* 158 (1):47–52.

Lowe, R., D. L. Pountney, P. H. Jensen, W. P. Gai, and N. H. Voelcker. 2004. Calcium (II) selectively induces alpha-synuclein annular oligomers via interaction with the C-terminal domain. *Protein Sci* 13 (12):3245–52.

Ly, T., and R. R. Julian. 2008. Protein-metal interactions of calmodulin and alpha-synuclein monitored by selective noncovalent adduct protein probing mass spectrometry. *J Am Soc Mass Spectrom* 19 (11):1663–72.

Maroteaux, L., J. T. Campanelli, and R. H. Scheller. 1988. Synuclein: a neuron-specific protein localized to the nucleus and presynaptic nerve terminal. *J Neurosci* 8 (8):2804–15.

Matsuoka, Y., M. Vila, S. Lincoln, A. McCormack, M. Picciano, J. LaFrancois, X. Yu, D. Dickson, W. J. Langston, E. McGowan, M. Farrer, J. Hardy, K. Duff, S. Przedborski, and D. A. Di Monte. 2001. Lack of nigral pathology in transgenic mice expressing human alpha-synuclein driven by the tyrosine hydroxylase promoter. *Neurobiol Dis* 8 (3):535–39.

Mattson, M. P. 2007. Calcium and neurodegeneration. *Aging Cell* 6 (3):337–50.

Molina-Holgado, F., R. C. Hider, A. Gaeta, R. Williams, and P. Francis. 2007. Metals ions and neurodegeneration. *Biometals* 20 (3–4):639–54.

Montgomery, E. B., Jr. 1995. Heavy metals and the etiology of Parkinson's disease and other movement disorders. *Toxicology* 97 (1–3):3–9.

Munishkina, L. A., A. L. Fink, and V. N. Uversky. 2008. Concerted action of metals and macromolecular crowding on the fibrillation of alpha-synuclein. *Protein Pept Lett* 15 (10):1079–85.

Neumann, M., S. Adler, O. Schluter, E. Kremmer, R. Benecke, and H. A. Kretzschmar. 2000. Alpha-synuclein accumulation in a case of neurodegeneration with brain iron accumulation type 1 (NBIA-1, formerly Hallervorden-Spatz syndrome) with widespread cortical and brainstem-type Lewy bodies. *Acta Neuropathol* 100 (5):568–74.

Nielsen, M. S., H. Vorum, E. Lindersson, and P. H. Jensen. 2001. Ca2+ binding to alpha-synuclein regulates ligand binding and oligomerization. *J Biol Chem* 276 (25):22680–84.

Oestreicher, E., G. J. Sengstock, P. Riederer, C. W. Olanow, A. J. Dunn, and G. W. Arendash. 1994. Degeneration of nigrostriatal dopaminergic neurons increases iron within the substantia nigra: a histochemical and neurochemical study. *Brain Res* 660 (1):8–18.

Ostrerova-Golts, N., L. Petrucelli, J. Hardy, J. M. Lee, M. Farer, and B. Wolozin. 2000. The A53T alpha-synuclein mutation increases iron-dependent aggregation and toxicity. *J Neurosci* 20 (16):6048–54.

Oyanagi, K., E. Kawakami, K. Kikuchi-Horie, K. Ohara, K. Ogata, S. Takahama, M. Wada, T. Kihira, and M. Yasui. 2006. Magnesium deficiency over generations in rats with special references to the pathogenesis of the Parkinsonism-dementia complex and amyotrophic lateral sclerosis of Guam. *Neuropathology* 26 (2):115–28.

Paik, S. R., J. H. Lee, D. H. Kim, C. S. Chang, and J. Kim. 1997. Aluminum-induced structural alterations of the precursor of the non-A beta component of Alzheimer's disease amyloid. *Arch Biochem Biophys* 344 (2):325–34.

Paik, S. R., H. J. Shin, and J. H. Lee. 2000. Metal-catalyzed oxidation of alpha-synuclein in the presence of Copper(II) and hydrogen peroxide. *Arch Biochem Biophys* 378 (2):269–77.

Paik, S. R., H. J. Shin, J. H. Lee, C. S. Chang, and J. Kim. 1999. Copper(II)-induced self-oligomerization of alpha-synuclein. *Biochem J* 340 (Pt 3):821–28.

Pifl, C., M. Khorchide, A. Kattinger, H. Reither, J. Hardy, and O. Hornykiewicz. 2004. Alpha-Synuclein selectively increases manganese-induced viability loss in SK-N-MC neuroblastoma cells expressing the human dopamine transporter. *Neurosci Lett* 354 (1):34–37.

Polymeropoulos, M. H., C. Lavedan, E. Leroy, S. E. Ide, A. Dehejia, A. Dutra, B. Pike, H. Root, J. Rubenstein, R. Boyer, E. S. Stenroos, S. Chandrasekharappa, A. Athanassiadou, T. Papapetropoulos, W. G. Johnson, A. M. Lazzarini, R. C. Duvoisin, G. Di Iorio, L. I. Golbe, and R. L. Nussbaum. 1997. Mutation in the alpha-synuclein gene identified in families with Parkinson's disease. *Science* 276 (5321):2045–47.

Pountney, D. L., R. Lowe, M. Quilty, J. C. Vickers, N. H. Voelcker, and W. P. Gai. 2004. Annular alpha-synuclein species from purified multiple system atrophy inclusions. *J Neurochem* 90 (2):502–12.

Pountney, D. L., N. H. Voelcker, and W. P. Gai. 2005. Annular alpha-synuclein oligomers are potentially toxic agents in alpha-synucleinopathy. Hypothesis. *Neurotox Res* 7 (1–2):59–67.

Rasia, R. M., C. W. Bertoncini, D. Marsh, W. Hoyer, D. Cherny, M. Zweckstetter, C. Griesinger, T. M. Jovin, and C. O. Fernandez. 2005. Structural characterization of copper (II) binding to alpha-synuclein: Insights into the bioinorganic chemistry of Parkinson's disease. *Proc Natl Acad Sci U S A* 102 (12):4294–99.

Riederer, P., E. Sofic, W. D. Rausch, B. Schmidt, G. P. Reynolds, K. Jellinger, and M. B. Youdim. 1989. Transition metals, ferritin, glutathione, and ascorbic acid in parkinsonian brains. *J Neurochem* 52 (2):515–20.

Rybicki, B. A., C. C. Johnson, J. Uman, and J. M. Gorell. 1993. Parkinson's disease mortality and the industrial use of heavy metals in Michigan. *Mov Disord* 8 (1):87–92.

Sandal, M., F. Valle, I. Tessari, S. Mammi, E. Bergantino, F. Musiani, M. Brucale, L. Bubacco, and B. Samori. 2008. Conformational equilibria in monomeric alpha-synuclein at the single-molecule level. *PLoS Biol* 6 (1):e6.

Schulz, J. B. 2007. Mechanisms of neurodegeneration in idiopathic Parkinson's disease. *Parkinsonism Relat Disord* 13 (Suppl 3):S306–8.

Spillantini, M. G., M. L. Schmidt, V. M. Lee, J. Q. Trojanowski, R. Jakes, and M. Goedert. 1997. Alpha-synuclein in Lewy bodies. *Nature* 388 (6645):839–40.

Sung, Y. H., C. Rospigliosi, and D. Eliezer. 2006. NMR mapping of copper binding sites in alpha-synuclein. *Biochim Biophys Acta* 1764 (1):5–12.

Tabner, B. J., S. Turnbull, O. M. El-Agnaf, and D. Allsop. 2002. Formation of hydrogen peroxide and hydroxyl radicals from A(beta) and alpha-synuclein as a possible mechanism of cell death in Alzheimer's disease and Parkinson's disease. *Free Radic Biol Med* 32 (11):1076–83.

Tamamizu-Kato, S., M. G. Kosaraju, H. Kato, V. Raussens, J. M. Ruysschaert, and V. Narayanaswami. 2006. Calcium-triggered membrane interaction of the alpha-synuclein acidic tail. *Biochemistry* 45 (36):10947–56.

Tanner, C. M. 1989. The role of environmental toxins in the etiology of Parkinson's disease. *Trends Neurosci* 12 (2):49–54.

Tanner, C. M., R. Ottman, S. M. Goldman, J. Ellenberg, P. Chan, R. Mayeux, and J. W. Langston. 1999. Parkinson disease in twins: an etiologic study. *J Amer Med Assoc* 281 (4):341–46.

Tiffany-Castiglion, E., and Y. Qian. 2001. Astroglia as metal depots: molecular mechanisms for metal accumulation, storage and release. *Neurotoxicology* 22 (5):577–92.

Trojanowski, J. Q., and V. M. Lee. 2002. Parkinson's disease and related synucleinopathies are a new class of nervous system amyloidoses. *Neurotoxicology* 23 (4–5):457–60.

Uitti, R. J., A. H. Rajput, B. Rozdilsky, M. Bickis, T. Wollin, and W. K. Yuen. 1989. Regional metal concentrations in Parkinson's disease, other chronic neurological diseases, and control brains. *Can J Neurol Sci* 16 (3):310–14.

Uversky, V. N. 2003. A protein-chameleon: conformational plasticity of alpha-synuclein, a disordered protein involved in neurodegenerative disorders. *J Biomol Struct Dyn* 21 (2):211–34.

———. 2007. Neuropathology, biochemistry, and biophysics of alpha-synuclein aggregation. *J Neurochem* 103 (1):17–37.

———. 2008. Alpha-synuclein misfolding and neurodegenerative diseases. *Curr Protein Pept Sci* 9 (5):507–40.

Uversky, V. N., A. S. Karnoup, D. J. Segel, S. Seshadri, S. Doniach, and A. L. Fink. 1998. Anion-induced folding of Staphylococcal nuclease: characterization of multiple equilibrium partially folded intermediates. *J Mol Biol* 278 (4):879–94.

Uversky, V. N., J. Li, K. Bower, and A. L. Fink. 2002. Synergistic effects of pesticides and metals on the fibrillation of alpha-synuclein: implications for Parkinson's disease. *Neurotoxicology* 23 (4–5):527–36.

Uversky, V. N., J. Li, and A. L. Fink. 2001a. Evidence for a partially folded intermediate in alpha-synuclein fibril formation. *J Biol Chem* 276 (14):10737–44.

———. 2001b. Metal-triggered structural transformations, aggregation, and fibrillation of human alpha-synuclein. A possible molecular NK between Parkinson's disease and heavy metal exposure. *J Biol Chem* 276 (47):44284–96.

Weinreb, P. H., W. Zhen, A. W. Poon, K. A. Conway, and P. T. Lansbury Jr. 1996. NACP, a protein implicated in Alzheimer's disease and learning, is natively unfolded. *Biochemistry* 35 (43):13709–15.

Wojda, U., E. Salinska, and J. Kuznicki. 2008. Calcium ions in neuronal degeneration. *IUBMB Life* 60 (9):575–90.

Wolozin, B., and N. Golts. 2002. Iron and Parkinson's disease. *Neuroscientist* 8 (1):22–32.

Yamin, G., C. B. Glaser, V. N. Uversky, and A. L. Fink. 2003. Certain metals trigger fibrillation of methionine-oxidized alpha-synuclein. *J Biol Chem* 278 (30):27630–35.

Yasui, M., T. Kihira, and K. Ota. 1992. Calcium, magnesium and aluminum concentrations in Parkinson's disease. *Neurotoxicology* 13 (3):593–600.

Youdim, M. B., D. Ben-Shachar, and P. Riederer. 1991. Iron in brain function and dysfunction with emphasis on Parkinson's disease. *Eur Neurol* 31 Suppl 1:34–40.

Zayed, J., G. Campanella, J. C. Panisset, S. Ducic, P. Andre, H. Masson, and M. Roy. 1990. [Parkinson disease and environmental factors]. *Rev Epidemiol Sante Publique* 38 (2):159–60.

Zayed, J., S. Ducic, G. Campanella, J. C. Panisset, P. Andre, H. Masson, and M. Roy. 1990. [Environmental factors in the etiology of Parkinson's disease]. *Can J Neurol Sci* 17 (3):286–91.

10 Zinc and p53 Misfolding

James S. Butler and Stewart N. Loh

CONTENTS

Introduction 193
Methodology and Experimental Approaches 194
Structure and Function of apoDBD 197
Folding Mechanisms of DBD and apoDBD 198
Zinc Binding, Zinc Loss, and Misfolding 200
Rescue of Misfolding by Metallochaperones 200
Zinc Bioavailability, p53, and Cancer 202
Conclusions and Clinical Perspective 203
References 204

INTRODUCTION

Up to 10% of human proteins bind zinc, making this family the most abundant class of metal-binding proteins in the metazoan proteome (Anzellotti and Farrell 2008; Maret 2008). Zinc plays a particularly dominant role in regulation of gene expression: Nearly one in two transcription factors binds Zn^{2+} (Andreini et al. 2006; Tupler, Perini, and Green 2001). The p53 tumor suppressor is chief among these in terms of importance to human health. P53 coordinates cell cycle arrest, apoptosis, and senescence pathways in response to DNA damage, oncogenic stress, and other genomic threats. Not surprisingly, loss of p53 function is associated with many forms of cancer. It was documented 20 years ago that half of human tumors harbor mutations in the p53 gene, establishing p53 as the single most frequently altered protein in cancer (Nigro et al. 1989). More recently, two studies sequenced over 20,000 protein-coding genes from 24 pancreatic tumors (Jones et al. 2008) and 22 glioblastoma multiforme tumors (Parsons et al. 2008) and identified p53 missense mutations as being among the most frequent genomic alterations that allow these cells to escape growth control. Each p53 monomer binds a single zinc ion, which is essential for DNA binding and transcription activation. As we will discuss, a substantial body of *in vitro* and *in vivo* evidence indicates that loss of Zn^{2+} as well as Zn^{2+}-misligation play major roles in loss of p53 function and likely contribute to development of cancer.

Like many transcription factors, p53 (393 amino acids) contains both structured and disordered elements and is comprised of distinct functional domains. The N-terminal domain (NTD; amino acids 1–61) possesses transactivation function and is intrinsically disordered. The core DNA-binding domain (DBD; amino acids 92–312) is responsible for site-specific DNA recognition. P53 binds DNA as a dimer-of-dimers, and the tetramerization domain is encoded by residues 325–356. The last

~30 residues of p53 are disordered and, being highly basic, bind nonspecifically to DNA. In this chapter we focus attention on the isolated DBD, as it binds the single zinc ion and has been the subject of most existing structural and biophysical studies of metal binding and loss. It should be noted, however, that much progress has been made toward solving the structure of full-length tetrameric p53 (Joerger and Fersht 2008) and elucidating its folding mechanism (Lubin, D.J., Butler, J.S., and Loh, S.N. 2010. Folding of tetrameric p53: oligomerization and tumorigenic mutations induce misfolding and loss of function. *J. Mol. Biol* 395:705–716.). We can look forward to future studies examining the effects of zinc binding and loss on the structure, function, and folding of the physiological form of p53.

Although DBD requires Zn^{2+} for DNA binding (Méplan, Richard, and Hainaut 2000a), it is not a zinc finger transcription factor. Both molecules bind Zn^{2+} in a tetrahedral fashion using Cys and His residues. In zinc fingers, two ligands are spaced one turn apart in a short α-helix, the other two ligands are in a small β-strand, and the two secondary structures are connected by a short loop. Zn^{2+} binding stabilizes the helical "finger," which docks into the major groove (often in multiple copies that spiral around the double helix). By contrast, the pairs of Zn^{2+}-binding residues in DBD are separated by ~60 amino acids and are located in large loops (C176/H179 in L2 and C238/C242 in L3; Figure 10.1a). L3 binds in the DNA minor groove (Cho et al. 1994) and interacts with L2. Major groove contacts are formed by the H2 helix and L1 loop.

METHODOLOGY AND EXPERIMENTAL APPROACHES

Preparing apoDBD. Metal-free DBD (apoDBD) is prepared by extracting Zn^{2+} from purified DBD. Care must be taken to minimize aggregation and misfolding during and subsequent to the zinc removal process, and the first priority is to use protein that is pure, monomeric, and functional. P53 and DBD are expressed at 18–20°C in BL21(DE3) *Escherichia coli* and purified using ion exchange chromatography as described by Bullock et al. (1997). Growth media rich in nutrients such as Luria-Bertani broth contain ample amounts of zinc, but it is essential to supplement minimal media with 0.1 mM $ZnCl_2$ to ensure proper expression. It is generally not necessary or advisable to add zinc to purification buffers. With the exception of severely destabilized mutants (e.g., R175H), most p53 variants bind metal tightly at low temperature and purify with one equivalent of bound Zn^{2+} (Butler and Loh 2003). Moreover, excess Zn^{2+} can cause DBD to aggregate *in vitro* (*vide infra*). A reducing agent (typically 2–5 mM DTT) is present during all purification and storage steps to ensure that cysteines remain in thiol form. We regularly recover 20 mg of pure wild-type (wt) DBD per liter of starting culture.

P53 retains its bound Zn^{2+} upon incubation at low temperature for days in the presence of excess EDTA (Lokshin et al. 2007; Butler and Loh 2007). To generate metal-free protein it is therefore necessary to destabilize or denature DBD so that Zn^{2+} is thermodynamically and kinetically accessible to chelating agents. Raising temperature greatly facilitates Zn^{2+} loss, but DBD and especially apoDBD aggregate above ~20°C. Although denaturants such as guanidine hydrochloride (GdnHCl) or urea can be used in place of temperature, refolding of DBD (and apoDBD) is a slow, error-prone process that must be performed at low μM concentration to

Zinc and p53 Misfolding

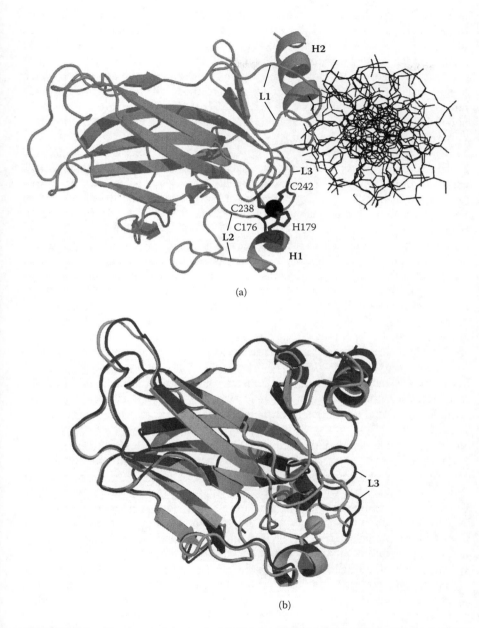

FIGURE 10.1 X-ray crystal structures of: (a) human DBD bound to DNA (Cho et al. 1994), (b) mouse DBD (gray) (W. C. Ho et al. 2006) and mouse apoDBD (black) (Kwon et al. 2008) in the absence of DNA. Zn^{2+} is shown as a sphere. Ligands to Zn^{2+} and secondary-structure elements surrounding the metal- and DNA-binding sites are indicated.

avoid irreversible aggregation. We find that transient exposure to acidic pH partially unfolds DBD and allows apoDBD to refold efficiently (Butler and Loh 2003). To remove Zn^{2+}, we reduce sample pH to 4.5 with acetic acid in the presence of 10 mM EDTA, on ice. After 1 min, pH is adjusted to 7.0 by rapidly adding buffer (e.g., 1 M sodium phosphate, HEPES, or bis-tris propane). The sample is immediately desalted to remove EDTA/Zn^{2+} complexes. Using this method, it is possible to completely remove Zn^{2+} from solutions of at least 200 µM DBD. Importantly, functional Zn^{2+}-bound DBD can be reconstituted from this material by carefully controlling Zn^{2+} availability as described later.

Reconstituting DBD from Zn^{2+} and apoDBD. In human cells, p53 is synthesized as the functional 1:1 complex with Zn^{2+}. The same holds true for recombinant DBD expressed in *E. coli*. However, reconstituting DBD from Zn^{2+} and apoDBD is a thorny process *in vitro*. As little as a 2.5-fold excess of $ZnCl_2$ is sufficient to precipitate native DBD within minutes, presumably by bridging multiple DBDs through shared metal-binding sites (Butler and Loh 2007). A similar misligation phenomenon is believed to cause apoDBD to misfold in the presence of Zn^{2+}, as discussed later. It is therefore clear that controlling Zn^{2+} bioavailability is critical for proper folding of p53. To that end, small-molecule chelators can assist *in vitro* reconstitution by acting as synthetic metallochaperones. The chelator should bind Zn^{2+} with greater affinity than the non-native sites, but lower affinity than the native site. It thereby serves as both sink and source of metal ions. Nitrilotriacetate (NTA; K_d = 17 nM) appears to satisfy that criterion. A ratio of 2.5:1 NTA:$ZnCl_2$ abolishes precipitation and misfolding, and allows apoDBD to bind a single zinc ion with the correct ligands (Butler and Loh 2007).

Experimental Considerations for Folding Studies. A common mechanism of p53 inactivation is mutation-induced destabilization of DBD. It is consequently of primary interest to measure thermodynamic stability. Chemical denaturants are preferred over thermal denaturation for these studies, as p53 tends to aggregate above 20°C. Unfolding from urea or GdnHCl is fully reversible; however, refolding is slow and complex and equilibration requires >10 h incubation at 10°C (Butler and Loh 2006; Bullock et al. 1997). The conformational state of DBD and apoDBD is typically monitored by fluorescence of the single Trp residue (W146). Fluorescence of native DBD is quenched with a Tyr-like maximum near 305 nm (Bullock et al. 1997). Unfolded DBD emits strongly at 355 nm. An isosbestic point is observed near 322 nm in urea denaturation experiments, indicating that the equilibrium transition is two-state (Bullock, Henckel, and Fersht 2000). ApoDBD exhibits nearly identical spectra, which suggests that the environment around W146 is not perturbed by Zn^{2+} loss (Butler and Loh 2007). Importantly, Trp fluorescence has proven to be valuable in identifying non-native species. Misfolded conformations (Butler and Loh 2007), including soluble aggregates (Friedler et al. 2003), are characterized by a maximum emission near 340 nm and a greater peak intensity than either native or unfolded DBD.

In addition to fluorescence methods, conformational stability of p53 has been measured by circular dichroism (CD) (Bell et al. 2002; Ishimaru et al. 2004; Butler and Loh 2007), differential scanning calorimetry (DSC) (Ang et al. 2006; Bullock et al. 1997), and thermal denaturation (Bullock, Henckel, and Fersht 2000). Techniques such as CD and DSC are required to study full-length p53 because three additional Trp residues in the unstructured NTD obscure the signal change from W146. DSC and variable-temperature CD experiments, however, result in irreversible aggregation so the observed melting temperature (T_m) is not a true quantity but an apparent one. Nevertheless, T_m correlates well with free energies measured by equilibrium experiments (Ang et al. 2006; Bullock, Henckel, and Fersht 2000), and T_m is useful to compare relative stabilities of p53 variants.

Although thermodynamic destabilization accounts for the deleterious effects of many tumorigenic mutations, kinetic folding studies have recently emphasized the role that kinetic traps and misfolding plays in loss of p53 function. Trp fluorescence is again the method of choice for kinetic experiments due to superior sensitivity. It is important to initiate folding by a double-jump sequence. Here, the protein is first diluted into ice-cold 7 M urea for 30 s, which is sufficient to fully unfold DBD yet not long enough to allow its 17 Pro residues (all *trans*) to isomerize to their equilibrium *cis/trans* distribution. The sample is then rapidly diluted into buffer, and folding begins from a denatured state containing native Pro isomers. Longer equilibration times in denaturing conditions reduce both the rate and yield of DBD folding (Butler and Loh 2005). Thus double-jump eliminates the Pro isomerization artifact and more closely resembles the unfolding-folding cycle in the cell. This artifact will no doubt be more pronounced in full-length p53, as it contains 43 Pro residues.

STRUCTURE AND FUNCTION OF APODBD

In order to understand the role that Zn^{2+} plays in the structure and function of p53, several groups have characterized apoDBD. Nuclear magnetic resonance (NMR) and fluorescence experiments reveal that apoDBD is folded and native-like (Butler and Loh 2003). Many cross peaks in 2D NMR spectra do not appear to shift upon Zn^{2+} removal. These resonances map to the L1 loop and the H1 helix contained within, and to regions of the β-sheet distant from the Zn^{2+}-binding site. Widespread chemical shift differences are observed in the H2 helix, L2 and L3 loops, and segments of the β-sheet scaffold proximal to the metal-binding site. These findings suggest that Zn^{2+} removal perturbs the conformation of the zinc-binding loops but does not affect more distant regions of the molecule. Compared to DBD, apoDBD exhibits 11-fold lower affinity for the consensus *gadd45* oligonucleotide but identical affinity for a nonconsensus sequence (Butler and Loh 2003). Apparent perturbation of the minor groove-binding L3 loop, coupled with retention of the major groove-recognition elements H2 and L1, were proposed to explain the loss of DNA-binding specificity.

A more recent molecular dynamics study of DBD and apoDBD yields additional structural insights (Duan and Nilsson 2006). L2 displays the greatest structural change between DBD and apoDBD. In addition, L3 is significantly farther away from the DNA in the apoDBD simulation, and the critical contact residue Arg248 is no longer able to form specific interactions with the minor groove. The simulations uncover other potentially relevant details. Zn^{2+} appears to interact electrostatically with the phosphate backbone, suggesting a direct role in DNA binding. The H1 helix is observed to transiently unfold and displace H179 from the Zn^{2+}-binding site. H1 unfolding may be the pathway for spontaneous Zn^{2+} release.

The x-ray crystal structure of oxidized mouse apoDBD was recently solved to 1.5 Å resolution (Kwon et al. 2008). Mouse and human DBD contain nine and ten Cys residues, respectively, and all are normally reduced. Crystals were grown in oxidizing conditions, and a single disulfide bond formed between the Zn^{2+}-coordinating residues Cys173 and Cys239 (equivalent to Cys176 and Cys242 in human DBD). This event may reflect oxidative changes that are known to occur (Hainaut and Milner 1993; Rainwater et al. 1995), although other cysteines may be oxidized as well. Oxidation of p53 abrogates its ability to recognize consensus DNA sequences (Parks, Bolinger, and Mann 1997) and thus functionally mimics Zn^{2+} loss in fully reduced DBD. The structure of mouse apoDBD is largely superimposable with that of mouse DBD (Figure 10.1b). C_α atoms overlap with a root mean square deviation of 0.57 Å when the L3 loop is excluded from comparison. By contrast, L3 adopts an alternate conformation in which Arg245 (equivalent to Arg248 in human DBD) is displaced. A recent survey of protein-DNA complexes in the protein data bank revealed that in many of these structures, Arg interacts with unusually narrow and negatively charged tracts of the minor groove (Kitayner, Rozenberg et al. 2010). This interaction appears to be critical for maintaining DNA binding specificity. It is noteworthy that Arg248 is out of position for minor groove binding in all existing apoDBD structures. The extent to which the structures of L2 and L3 are affected by the nascent cross-link between them remains uncertain. Nonetheless, x-ray, computational, and NMR results agree that Zn^{2+} plays a critical role in orienting L3 for minor groove recognition.

FOLDING MECHANISMS OF DBD AND apoDBD

How do tumorigenic mutations cause p53 to lose function? Work from the Fersht group reveals that mutations exert their effects by at least two mechanisms: alteration of key DNA contact residues and thermodynamic destabilization (Bullock, Henckel, and Fersht 2000; Bullock et al. 1997). A growing body of evidence, however, argues that the details of the folding and unfolding reactions play a significant role in the inactivation process. Elucidation of folding and unfolding pathways has provided insight into how Zn^{2+} binding and loss affect p53 function in ways not previously recognized.

The kinetic folding mechanism of DBD is unusually complex. Fluorescence-monitored folding traces are minimally fit by four exponentials plus a sub-ms burst phase (Butler and Loh 2005). Observable time constants range from ~10 s to several hours for the slowest phase. These data can be modeled by a sequential pathway involving multiple intermediates, parallel pathways in which no intermediates are populated, or a combination of the two. Interrupted folding experiments (Schmid 1983), which specifically monitor formation of the native state, rule out the purely sequential model by demonstrating that native molecules form in several phases. Available evidence indicates that DBD folding is minimally fit by a parallel, fast- and slow-track mechanism (Figure 10.2).

The most noteworthy aspect of the folding mechanism is the presence of multiple kinetic traps. For example, within the first milliseconds of folding, molecules partition into fast and slow tracks. The physical basis of this choice remains unknown, but once a molecule enters a track it is committed to remain in that pathway. Other kinetic traps are present within both fast- and slow-folding channels (intermediates

Zinc and p53 Misfolding

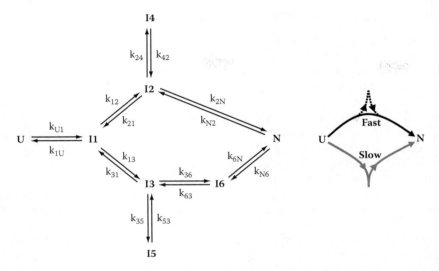

FIGURE 10.2 Proposed folding mechanism of DBD (left), illustrating the flux of molecules through each folding track (right). Fast and slow channels are the upper and lower pathways at left, respectively, and are colored black and gray on the right. Elementary rate constants that reproduce observed folding rates are listed in Butler and Loh (2007). The folding mechanism of apoDBD is essentially identical.

I4 and I5, respectively) and these are believed to be off-pathway aggregation-prone conformations (Figure 10.2). About 25% of molecules choose the fast track, avoid off-pathway traps, and fold to the native state within seconds. The remainder takes the slow pathway, falls into misfolding traps, or both. The net result is that a fraction of molecules become ensnared in a nonfunctional state with each unfolding-folding cycle. Strikingly, several hot-spot tumorigenic mutations examined do not affect the DBD folding mechanism; they simply accelerate unfolding. This result provides an explanation for why mildly destabilizing mutations inactivate p53. They may facilitate misfolding by causing DBD to cycle unusually rapidly between folded and unfolded states.

To what extent does zinc binding influence the folding landscape of p53? In some cases, metals or other cofactors accelerate folding, which suggests that they bind to the transition state ensemble (Pandit et al. 2007; Bushmarina et al. 2006) or guide the initial conformational search by binding to the unfolded state (Higgins, Muralidhara, and Wittung-Stafshede 2005). Cofactors can also hinder folding by making non-native contacts with unfolded or partially folded states. Zinc removal has little effect on DBD-folding kinetics (Butler and Loh 2007). A somewhat greater percentage of apoDBD molecules fold by the fast track but the essential mechanism remains unchanged. The observed kinetic traps are therefore not caused by metal binding or misligation; kinetic complexity appears to be an intrinsic feature of the DBD-folding landscape. ApoDBD binds Zn^{2+} late in the folding reaction, beyond the transition state, and likely after it has attained its native conformation.

ZINC BINDING, ZINC LOSS, AND MISFOLDING

The mechanisms by which Zn^{2+} binding and release affect p53 function extend beyond direct modulation of the DNA-binding interface. Zinc dissociation as well as improper zinc binding can lead to severe misfolding. The two misfolding mechanisms are distinct but both result in aggregated, nonfunctional p53.

Zinc loss and destabilizing mutations are similar in that they both destabilize DBD and exert this effect by accelerating the unfolding rate (Butler and Loh 2007). Zinc removal decreases stability of WT DBD from 10 kcal mol^{-1} to 6 kcal mol^{-1} and increases the unfolding rate (k_{unf}) by as much as 100-fold. At 10°C, $k_{unf} \sim 10^{-4}$ s^{-1}; apoDBD thus cycles between folded and unfolded states very slowly. Misfolding is consequently not problematic at low temperatures. However, k_{unf} is strongly temperature dependent and it increases by nearly 10^4-fold at 37°C (Butler and Loh 2006). ApoDBD cycles between states much more frequently at physiological temperature, and the protein aggregates within minutes at low μM concentration.

In addition to the off-pathway aggregation mechanism mentioned previously, native apoDBD appears to promote aggregation of zinc-bound DBD by a nucleation-growth process. When apoDBD and DBD are mixed in a 1:3 ratio (mimicking the condition where one of the monomers in a p53 tetramer has lost its Zn^{2+}), the turbidity of the solution increases much faster than can be explained by the sum of independent apoDBD and DBD aggregation rates (Butler and Loh 2003). Neither apoDBD nor DBD precipitate individually at those concentrations, but when they are mixed, precipitation occurs within minutes. Subdenaturing concentrations of urea eliminate this seeding effect, suggesting that native apoDBD is the species that is responsible for converting soluble DBD to the aggregated form. These results imply that if one subunit of the p53 tetramer loses metal, it can rapidly lead to aggregation of functional DBDs in the same tetramer as well as in other tetramers. This cascade may help explain the dominant-negative behavior exhibited by many tumorigenic mutations.

What factors influence zinc binding and loss? Does metal-free p53 exist in the cell? The answers are tied to the thermodynamic stability of p53. Native DBD binds Zn^{2+} tightly at low temperatures. DBD loses one-half equivalent of Zn^{2+} only after incubating for 72 h in the presence of excess EDTA at 10°C (Butler and Loh 2007). The off-rate is similar to k_{unf}, which suggests that Zn^{2+} release may be limited by global unfolding. Consistent with that hypothesis, Zn^{2+} dissociates from DBD within minutes at 37°C, which is again similar to the DBD-unfolding rate at this temperature. A significant observation is that of several tumorigenic mutations tested, all were found to destabilize DBD but not apoDBD (Butler and Loh 2003). Thus these mutations simultaneously promote Zn^{2+} release as well as preferentially stabilize apoDBD relative to DBD. The logical conclusion is that a significant fraction of p53 may exist in the cell as the metal-free form, especially when tumorigenic mutations are present.

RESCUE OF MISFOLDING BY METALLOCHAPERONES

Free Zn^{2+} causes native DBD to precipitate when present in as low as a twofold excess (Figure 10.3). Zinc-induced misfolding is not limited to native DBD. As a dramatic example, addition of $ZnCl_2$ completely arrests folding of apoDBD (Figure 10.4). No

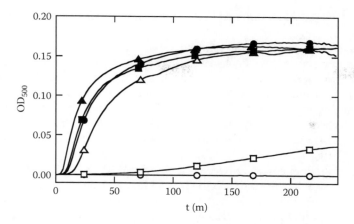

FIGURE 10.3 Zinc-induced aggregation of native DBD and rescue by NTA. Turbidity is monitored by light scatter at 500 nm (10°C). DBD concentration is 2 µM. ZnCl$_2$ is added at time zero at the following concentrations: 4 µM (open squares), 10 µM (open triangles), 14 µM (closed circles), 18 µM (closed squares), and 40 µM (closed triangles). DBD does not aggregate in the presence of 74 µM ZnCl$_2$ and 186 µM NTA (open circles). (Reprinted from Butler, J. S., and S. N. Loh, *Biochemistry*, 46, 2630–39, 2007. With permission.)

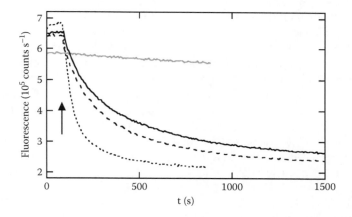

FIGURE 10.4 Zinc-induced arrest of DBD folding and rescue by metal chelators. DBD (0.75 µM) is refolded in the presence of 16 µM ZnCl$_2$ (10°C). After 80 s of stalled folding the following is added: NTA (solid line), EGTA (long dashed line), EDTA (short dashed line), or nothing (gray line). Concentration of all chelators is 0.5 mM. (Reprinted from Butler, J. S., and S. N. Loh, *Biochemistry*, 46, 2630–39, 2007. With permission.)

precipitation is observed, however, and folding resumes immediately after addition of EDTA or EGTA. A straightforward explanation for both of these results is that Zn^{2+} coordinates to nonphysiological Cys and or His residues. In the case of refolding DBD, Zn^{2+} traps the protein in partially folded, nonnative conformations. For folded DBD, misligation appears to cross-link multiple molecules in an essentially irreversible fashion: Subsequent addition of EDTA does not dissolve the aggregates.

Figure 10.4 suggests that the rate and yield of DBD refolding can be enhanced by natural or chemical metallochaperones. The concept is to employ a chelator that binds Zn^{2+} with greater affinity than the nonnative sites, but with lower affinity than the native site. The chelator can protect against misfolding by sequestering Zn^{2+} during the critical early stages of folding, where the lack of well-defined structure allows for a multitude of nonnative coordination possibilities. The chelator can then transfer zinc to apoDBD once the high-affinity physiological pocket is formed. EDTA binds Zn^{2+} too tightly ($K_d \sim 0.1$ pM), and releases Zn^{2+} too slowly ($k_{off} \sim 10^{-5}$), to facilitate zinc transfer on a biologically relevant timescale. NTA ($K_d \sim 10$ nM) rescues p53 folding to an extent comparable to EDTA and EGTA (Figure 10.4). Unlike EDTA, however, NTA readily donates Zn^{2+} to native apoDBD and functional DBD forms spontaneously within seconds (Butler and Loh 2007). Importantly, once DBD is regenerated, NTA has the requisite Zn^{2+} affinity to prevent metal-induced aggregation (Figure 10.3).

ZINC BIOAVAILABILITY, p53, AND CANCER

A significant percentage of the world population suffers some degree of zinc deficiency (Haase, Overbeck, and Rink 2008). Zinc uptake decreases with age; thus the elderly face simultaneous—and possibly linked—threats of zinc deficiency and cancer. The connection between intracellular Zn^{2+} concentration, p53 activity, and cancer has been explored by numerous *in vivo* zinc depletion and supplementation studies. Treating cultured cells with the cell-permeant metal chelator TPEN causes p53 to accumulate in a misfolded form that is unable to bind DNA (Méplan, Richard, and Hainaut 2000a). Removal of the chelator or addition of exogenous Zn^{2+} restores proper folding and DNA-binding activity (Méplan, Richard, and Hainaut 2000b).

Reduced cellular Zn^{2+} is associated with prostate and other cancers (Franklin and Costello 2007; Murakami and Hirano 2008). Depleting Zn^{2+} from the growth serum causes nuclear p53 protein levels to increase in a variety of cultured cell types (Clegg et al. 2005) including prostate (Yan et al. 2008), hepatoblastoma (Alshatwi et al. 2006; Reaves et al. 2000), lung fibroblast (E. Ho, Courtemanche, and Ames 2003), glioma (E. Ho and Ames 2002), aortic endothelial (Fanzo et al. 2002), and bronchial epithelial cells (Fanzo et al. 2001). As is the case with chelation-induced Zn^{2+} depletion, much of the amassed p53 appears to be nonfunctional: P53 DNA binding decreases and there is evidence that apoptotic and DNA repair pathways are not properly activated (Yan et al. 2008; Clegg et al. 2005). Similar findings were reported in animal studies of esophageal and forestomach cancers (Fong, Jiang, and Farber 2006; Fong et al. 2003). These results are consistent with the view that zinc loss causes p53 to misfold and aggregate in the cell.

Mutation-induced zinc loss has also been implicated in several forms of cancer. The most compelling evidence comes from sequence analysis of *TP53* genes from patient tumors. As mentioned, p53 missense mutations are detected in a large percentage of tumors regardless of origin. Significantly, mutations in the L2 and L3 loops are associated with inferior prognosis. This correlation has been noted for cancers of the prostate (Russo et al. 2002, 2005), breast (Van Slooten et al. 1999; Gentile et al. 1999), non-small-cell lung (Skaug et al. 2000), and esophagus (Kihara et al. 2000). The majority of these mutations have not been tested for metal binding and it is unknown whether zinc loss is the primary mechanism for impairment of p53 function. However, it has been shown that the C176S Zn^{2+}-binding mutant accumulates in nuclei of neuroblastoma cells and is able to bind DNA but not activate transcription (Wolff et al. 2001).

Consistent with *in vitro* aggregation results, unusually high intracellular Zn^{2+} also causes p53 to accumulate in the cell (Rudolf and Cervinka 2008; Fanzo et al. 2001, 2002). It is unclear whether this protein is folded correctly or incorrectly, and Zn^{2+} may affect p53 expression indirectly (e.g., via NF-κB). Thus the extent to which increased intracellular zinc concentration causes p53 to misfold remains unresolved.

Nonetheless, the balance of evidence suggests that the cell must tightly regulate Zn^{2+} bioavailability in order to maintain proper p53 function. The metallothionein (MT) family of metal-binding proteins is likely to figure prominently in this process. MT binds up to seven Zn^{2+} with K_d ranging from 10 nM to 1 pM (Krezel and Maret 2007). P53 binds MT directly (Ostrakhovitch et al. 2006). Cellular p53/MT complexes contain mostly metal-free MT, suggesting that MT may load Zn^{2+} onto apo-p53 in a manner akin to NTA. Several observations support this putative metallochaperone function. Moderate overexpression of MT in cultured cells enhances p53 transcriptional activity (Méplan, Richard, and Hainaut 2000b). In addition, p53 up-regulates expression of MT after exposure to metals (Ostrakhovitch et al. 2007; Rudolf and Cervinka 2008). Increased MT levels probably discourage metal-induced misfolding of p53 and other proteins as well protect the cell against oxidative stress (Bell and Vallee 2009). Ironically but not unexpectedly, extreme overexpression of recombinant MT (Méplan, Richard, and Hainaut 2000b) or up-regulation of endogenous MT (Puca et al. 2008, 2009) cause p53 to misfold and lose ability to activate transcription. It appears that a large surplus of MT results in excessive Zn^{2+} abstraction from p53 and misfolding by the mechanisms described earlier.

CONCLUSIONS AND CLINICAL PERSPECTIVE

Many tumorigenic p53 mutations enhance zinc loss, either directly (by perturbing structure surrounding the metal binding site) or indirectly (through loss of protein stability). Zinc-free p53 is doubly dysfunctional. It can no longer distinguish cognate DNA sequences from nonconsensus sequences and it exhibits a pronounced tendency to misfold and aggregate. Use of natural or synthetic metallochaperones to facilitate proper zinc binding and prevent misligation is therefore a viable clinical approach. Such a compound should bind Zn^{2+} tightly enough, or be present in sufficient abundance, so that the free Zn^{2+} concentration is below K_d of the misligation interaction (~10^{-9} M). It should also release Zn^{2+} fast enough to deliver it to the physiological binding site in a timely manner. In our view it is doubtful that a single small-molecule chelator can strike that balance *in vivo*, although the chemical chelator 1,10-phenanthroline (K_d ~ 10^{-9} M) has been shown to activate p53 in mouse cells (Sun et al. 1997). What is likely needed is a Zn^{2+} buffering agent that binds multiple metal ions with a range of affinities. In this way it can serve as both a sink and source of metal ions as needed. Proteins such as MT, synthetic compounds that contain multiple metal-binding sites, or a combination of chemical chelators may offer the best solution. A long-standing therapeutic goal has been to develop small molecules that bind to p53 and stabilize the native conformation (Boeckler et al. 2008; Joerger and Fersht 2007; Joerger, Ang, and Fersht 2006). Such molecules will help prevent zinc loss as well, and this strategy remains a priority for treating p53-related cancers.

REFERENCES

Alshatwi, A. A., C. T. Han, N. W. Schoene, and K. Y. Lei. 2006. Nuclear accumulations of p53 and Mdm2 are accompanied by reductions in c-Abl and p300 in zinc-depleted human hepatoblastoma cells. *Exp Biol Med (Maywood)* 231 (5):611–18.

Andreini, C., L. Banci, I. Bertini, and A. Rosato. 2006. Counting the zinc-proteins encoded in the human genome. *J Proteome Res* 5 (1):196–201.

Ang, H. C., A. C. Joerger, S. Mayer, and A. R. Fersht. 2006. Effects of common cancer mutations on stability and DNA binding of full-length p53 compared with isolated core domains. *J Biol Chem* 281:21934–41.

Anzellotti, A. I., and N. P. Farrell. 2008. Zinc metalloproteins as medicinal targets. *Chem Soc Rev* 37 (8):1629–51.

Bell, S. G., and B. L. Vallee. 2009. The metallothionein/thionein system: an oxidoreductive metabolic zinc link. *Chembiochem* 10 (1):55–62.

Bell, S., C. Klein, L. Müller, S. Hansen, and J. Buchner. 2002. P53 contains large unstructured regions in its native state. *J Mol Biol* 322:917–27.

Boeckler, F. M., A. C. Joerger, G. Jaggi, T. J. Rutherford, D. B. Veprintsev, and A. R. Fersht. 2008. Targeted rescue of a destabilized mutant of p53 by an in silico screened drug. *Proc Natl Acad Sci U S A* 105 (30):10360–65.

Bullock, A. N., J. Henckel, B. S. DeDecker, C. M. Johnson, P. V. Nikolova, M. R. Proctor, D. P. Lane, and A. R. Fersht. 1997. Thermodynamic stability of wild-type and mutant p53 core domain. *Proc Natl Acad Sci U S A* 94:14338–42.

Bullock, A. N., J. Henckel, and A. R. Fersht. 2000. Quantitative analysis of residual folding and DNA binding in mutant p53 core domain: definition of mutant states for rescue in cancer therapy. *Oncogene* 19:1245–56.

Bushmarina, N. A., C. E. Blanchet, G. Vernier, and V. Forge. 2006. Cofactor effects on the protein folding reaction: acceleration of α-lactalbumin refolding by metal ions. *Protein Sci* 15:659–71.

Butler, J. S., and S. N. Loh. 2003. Structure, function, and aggregation of the zinc-free form of the p53 DNA binding domain. *Biochemistry* 42:2396–2403.

———. 2005. Kinetic partitioning during folding of the p53 DNA binding domain. *J Mol Biol* 350:906–18.

———. 2006. Folding and misfolding mechanisms of the p53 DNA binding domain at physiological temperature. *Protein Sci* 15:2457–65.

———. 2007. Zn^{2+}-dependent misfolding of the p53 DNA binding domain. *Biochemistry* 46:2630–39.

Cho, Y., S. Gorina, P. D. Jeffrey, and N. P. Pavletich. 1994. Crystal structure of a p53 tumor suppressor-DNA complex: Understanding tumorigenic mutations. *Science* 265:346–55.

Clegg, M. S., L. A. Hanna, B. J. Niles, T. Y. Momma, and C. L. Keen. 2005. Zinc deficiency-induced cell death. *IUBMB Life* 57 (10):661–69.

Duan, J., and L. Nilsson. 2006. Effects of Zn^{2+} on DNA recognition and stability of the p53 DNA binding domain. *Biochemistry* 45:7483–92.

Fanzo, J. C., S. K. Reaves, L. Cui, L. Zhu, and K. Y. Lei. 2002. P53 protein and p21 mRNA levels and caspase-3 activity are altered by zinc status in aortic endothelial cells. *Am J Physiol Cell Physiol* 283 (2):C631–38.

Fanzo, J. C., S. K. Reaves, L. Cui, L. Zhu, J. Y. Wu, Y. R. Wang, and K. Y. Lei. 2001. Zinc status affects p53, gadd45, and c-fos expression and caspase-3 activity in human bronchial epithelial cells. *Am J Physiol Cell Physiol* 281 (3):C751–57.

Fong, L. Y., H. Ishii, V. T. Nguyen, A. Vecchione, J. L. Farber, C. M. Croce, and K. Huebner. 2003. P53 deficiency accelerates induction and progression of esophageal and forestomach tumors in zinc-deficient mice. *Cancer Res* 63 (1):186–95.

Fong, L. Y., Y. Jiang, and J. L. Farber. 2006. Zinc deficiency potentiates induction and progression of lingual and esophageal tumors in p53-deficient mice. *Carcinogenesis* 27 (7):1489–96.

Franklin, R. B., and L. C. Costello. 2007. Zinc as an anti-tumor agent in prostate cancer and in other cancers. *Archi Biochem Biophys* 463 (2):211–17.

Friedler, A., D. B. Veprintsev, L. O. Hansson, and A. R. Fersht. 2003. Kinetic instability of p53 core domain mutants. *J. Biol. Chem.* 278:24108–12.

Gentile, M., M. Bergman Jungestrom, K. E. Olsen, P. Soderkvist, and S. Wingren. 1999. p53 and survival in early onset breast cancer: analysis of gene mutations, loss of heterozygosity and protein accumulation. *Eur J Cancer* 35 (8):1202–7.

Haase, H., S. Overbeck, and L. Rink. 2008. Zinc supplementation for the treatment or prevention of disease: current status and future perspectives. *Exp Gerontol* 43 (5):394–408.

Hainaut, P., and J. Milner. 1993. Redox modulation of p53 conformation and sequence-specific DNA binding *in vitro*. *Cancer Res* 53:4469–73.

Higgins, C. L., B. K. Muralidhara, and P. Wittung-Stafshede. 2005. How do cofactors modulate protein folding? *Protein Peptide Lett* 12 (2):165–70.

Ho, E., and B. N. Ames. 2002. Low intracellular zinc induces oxidative DNA damage, disrupts p53, NFkappa B, and AP1 DNA binding, and affects DNA repair in a rat glioma cell line. *Proc Natl Acad Sci U S A* 99 (26):16770–75.

Ho, E., C. Courtemanche, and B. N. Ames. 2003. Zinc deficiency induces oxidative DNA damage and increases p53 expression in human lung fibroblasts. *J Nutr* 133 (8):2543–48.

Ho, W. C., C. Luo, K. Zhao, X. Chai, M. X. Fitzgerald, and R. Marmorstein. 2006. High-resolution structure of the p53 core domain: implications for binding small-molecule stabilizing compounds. *Acta Crystallogr D* 62 (Pt 12):1484–93.

Ishimaru, D., L. M. Lima, L. F. Maia, P. M. Lopez, A. P. Ano Bom, A. P. Valente, and J. L. Silva. 2004. Reversible aggregation plays a crucial role in the folding landscape of p53 core domain. *Biophys J* 87:2691–2700.

Joerger, A. C., H. C. Ang, and A. R. Fersht. 2006. Structural basis for understanding oncogenic p53 mutations and designing rescue drugs. *Proc Natl Acad Sci U S A* 103 (41):15056–61.

Joerger, A. C., and A. R. Fersht. 2007. Structure-function-rescue: the diverse nature of common p53 cancer mutants. *Oncogene* 26 (15):2226–42.

———. 2008. Structural biology of the tumor suppressor p53. *Annu Rev Biochem* 77:557–82.

Jones, S., X. Zhang, D. W. Parsons, J. C. Lin, R. J. Leary, P. Angenendt, P. Mankoo, H. Carter, H. Kamiyama, A. Jimeno, S. M. Hong, B. Fu, M. T. Lin, E. S. Calhoun, M. Kamiyama, K. Walter, T. Nikolskaya, Y. Nikolsky, J. Hartigan, D. R. Smith, M. Hidalgo, S. D. Leach, A. P. Klein, E. M. Jaffee, M. Goggins, A. Maitra, C. Iacobuzio-Donahue, J. R. Eshleman, S. E. Kern, R. H. Hruban, R. Karchin, N. Papadopoulos, G. Parmigiani, B. Vogelstein, V. E. Velculescu, and K. W. Kinzler. 2008. Core signaling pathways in human pancreatic cancers revealed by global genomic analyses. *Science* 321 (5897):1801–6.

Kihara, C., T. Seki, Y. Furukawa, H. Yamana, Y. Kimura, P. van Schaardenburgh, K. Hirata, and Y. Nakamura. 2000. Mutations in zinc-binding domains of p53 as a prognostic marker of esophageal-cancer patients. *Jpn J Cancer Res* 91 (2):190–98.

Kitayner, M., Rozenberg, H., Rohs, R., Suad, O., Rabinovitch, D., Honig, B., and Shakked, Z. 2010. Diversity in DNA recognition by p53 revealed by crystal structures with Hoogsteen base pairs, *Nat Struct Mol Biol* 17, 423–429.

Krezel, A., and W. Maret. 2007. Dual nanomolar and picomolar Zn(II) binding properties of metallothionein. *J Am Chem Soc* 129 (35):10911–21.

Kwon, E., D. Y. Kim, S. W. Suh, and K. K. Kim. 2008. Crystal structure of the mouse p53 core domain in zinc-free state. *Proteins* 70 (1):280–83.

Lokshin, M., Y. Li, C. Gaiddon, and C. Prives. 2007. P53 and p73 display common and distinct requirements for sequence specific binding to DNA. *Nucleic Acids Res* 35 (1):340–52.

Maret, W. 2008. Metallothionein redox biology in the cytoprotective and cytotoxic functions of zinc. *Exp Gerontol* 43 (5):363–69.

Méplan, C., M.-J. Richard, and P. Hainaut. 2000a. Redox signalling and transition metals in the control of the p53 pathway. *Biochem Pharmacol* 59:25–33.

———. 2000b. Metallogregulation of the tumor supressor protein p53: zinc mediates the renaturation of p53 after exposure to metal chelators *in vitro* and in intact cells. *Oncogene* 19:5227–36.

Murakami, M., and T. Hirano. 2008. Intracellular zinc homeostasis and zinc signaling. *Cancer Sci* 99 (8):1515–22.

Nigro, J. M., S. J. Baker, A. C. Presinger, J. M. Jessup, R. Hostetter, K. Cleary, S. H. Bigner, N. Davidson, S. Baylin, P. Devilee, and B. Vogelstein. 1989. Mutations in the p53 gene occur in diverse human tumour types. *Nature* 342:705–8.

Ostrakhovitch, E. A., P. E. Olsson, S. Jiang, and M. G. Cherian. 2006. Interaction of metallothionein with tumor suppressor p53 protein. *FEBS Lett* 580 (5):1235–38.

Ostrakhovitch, E. A., P. E. Olsson, J. von Hofsten, and M. G. Cherian. 2007. P53 mediated regulation of metallothionein transcription in breast cancer cells. *J Cell Biochem* 102 (6):1571–83.

Pandit, A. D., B. A. Krantz, R. S. Dothager, and T. R. Sosnick. 2007. Characterizing protein folding transition States using Psi-analysis. *Method Mol Biol* 350:83–104.

Parks, D., R. Bolinger, and K. Mann. 1997. Redox state regulates binding of p53 to sequence-specific DNA, but not to non-specific or mismatched DNA. *Nucleic Acids Res* 25:1289–95.

Parsons, D. W., S. Jones, X. Zhang, J. C. Lin, R. J. Leary, P. Angenendt, P. Mankoo, H. Carter, I. M. Siu, G. L. Gallia, A. Olivi, R. McLendon, B. A. Rasheed, S. Keir, T. Nikolskaya, Y. Nikolsky, D. A. Busam, H. Tekleab, L. A. Diaz Jr., J. Hartigan, D. R. Smith, R. L. Strausberg, S. K. Marie, S. M. Shinjo, H. Yan, G. J. Riggins, D. D. Bigner, R. Karchin, N. Papadopoulos, G. Parmigiani, B. Vogelstein, V. E. Velculescu, and K. W. Kinzler. 2008. An integrated genomic analysis of human glioblastoma multiforme. *Science* 321 (5897):1807–12.

Puca, R., L. Nardinocchi, G. Bossi, A. Sacchi, G. Rechavi, D. Givol, and G. D'Orazi. 2009. Restoring wtp53 activity in HIPK2 depleted MCF7 cells by modulating metallothionein and zinc. *Exp Cell Res* 315:67–75.

Puca, R., L. Nardinocchi, H. Gal, G. Rechavi, N. Amariglio, E. Domany, D. A. Notterman, M. Scarsella, C. Leonetti, A. Sacchi, G. Blandino, D. Givol, and G. D'Orazi. 2008. Reversible dysfunction of wild-type p53 following homeodomain-interacting protein kinase-2 knockdown. *Cancer Research* 68 (10):3707–14.

Rainwater, R., D. Parks, M.E. Anderson, and P. Tegtmeyer. 1995. Role of cysteine residues in regulation of p53 function. *Mol Cell Biol* 15:3892–3903.

Reaves, S. K., J. C. Fanzo, J. Y. Arima, J. Wu, Y. R. Wang, and K. Y. Lei. 2000. Expression of the p53 tumor suppressor gene is up-regulated by depletion of intracellular zinc in HepG2 cells. *J Nutr* 130:1688–94.
Rudolf, E., and M. Cervinka. 2008. External zinc stimulates proliferation of tumor Hep-2 cells by active modulation of key signaling pathways. *J Trace Elem Med Bio* 22 (2):149–61.
Russo, A., V. Bazan, B. Iacopetta, D. Kerr, T. Soussi, and N. Gebbia. 2005. The TP53 colorectal cancer international collaborative study on the prognostic and predictive significance of p53 mutation: influence of tumor site, type of mutation, and adjuvant treatment. *J Clin Oncol* 23 (30):7518–28.
Russo, A., M. Migliavacca, I. Zanna, M. R. Valerio, M. A. Latteri, N. Grassi, G. Pantuso, S. Salerno, G. Dardanoni, I. Albanese, M. La Farina, R. M. Tomasino, N. Gebbia, and V. Bazan. 2002. P53 mutations in L3-loop zinc-binding domain, DNA-ploidy, and S phase fraction are independent prognostic indicators in colorectal cancer: a prospective study with a five-year follow-up. *Cancer Epidem Biomar* 11 (11):1322–31.
Schmid, F. X. 1983. Mechanism of folding of ribonuclease A. Slow refolding is a sequential reaction via structural intermediates. *Biochemistry* 22:4690–96.
Skaug, V., D. Ryberg, E. H. Kure, M. O. Arab, L. Stangeland, A. O. Myking, and A. Haugen. 2000. P53 mutations in defined structural and functional domains are related to poor clinical outcome in non-small cell lung cancer patients. *Clin Cancer Res* 6 (3):1031–37.
Sun, Y., J. Bian, Y. Wang, and C. Jacobs. 1997. Activation of p53 transcriptional activity by 1,10-phenanthroline, a metal chelator and redox sensitive compound. *Oncogene* 14:385–93.
Tupler, R., G. Perini, and M. R. Green. 2001. Expressing the human genome. *Nature* 409:832–33.
van Slooten, H. J., M. J. van De Vijver, A. L. Borresen, J. E. Eyfjord, R. Valgardsdottir, S. Scherneck, J. M. Nesland, P. Devilee, C. J. Cornelisse, and J. H. van Dierendonck. 1999. Mutations in exons 5–8 of the p53 gene, independent of their type and location, are associated with increased apoptosis and mitosis in invasive breast carcinoma. *J Pathol* 189 (4):504–13.
Wolff, A., A. Technau, C. Ihling, K. Technau-Ihling, R. Erber, F. X. Bosch, and G. Brandner. 2001. Evidence that wild-type p53 in neuroblastoma cells is in a conformation refractory to integration into the transcriptional complex. *Oncogene* 20 (11):1307–17.
Yan, M., Y. Song, C. P. Wong, K. Hardin, and E. Ho. 2008. Zinc deficiency alters DNA damage response genes in normal human prostate epithelial cells. *J Nutr* 138 (4):667–73.

11 The Octarepeat Domain of the Prion Protein and Its Role in Metal Ion Coordination and Disease

Glenn L. Millhauser

CONTENTS

Introduction 209
The Octarepeat Domain 211
Zinc Uptake in the Octarepeat Domain 214
Octarepeat Expansion Disease 217
Future Perspective 221
References 221

INTRODUCTION

The octarepeat (OR) domain of the prion protein (PrP) is deceptively simple in sequence and structure, yet appears to carry out a remarkable copper regulatory role (see reviews in Millhauser 2004, 2007). The domain is composed of a repeating eight-amino acid sequence, with glycine representing 50% of its amino acid content, and is devoid of secondary or tertiary structure in the absence of copper. Despite its lack of fold or sequence complexity, it takes up multiple divalent copper ions (Cu^{2+}) with remarkable selectivity over other metal ions (Whittal et al. 2000), binds with profound negative cooperativity (Walter, Chattopadhyay, and Millhauser 2006), exhibits Cu^{2+} affinity well matched to micromolar copper levels in extracellular fluids (Kramer et al. 2001), and alters copper's intrinsic redox properties (Bonomo et al. 2000). Moreover, insert expansions increasing the OR domain length by one to nine repeats, but without alterations of the canonical repeat sequence, are linked to inherited prion disease (Croes et al. 2004; Kong et al. 2004).

The cellular form of the prion protein, PrP^C, is a normal constituent of the brain and is found at high levels at synaptic junctions (Vassallo and Herms 2003). The transmissible spongiform encephalopathies (TSEs), or prion diseases, are infectious neurodegenerative disorders that arise from misfolding of PrP^C to the scrapie conformer PrP^{Sc} (scrapie form). The TSEs affect a wide range of mammalian species and include mad cow disease (BSE), scrapie in goats and sheep, chronic wasting disease in deer and elk, and the human diseases kuru, Gerstmann-Sträussler-Scheinker syndrome, and Creutzfeldt-Jacob disease (Prusiner 1998, 2001).

The prion diseases involve accumulation of PrP^{Sc}, often in the form of amyloid, and are therefore grouped with Alzheimer's and Parkinson's disease. Most interest to date has focused on the cellular processes that lead to misfolding and the mechanisms by which PrP^{Sc} damages cellular structures. However, over the last decade a new branch of prion research has emerged with specific focus on the function of PrP^{C} in healthy cells. This new research avenue promises to not only reveal new molecular features underlying neuron function, but also will provide insight into potential loss of function upon PrP^{C} misfolding, which may be critical for understanding etiology of neurodegeneration.

Although the specific function of PrP^{C} is not known, a growing body of information links the protein to copper regulation in the central nervous system (CNS). PrP^{C} is an extracellular protein bound to cell surfaces through a glycophosphatidylinositol (GPI) linker. Exchangeable copper, typically in the Cu^{2+} oxidation state, has been identified in the extracellular regions of the CNS, including the cerebral spinal fluid (CSF), extracellular and synaptic fluids (Que, Domaille, and Chang 2008). Copper concentrations in these regions vary from 0.3 μM to perhaps 500 μM (Que, Domaille, and Chang 2008). As established by our lab and others, these concentrations exceed the dissociation constants for PrP^{C} copper binding, thus providing compelling evidence that the protein is copper bound *in vivo* (Walter, Chattopadhyay, and Millhauser 2006). PrP^{C} confers protection of neurons against oxidative assaults, as demonstrated in both cell culture (Rachidi et al. 2003) and animal experiments (Klamt et al. 2001). Copper binds to the PrP promoter, increases PrP^{C} expression, and stimulates PrP^{C} cycling through endocytosis.

Copper is vital for cellular function in the CNS. Copper concentrations in the brain are higher than in any another other tissue of the body (Que, Domaille, and Chang 2008). The brain uses more oxygen than any other organ, and many essential respiratory proteins function through copper's ability to cycle between the Cu^{+} and Cu^{2+} redox states. At the same time, uncomplexed copper facilitates the production of reactive oxygen species (ROS) and is therefore toxic to neurons. Interestingly, a recent systems biology study identified metal ion regulation as one of three dominant gene modules altered by prion disease (Hwang et al. 2009).

Most of the aforementioned issues have been recently covered elsewhere. Available are detailed discussions of potential PrP^{C} functions (Aguzzi, Baumann, and Bremer 2008), influences of copper on conversion to PrP^{Sc} (Bocharova et al. 2005; Cox, Pan, and Singh 2006), molecular features of the various copper sites (Millhauser 2007), binding energetics (Hodak et al. 2009; Walter, Chattopadhyay, and Millhauser 2006; Pushie and Rauk 2003), and so forth. My goal in this chapter is not to revisit these issues but, instead, to focus on the remarkable features of the octarepeat domain, the dynamic region in PrP responsible for taking up most of the Cu^{2+} equivalents. Unlike other well-defined metal coordination sites, the OR domain adapts its coordination mode depending on the precise ratio of copper to protein (Chattopadhyay et al. 2005). And even though the OR domain presents multiple histidine residues, it nevertheless excludes most other metal ions other than copper, and takes up zinc only under very specific conditions (Walter et al. 2007). Finally, expansions of the

OR domain are linked to inherited prion disease (Croes et al. 2004; Kong et al. 2004). We recently demonstrated an OR length threshold that correlates with early-onset disease and concomitant alterations in copper uptake (Stevens et al. 2009).

Next I begin with a general discussion of the OR domain, its sequence, copper coordination modes, and affinity. Then I will examine how the OR domain responds when challenged with both Cu^{2+} and Zn^{2+}, the dominant extracellular metal ions in the CNS. Finally, I consider recent investigations in OR expansion prion disease and future directions.

THE OCTAREPEAT DOMAIN

Figure 11.1a shows a schematic of the human prion protein with locations of secondary structure in the folded C-terminal domain, and identification of the disulfide bond, the GPI anchor, and N-linked glycans. The N-terminal portion, approximately up through residue 120, is largely unstructured. Segment 60–91, within this N-terminal portion, is the octarepeat domain, composed of tandem repeats of the fundamental eight-residue sequence PHGGGWGQ. Copper binding has been identified in both the N-terminal and C-terminal PrP domain (Van Doorslaer et al. 2001), but current consensus suggests that relevant Cu^{2+} sites are exclusively within the N-terminal half of PrP (Walter et al. 2009). These sites involve His residues in the OR domain, and His96 and His111. At the His96 and His111 sites, equatorial Cu^{2+}

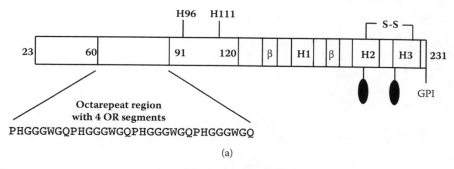

FIGURE 11.1 Sequence and structural features of the prion protein. (a) The cellular form of the prion protein, PrP^C, has a flexible N-terminal domain, up through residue 120, and a structured C-terminal domain composed of three α-helices (H1–H3) and two β-strands. The octarepeat domain spanning residues 60 to 91 is composed of four tandem PHGGGWGQ segments. Also shown are the disulfide bond, two N-linked glycans (ovals), and the C-terminal GPI anchor. (b) Three representative OR sequences showing conservation of the HGG copper-binding residues after each proline. Variations are minor and include Gly→Ser at two locations in mouse, and an extra repeat in cow.

coordination is from the histidine imidazole and exocyclic nitrogen, as well as backbone nitrogens from the two residues immediately preceding the histidine (Burns et al. 2003). There is also evidence for copper uptake at the PrP N-terminus, but this is not yet firmly established.

As seen in Figure 11.1b, the OR domain is highly conserved among mammalian species (Wopfner et al. 1999). Three representative sequences are shown—human, mouse, and cow. Comparison of these sequences reveals the typical variations observed among OR domains. In some rodent sequences, Gly to Ser variations at the third glycine, within the PHGGGWGQ repeat, are common. On the other hand, cows and several other species, possess an extra repeat with a nine-residue PHGGGGWGQ module. It is noteworthy that for all species, each OR His is rigorously followed by two Gly residues, regardless of any other sequence variations.

The structurally dynamic nature of the OR domain is demonstrated in Figure 11.2. As determined primarily by electron paramagnetic resonance (EPR) (Chattopadhyay et al. 2005), nuclear magnetic resonance (NMR) (Valensin et al. 2004), and crystallography (Burns et al. 2002), Cu^{2+} binding is well described by three fundamental coordination modes. At low copper concentration, corresponding to 1:1 copper to protein, the OR domain coordinates exclusively through His side chain imidazole groups. Referred to as "component 3," this coordination mode involves three or four histidines and is stable below pH = 6.5. Binding competition studies using a range of competitor species show that component 3 dissociation constant is approximately 0.1 nM (Walter, Chattopadhyay, and Millhauser 2006). Component 2 coordination arises at 2:1 copper to protein and involves two adjacent repeats for each Cu^{2+} ion.

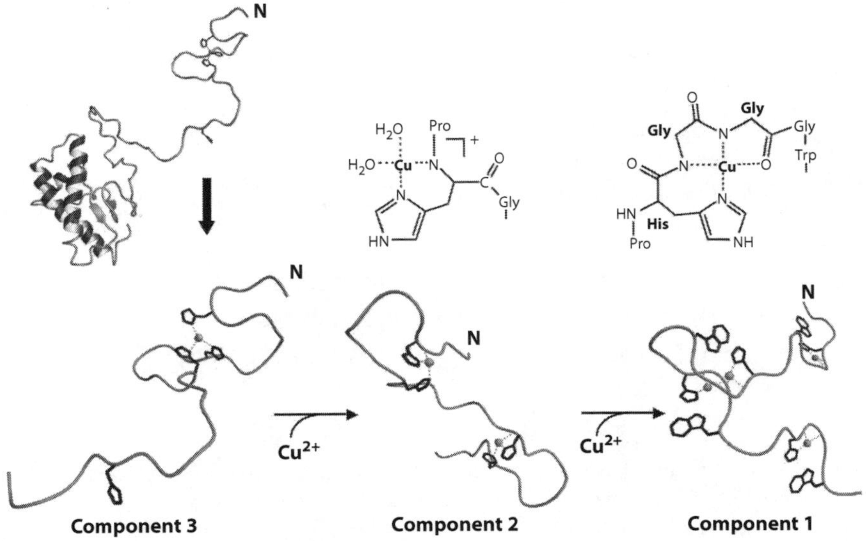

FIGURE 11.2 The octarepeat domain takes up copper (Cu^{2+}) with three different coordination modes. At 1.0 equivalent, copper interacts strictly with His side chain imidazoles. The OR domain saturates at 4.0 equivalents of Cu^{2+} and coordinates through the His imidazole and sequential backbone amide nitrogens.

Equatorial interactions are provided by the His residues, using both imidizole and exocyclic nitrogens, as well as water molecules. The complex is further stabilized by axial coordination from a His imidazole on an adjacent repeat. The dissociation constant for this complex is approximately 12 µM, and therefore exhibits a much weaker copper affinity than component 3 (Walter, Chattopadhyay, and Millhauser 2006).

At full copper saturation, the OR domain takes up four Cu^{2+} equivalents. Coordinating residues in this component 1 motif are within the HGGGW pentapeptide sequence (Figure 11.2). As determined by both EPR and X-ray crystallography, copper binding in the equatorial plane involves the His imidazole, deprotonated backbone nitrogens from the first and second glycines immediately following the His, and the carbonyl of the second His (Aronoff-Spencer et al. 2000; Burns et al. 2002). The crystal structure further reveals an axial water molecule that forms a hydrogen bond to the indole NH of the tryptophan. This is an unusual Cu^{2+} site not previously observed in copper-binding proteins. In most Cu^{2+} sites comprised of His and backbone amide nitrogens, coordination is on the N-terminal side of the anchoring histidine, as observed for component 2. This mode is likely favored by the six-member ring configuration formed by the Cu^{2+} binding between the two His nitrogens. Component 1 coordination takes place on the C-terminal side of the His, with the His-Gly-Cu^{2+} forming a seven-member ring. The low energy configuration of this coordination site is supported by theoretical studies, which suggest that the conserved prolines serve as natural break points directing Cu^{2+} away from the N-terminal side of His (Pushie and Rauk 2003). (The proline immide contains a tertiary nitrogen that cannot coordinate Cu^{2+}.) Binding affinity, determined by both fluorescence and competition studies, displays a dissociation constant of approximately 7.0 µM, similar to that found for component 2 coordination (Walter, Chattopadhyay, and Millhauser 2006).

Copper metalloproteins often possess very high copper affinities (low K_d values). For example, copper superoxide dismutase exhibits a K_d of approximately 10^{-14} M, depending on pH and specific copper site (Hirose et al. 1982). The substantially higher affinity found for component 3 coordination, relative to component 1, therefore suggests that this is a physiologically relevant copper-bound state. However, the component 1 dissociation constant approximately matches the concentration of extracellular copper and supports the concept that the PrP OR domain is tuned to take up exchangeable copper (Kramer et al. 2001). Consideration of the conserved features in the sequences in Figure 11.2 points to physiological relevance of both coordination modes (Millhauser 2007). Component 3 depends on the number of histidines, and requires four or more. The intervening residues between histidines are not important, as long as the polypeptide backbone is sufficiently flexible. As seen in the alignment (Figure 11.2), the histidine count is rigorously conserved, with all wild-type sequences possessing four or five His residues. Component 1 coordination, which takes up a copper equivalent in each repeat segment, requires a PHGGXW sequence, where X is Gly, Ser, or Gly-Gly. Again, this is rigorously conserved. As we have suggested previously, it appears that the ability of the OR domain to transition among coordination modes, depending on copper concentration, plays an important role in PrP function.

Copper affinity in the OR domain decreases with each added equivalent. As such, the domain exhibits strong negative cooperativity. This is in contrast to early suggestions of strong positive cooperativity, with a Hill coefficient of 3.4, similar to how hemoglobin takes up oxygen. Cooperativity often suggests molecular communication among binding sites, thereby influencing subsequent binding events. In the case of hemoglobin, for example, O_2 binding within each protein subunit leads to a conformational change that alters the interface with neighboring subunits (Voet and Voet 1990). With the OR domain, the specific mechanism underlying the observed negative cooperativity is not known. However, insight can be gained by considering chemical changes associated with copper uptake. At low Cu^{2+} occupancy, favoring component 3 coordination, four His residues are available, with no limiting conformational restrictions from the backbone. Higher Cu^{2+} occupancy, with a lower His to Cu^{2+} ratio, necessarily involves backbone nitrogens for equatorial coordination. Although this is energetically more favorable than a simple two-state equilibrium between free aquo copper and component 3, there is an energetic cost for deprotonating the backbone amide nitrogens (Kozlowski et al. 1999; Sigel and Martin 1982; Sundberg and Martin 1974; Pushie and Rauk 2003). There are a number of factors that contribute to this energetic cost. As expected, there is a strong pH dependence that disfavors both component 2 and component 1 coordination below pH 6.5. Also, electrostatics are important with localized positive charge decreasing the amide proton pK. Electrochemical experiments demonstrate that copper uptake in histidine containing peptides begins with coordination to the imidazole (Kozlowski et al. 1999). The localized positive charge from the Cu^{2+} center facilitates deprotonation of nearby backbone amides, which then become available for coordination (Sundberg and Martin 1974). EPR dipolar couplings reveal close copper-copper contacts, which suggests that component 1 coordination leads to compactness of the OR domain (Chattopadhyay et al. 2005). Finally, there is evidence of interaction between the OR domain and the folded PrP C-terminal domain (Walter et al. 2009). Thus while the global energetics clearly favor negative cooperativity, we are only beginning to understand the molecular mechanism and organization of the OR domain in its various copper-bound states.

ZINC UPTAKE IN THE OCTAREPEAT DOMAIN

Although most physiological studies on metal regulation link PrP^C to copper, zinc is also implicated. Zinc (Zn^{2+}), like Cu^{2+}, stimulates PrP^C endocytosis (Pauly and Harris 1998), and selectivity studies demonstrate that both metal ions bind to the OR domain (Whittal et al. 2000). In membrane-anchored OR constructs, zinc facilitates intermolecular interactions with greater efficacy than copper (Kenward, Bartolotti, and Burns 2007). In addition to binding Zn^{2+}, PrP^C is also found to alter Zn^{2+} localization within the cell, increase resistance to Zn^{2+} toxicity, and contribute to metallothionein expression (Rachidi et al. 2009). The zinc concentration in the brain and, specifically within the synapse, is maintained at a higher level than copper (Que, Domaille, and Chang 2008). For example, whereas synaptic Cu^{2+} concentrations range from 10 to 200 µM, Zn^{2+} concentrations are 60–300 µM. These values are approximations, but certainly motivate a broader examination of Zn^{2+} uptake and the interplay with Cu^{2+}.

The higher zinc concentrations in the extracellular space might argue that Zn^{2+}, not Cu^{2+}, is the natural PrP^C ligand *in vivo* (Watt and Hooper 2003). Resolution of this issue is certainly critical for assessing the role of PrP^C in the body. Because Cu^{2+} exhibits several distinct binding modes, one cannot simply compare dissociation constants to evaluate the relative concentrations of PrP^C-Cu^{2+} and PrP^C-Zn^{2+} species. We addressed this issue with direct competition experiments and our methods for decomposition of EPR spectra (Walter et al. 2007). Figure 11.3a shows how the OR domain (50 μM) with a single equivalent of Cu^{2+} responds when challenged with increasing concentrations of Zn^{2+}. The top curve marks the total Cu^{2+} bound and shows that there is no observable variation. Thus even when Zn^{2+} is at the maximum physiologic concentration (approximately 300 μM), and six times the Cu^{2+} concentration, it is not able to displace copper. However, evaluation of the specific copper-binding mode reveals remarkable trend, as shown in the lower curves of Figure 11.3a. With no added Zn^{2+}, component 3 coordination dominates, as expected. However, with increasing zinc, there is a progressive shift from component 3 to component 1 coordination. These data show that the OR domain reorganizes in order to accommodate both metal ions.

This trend is further illustrated in Figure 11.3b, where we examine coordination modes with and without 300 μM Zn^{2+}. Although we observe a maximum of component 3 coordination at 1.0 equivalent in both cases, the concentration of component 3 is greatly suppressed in the presence of zinc. Analysis of these binding curves finds that the PrPC-Zn^{2+} binding constant is approximately 10^{-4} M, which is much weaker than any of the binding modes evaluated for copper. We performed equivalent experiments on Syrian hamster PrP and found similar trends—no release

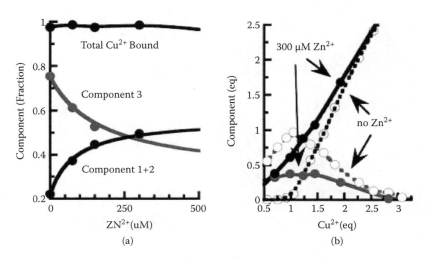

FIGURE 11.3 Zinc (Zn^{2+}) does not displace Cu^{2+}, but it does alter copper coordination modes. (a) With 1.0 equivalent of Cu^{2+}, added Zn^{2+} causes a shift from component 3 coordination to component 1. (b) Titration of Cu^{2+} with zero Zn^{2+} (dashed lines, open circles) and 300 μM Zn^{2+} (solid lines and circles). The higher zinc concentration suppresses component 3 coordination.

of Cu^{2+} regardless of the Zn^{2+} concentration, but a significant suppression of the multihistidine component 3 coordination mode (Walter et al. 2007). Direct Zn^{2+} coordination was further examined using a chemical modification scheme that labels noncoordinated imidazoles. These experiments showed that the OR domain saturates with a single equivalent of Zn^{2+}, with coordination strictly through histidine imidazoles, and a dissociation constant of approximately 200 µM (consistent with the equilibrium constant identified by EPR binding curves). Therefore Zn^{2+} coordinates to the OR domain in a fashion that is equivalent to component 3 identified for Cu^{2+}, although with lower affinity.

These findings demonstrate that it is not appropriate to consider PrP^C as either a zinc- or copper-binding protein, but instead as a protein that accommodates coordination of both metal ions. High zinc levels compete for Cu^{2+} component 3 coordination and alter the distribution of copper modes to favor component 1. This is summarized in Figure 11.4, where relevant Cu^{2+} and Zn^{2+} concentrations are indicated. (For clarity of the copper-binding modes, only component 1 and component 3 coordination are shown.) At micromolar Zn^{2+} levels, PrP^C interacts primarily with copper, following the scheme described in Figure 11.2. However, at Zn^{2+} levels above 100 µM, and low, subnanomolar Cu^{2+} levels, zinc is the dominant bound species. With progressively increasing Cu^{2+} concentrations, PrP^C takes up copper first in the non-OR binding sites (His96 and His111) and then finally in the OR domain where there emerges a coexistence between Zn^{2+}-bound and Cu^{2+}-bound species. The higher overall concentrations found for zinc relative to copper in CSF and synaptic fluid suggest to us that the underlined species in Figure 11.4 are the most relevant *in vivo*.

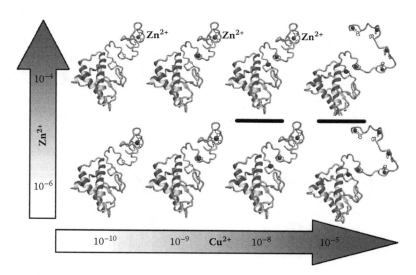

FIGURE 11.4 Models representing Cu^{2+} and Zn^{2+} binding in the N-terminal domain of PrP. Cu^{2+} is represented by spheres, except where labeled as Zn^{2+} (top row shows species with multi-His coordination). The bottom row shows the canonical structures identified for copper. Elevated zinc, top row, displaces component 3 copper. However, high copper and zinc lead to coexistence of the two underlined species that are likely relevant *in vivo*.

Our results further underscore the remarkable metal ion–binding capabilities of the OR domain and, in doing so, address several fundamental issues regarding putative PrP^C function. First, it is clear that, despite high levels of zinc *in vivo,* PrP^C is capable of taking up copper through coordination modes characterized by our lab and others. Consequently, emerging genetic and biochemical studies linking PrP^C to copper regulation or signaling are well founded and merit continued study. Next, component 3 coordination, in particular, raises issues regarding possible redox activity at the copper center. Several papers suggest that copper-bound PrP^C is either a superoxide dismutase or copper reductase (Davies and Brown 2008), and multiple imidazole coordination may facilitate enzymatic activity at the copper center (Miura et al. 2005). But zinc may largely displace component 3 copper coordination, thus quenching copper redox cycling. Finally, as described earlier, high levels of both copper and zinc stimulate PrP^C endocytosis. For copper, one can imagine that transition from component 3 to component 1 coordination might serve as a molecular trigger. But with zinc, there is only a single binding mode, with features similar to component 3. How then would zinc cause a conformational change in the OR domain? The species represented in Figure 11.4 provide insight. At low Zn^{2+}, but intermediate Cu^{2+}, the OR domain is dominated by its copper-bound form in component 3 coordination. However, when challenged with high Zn^{2+} levels, the domain responds by forming an equilibrium mixture of the zinc-bound species and the component 1 copper-bound species (indicated by the thick underlines). The result is that zinc drives a change in copper coordination, similar to the change observed by changing copper concentrations alone. PrP^C endocytosis requires concentrations for both Zn^{2+} and Cu^{2+} of 100 μM or greater, which lies just beyond the dissociation constants for both zinc and component 1 copper.

OCTAREPEAT EXPANSION DISEASE

Although the structure of PrP^{Sc} is not known, most experiments suggest that the misfolded segment leading to formation of the infectious prion is localized to the C-terminal domain. Treatment of PrP^{Sc} with proteinase K results in removal of the segment on the N-terminal side of residue 90, but without loss of infectivity (McKinley, Bolton, and Prusiner 1983). Moreover, all point mutations that segregate with inherited prion disease in humans are beyond residue 90 (Mead 2006). Consequently, the OR domain is outside of the region in PrP that misfolds to produce PrP^{Sc}. However, the N-terminal domain, including the octarepeats, seems to modulate the course of disease. Transgenic animals that produce a truncated PrP^C lacking residues 32–93 develop spongiform degeneration, but with longer incubation times and reduced tissue pathology (Flechsig et al. 2000).

The potential role of the OR domain is further underscored by a rare form of inherited disease resulting from domain expansion, where individuals carry from one to nine additional octarepeat insertions. Statistical analysis suggests that the number of inserts determines disease progression, with a remarkable threshold just beyond four inserts (Croes et al. 2004; Kong et al. 2004). With one to four extra octarepeats, the average onset age is above 60 years, whereas five to nine extra octarepeats results in early-onset prion disease with starting between 30 and 40 years of age, a difference of almost three decades. Earlier studies examined whether the expanded OR domain increased the rate of PrP^{Sc} formation, but the results have been inconsistent. For example, chimeric Sup35 yeast proteins, in which octarepeats replaced the endogenous repeat sequences, do give shorter fibril formation times for longer OR segments (although not in the presence of copper) (Dong et al. 2007). In addition,

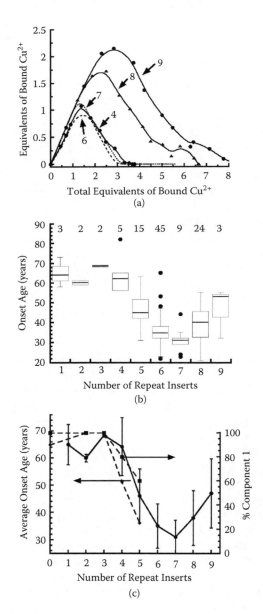

fusion constructs of glutathione-S transferase with PrP OR domains exhibit enhanced multimerization with longer repeat segments (Leliveld et al. 2006). While supportive of a kinetic role, neither of these studies identifies biophysical changes at the OR length threshold that correlates with early-onset disease. Moreover, recombinant mouse PrP with repeat inserts exhibits decreased amyloid formation (Leliveld, Stitz, and Korth 2008).

The previous sections demonstrate how sensitive the OR domain is to precise copper concentration. Component 3 coordination is thermodynamically stable, but readily transitions to component 1 coordination, at higher copper levels. Given the profound influence of OR domain length in expansion disease, we explored whether OR response to copper is altered by insertion number (Stevens et al. 2009). Figure 11.5a shows copper uptake in a series of OR peptide constructs of varying length. We used EPR spectral decomposition to track component 3 coordination as a function of added copper (in equivalents). The total number of repeats is labeled for each curve. With four to seven repeats (i.e., zero to three inserts), the behavior is much like wild type, reaching a maximum of component 3 at 1.0–1.5 equivalents. However, the eight-OR construct (four inserts) exhibits persistent component 3 coordination that reaches a maximum at approximately 2.0–2.5 equivalents. The maximum observed for the nine-OR construct is shifted to even higher copper concentration. Equivalent trends were observed with full-length recombinant protein, where we compared wild-type with mutant PrP containing five OR inserts.

These data show that expanded OR domains, with four or five inserts, resist transitioning from component 3 to component 1 coordination. For example, whereas 3.0 equivalents of copper would normally lead to almost exclusive component 1 coordination, expanded OR domains with more than four inserts remain predominantly as component 3. This is explained, in part, by considering the number of available His imidazoles and the coordination requirements for component 3. As demonstrated in our previous work, component 3 coordination requires approximately four side chain imidazoles from adjacent octarepeat segments (Chattopadhyay et al. 2005). If four His residues are required, then OR domains with up to seven repeats may take up only a single Cu^{2+} in the component 3 coordination mode, as observed. However, eight total repeats is just at the threshold that allows for two equivalents of component 3 coordination. Beyond stoichiometric arguments, we also find that expanded OR domains of eight or more repeats also exhibit an increase in component 3 affinity, with the dissociation constant lower by a factor of 10 relative to wild type.

FIGURE 11.5 (Opposite) Biophysical studies of OR expansions and the relationship to inherited prion disease. (a) Component 3 coordination from copper titrations of OR domains with four to nine total repeats. Note that beyond seven total repeats (three inserts), the OR domain requires higher Cu^{2+} concentrations to transition out of component 3 coordination. (b) Parallel box plot representing onset age for all documented cases of prion disease due to OR expansion. The number of individuals in each group is given at the top. Note that the onset age drops dramatically beyond four inserts. (c) Comparison of mean onset age (solid line with error bars) and OR length-dependent changes in copper uptake. The dashed lines show residual component 1 coordination at 3.0 (black, diamonds) and 4.0 (gray, squares) equivalents. These data show that loss of component 1 coordination, as a function of OR length, correlates strongly with early-onset prion disease.

To compare our findings to onset age in inherited OR expansion disease, we pooled all published clinical cases and performed a detailed statistical analysis, as represented in the box plot of Figure 11.5b (Stevens et al. 2009). These data are from approximately 110 individuals. Median onset age versus insert count show a clear clustering with up to four inserts showing late onset (>60 years) and more than five inserts showing early onset (<40 years). OR domains with five inserts are intermediate. (Individuals with nine inserts give a higher median, but with only three cases, this is not statistically significant.) The observed clustering into two groups distinguished by mean onset age is supported by an overall F test, which finds two distinct groups (p-value 2.8 e-14).

Figure 11.5c compares the average onset age and standard deviation, as a function of OR length, to Cu^{2+}-binding properties. As developed previously, the longest OR expansions favor component 3 coordination and resist component 1. Thus component 1 coordination serves as a convenient measure of altered Cu^{2+}-binding properties. Figure 11.5c shows the relative population of component 1 coordination for each OR construct, as derived from our copper titrations (at two fixed Cu^{2+} concentrations), superimposed on the average age of onset. For wild type and expansions involving up to three inserts, component 1 coordination is dominant for both 3.0 and 4.0 equivalents Cu^{2+}. However, at four and five inserts, respectively, the population of component 1 coordination drops precipitously. For example, at 3.0 equivalents Cu^{2+}, component 1 coordination is nearly 100% for three inserts and drops to approximately 25% for five inserts.

For OR domain expansion disease, these data demonstrate a striking relationship between onset age and altered Cu^{2+} binding. Expansion beyond four OR inserts inhibits transition to component 1 coordination and thus points to altered copper binding in lowering the onset age for prion disease. At this juncture it is not clear how loss of component 1 coordination accelerates prion disease, but there are several possible mechanisms. First, component 1 coordination may suppress redox cycling and loss of this coordination mode may therefore enhance redox stress. Another possibility is that component 1 coordination is required for efficient docking with companion proteins at the cell surface. Several recent studies suggest that PrP^C makes tight associations with other membrane-associated proteins, such as the low-density lipoprotein receptor related-protein 1 (LRP1) (Taylor and Hooper 2007). Even subtle changes in the linker segment between the OR and C-terminal globular domains lead to neurodegeneration, and can even halt development in the embryonic stage (Baumann et al. 2007; Li et al. 2007). Finally, component 1 may protect against conversion to amyloid forms of PrP^{Sc}. Recent studies do suggest that Cu^{2+} inhibits the *in vitro* formation of synthetic prions (Bocharova et al. 2005; Cox, Pan, and Singh 2006). Interestingly, many cases of OR expansion disease, with longer insert lengths, show expansive amyloid observed in GSS.

The role of the octarepeats in prion disease is enigmatic. Although the OR domain is not part of the protease-resistant scrapie particle, it nevertheless modulates disease progression. These studies on OR expansion disease point to contributing factors in prion neurodegeneration, and suggest either loss of PrP^C copper protein function or loss of copper-mediated protection against conversion to PrP^{Sc}.

FUTURE PERSPECTIVE

The fundamental question remains: What is the normal function of the prion protein? In my view, there is little doubt that the octarepeat domain, with its ability to take up both copper and zinc, will play a key role in PrP^C function, and perhaps loss of function in disease. Biophysical studies are now reaching a mature phase where we understand binding modes, energetics, and the effects of certain mutations. The next phase is to evaluate function at the cellular level with a focus on how PrP^C influences metal ion partitioning and mobility and, in turn, how metals modulate PrP's movement and its interactions with companion proteins. It is only within these contexts that we will be able to critically evaluate the role of this ubiquitous and remarkable protein.

REFERENCES

Aguzzi, A., F. Baumann, and J. Bremer. 2008. The prion's elusive reason for being. *Annu Rev Neurosci* 31:439–77.

Aronoff-Spencer, E., C. S. Burns, N. I. Avdievich, G. J. Gerfen, J. Peisach, W. E. Antholine, H. L. Ball, F. E. Cohen, S. B. Prusiner, and G. L. Millhauser. 2000. Identification of the Cu^{2+} binding sites in the N-terminal domain of the prion protein by EPR and CD spectroscopy. *Biochemistry* 39:13760–71.

Baumann, F., M. Tolnay, C. Brabeck, J. Pahnke, U. Kloz, H. H. Niemann, M. Heikenwalder, T. Rulicke, A. Burkle, and A. Aguzzi. 2007. Lethal recessive myelin toxicity of prion protein lacking its central domain. *EMBO J* 26 (2):538–47.

Bocharova, O. V., L. Breydo, V. V. Salnikov, and I. V. Baskakov. 2005. Copper(II) inhibits *in vitro* conversion of prion protein into amyloid fibrils. *Biochemistry* 44 (18):6776–87.

Bonomo, R. P., G. Impellizzeri, G. Pappalardo, E. Rizzarelli, and G. Tabbi. 2000. Copper(II) binding modes in the prion octarepeat PHGGGWGQ: a spectroscopic and voltammetric study. *Chem Eur J* 6:4195–4202.

Burns, C. S., E. Aronoff-Spencer, C. M. Dunham, P. Lario, N. I. Avdievich, W. E. Antholine, M. M. Olmstead, A. Vrielink, G. J. Gerfen, J. Peisach, W. G. Scott, and G. L. Millhauser. 2002. Molecular features of the copper binding sites in the octarepeat domain of the prion protein. *Biochemistry* 41:3991–4001.

Burns, C. S., E. Aronoff-Spencer, G. Legname, S. B. Prusiner, W. E. Antholine, G. J. Gerfen, J. Peisach, and G. L. Millhauser. 2003. Copper coordination in the full-length, recombinant prion protein. *Biochemistry* 42 (22):6794–6803.

Chattopadhyay, M., E. D. Walter, D. J. Newell, P. J. Jackson, E. Aronoff-Spencer, J. Peisach, G. J. Gerfen, B. Bennett, W. E. Antholine, and G. L. Millhauser. 2005. The octarepeat domain of the prion protein binds Cu(II) with three distinct coordination modes at pH 7.4. *J Am Chem Soc* 127 (36):12647–56.

Cox, D. L., J. Pan, and R. R. Singh. 2006. A mechanism for copper inhibition of infectious prion conversion. *Biophys J* 91 (2):L11–13.

Croes, E. A., J. Theuns, J. J. Houwing-Duistermaat, B. Dermaut, K. Sleegers, G. Roks, M. van den Broeck, B. van Harten, J. C. van Swieten, M. Cruts, C. van Broeckhoven, and C. M. van Duijn. 2004. Octapeptide repeat insertions in the prion protein gene and early onset dementia. *J Neurol Neurosur Ps* 75 (8):1166–70.

Davies, P., and D. R. Brown. 2008. The chemistry of copper binding to PrP: is there sufficient evidence to elucidate a role for copper in protein function? *Biochem J* 410 (2):237–44.

Dong, J., J. D. Bloom, V. Goncharov, M. Chattopadhyay, G. L. Millhauser, D. G. Lynn, T. Scheibel, and S. Lindquist. 2007. Probing the role of PrP repeats in conformational conversion and amyloid assembly of chimeric yeast prions. *J Biol Chem* 282 (47):34204–12.

Flechsig, E., D. Shmerling, I. Hegyi, A. J. Raeber, M. Fischer, A. Cozzio, C. von Mering, A. Aguzzi, and C. Weissmann. 2000. Prion protein devoid of the octapeptide repeat region restores susceptibility to scrapie in PrP knockout mice. *Neuron* 27 (2):399–408.

Hirose, J., T. Ohhira, H. Hirata, and Y. Kidani. 1982. The pH dependence of apparent binding constants between apo-superoxide dismutase and cupric ions. *Arch Biochem Biophys* 218 (1):179–86.

Hodak, M., R. Chisnell, W. Lu, and J. Bernholc. 2009. Functional implications of multistage copper binding to the prion protein. *Proc Natl Acad Sci U S A* 19 (20):8866–75.

Hwang, D., I. Y. Lee, H. Yoo, N. Gehlenborg, J. H. Cho, B. Petritis, D. Baxter, R. Pitstick, R. Young, D. Spicer, N. D. Price, J. G. Hohmann, S. J. Dearmond, G. A. Carlson, and L. E. Hood. 2009. A systems approach to prion disease. *Mol Syst Biol* 5:252.

Kenward, A. G., L. J. Bartolotti, and C. S. Burns. 2007. Copper and zinc promote interactions between membrane-anchored peptides of the metal binding domain of the prion protein. *Biochemistry* 46 (14):4261–71.

Klamt, F., F. Dal-Pizzol, M. L. Conte, D. A. Frota Jr., R. Walz, M. E. Andrades, E. Gomes, D. A Silva, R. R. Brentani, I. Izquierdo, and J. C. F. Moreira. 2001. Imbalance of antioxidant defense in mice lacking cellular prion protein. *Free Radical Bio Med* 30:1137–44.

Kong, Q., W.K. Surewicz, R. B. Petersen, W. Zou, S.G. Chen, P. Gambetti, P. Parchi, S. Capellari, L. Goldfarb, P. Montagna, E. Lugaresi, P. Piccardo, and B. Ghetti. 2004. Inherited prion diseases. In *Prion biology and diseases,* ed. S. B. Prusiner. Cold Spring Harbor, NY: Cold Spring Harbor Library Press.

Kozlowski, H., W. Bal, M. Dyba, and T. Kowalik-Jankowska. 1999. Specific structure-stability relations in metallopeptides. *Coordin Chem Rev* 184:319–46.

Kramer, M. L., H. D. Kratzin, B. Schmidt, A. Romer, O. Windl, S. Liemann, S. Hornemann, and H. Kretzschmar. 2001. Prion protein binds copper within the physiological concentration range. *J Biol Chem* 276:16711–19.

Leliveld, S. R., R. T. Dame, G. J. Wuite, L. Stitz, and C. Korth. 2006. The expanded octarepeat domain selectively binds prions and disrupts homomeric prion protein interactions. *J Biol Chem* 281 (6):3268–75.

Leliveld, S. R., L. Stitz, and C. Korth. 2008. Expansion of the octarepeat domain alters the misfolding pathway but not the folding pathway of the prion protein. *Biochemistry* 47 (23):6267–78.

Li, A., H. M. Christensen, L. R. Stewart, K. A. Roth, R. Chiesa, and D. A. Harris. 2007. Neonatal lethality in transgenic mice expressing prion protein with a deletion of residues 105–125. *EMBO J* 26 (2):548–58.

McKinley, M. P., D. C. Bolton, and S. B. Prusiner. 1983. A protease-resistant protein is a structural component of the scrapie prion. *Cell* 35 (1):57–62.

Mead, S. 2006. Prion disease genetics. *Eur J Hum Genet* 14 (3):273–81.

Millhauser, G. L. 2004. Copper binding in the prion protein. *Acc Chem Res* 37 (2):79–85.

———. 2007. Copper and the prion protein: methods, structures, function, and disease. *Annu Rev Phys Chem* 58:299–320.

Miura, T., S. Sasaki, A. Toyama, and H. Takeuchi. 2005. Copper reduction by the octapeptide repeat region of prion protein: pH dependence and implications in cellular copper uptake. *Biochemistry* 44 (24):8712–20.

Pauly, P. C, and D. A. Harris. 1998. Copper stimulates endocytosis of the prion protein. *J Biol Chem* 273:33107–19.

Prusiner, S. B. 1998. Prions. *Proc Natl Acad Sci U S A* 95:13363–83.

———. 2001. Shattuck lecture—neurodegenerative diseases and prions. *N Engl J Med* 344 (20):1516–26.

Pushie, M. J., and A. Rauk. 2003. Computational studies of Cu(II)[peptide] binding motifs: Cu[HGGG] and Cu[HG] as models for Cu(II) binding to the prion protein octarepeat region. *J Biol Inorg Chem* 8 (1–2):53–65.

Que, E. L., D. W. Domaille, and C. J. Chang. 2008. Metals in neurobiology: probing their chemistry and biology with molecular imaging. *Chem Rev* 108 (5):1517–49.

Rachidi, W., F. Chimienti, M. Aouffen, A. Senator, P. Guiraud, M. Seve, and A. Favier. 2009. Prion protein protects against zinc-mediated cytotoxicity by modifying intracellular exchangeable zinc and inducing metallothionein expression. *J Trace Elem Med Biol* 23 (3):214–23.

Rachidi, W., D. Vilette, P. Guiraud, M. Arlotto, J. Riondel, H. Laude, S. Lehmann, and A. Favier. 2003. Expression of prion protein increases cellular copper binding and antioxidant enzyme activities but not Copper delivery. *J Biol Chem* 278 (11):9064–72.

Sigel, H., and R. B. Martin. 1982. Coordinating properties of the amide bond—stability and structure of metal-ion complexes of peptides and related ligands. *Chem Rev* 82 (4):385–426.

Stevens, D. J., E. D. Walter, A. Rodriguez, D. Draper, P. Davies, D. R. Brown, and G. L. Millhauser. 2009. Early onset prion disease from octapeptide expansion correlates with copper binding properties. *PLoS Pathog* 5 (4):e1000390.

Sundberg, R. J., and R. B. Martin. 1974. Interactions of histidine and other imidazole derivatives with transition metal ions in chemical and biological systems. *Chem Rev* 74:471–517.

Taylor, D. R., and N. M. Hooper. 2007. The low-density lipoprotein receptor-related protein 1 (LRP1) mediates the endocytosis of the cellular prion protein. *Biochem J* 402 (1):17–23.

Valensin, D., M. Luczkowski, F. M. Mancini, A. Legowska, E. Gaggelli, G. Valensin, K. Rolka, and H. Kozlowski. 2004. The dimeric and tetrameric octarepeat fragments of prion protein behave differently to its monomeric unit. *Dalton Trans* (9):1284–93.

van Doorslaer, S., G. M. Cereghetti, R. Glockshuber, and A. Schweiger. 2001. Unraveling the Cu^{2+} binding sites in the C-terminal domain of the murine prion protein: a pulse EPR and ENDOR study. *J Phys Chem* 105:1631–39.

Vassallo, N., and J. Herms. 2003. Cellular prion protein function in copper homeostasis and redox signalling at the synapse. *J Neurochem* 86 (3):538–44.

Voet, D., and J. G. Voet. 1990. *Biochemistry*. New York: Wiley.

Walter, E. D., M. Chattopadhyay, and G. L. Millhauser. 2006. The affinity of copper binding to the prion protein octarepeat domain: evidence for negative cooperativity. *Biochemistry* 45 (43):13083–92.

Walter, E. D., D. J. Stevens, A. R. Spevacek, M. P. Visconte, A. Dei Rossi, and G. L. Millhauser. 2009. Copper binding extrinsic to the octarepeat region in the prion protein. *Curr Protein Pept Sci* 10 (5):529–35.

Walter, E. D., D. J. Stevens, M. P. Visconte, and G. L. Millhauser. 2007. The prion protein is a combined zinc and copper binding protein: Zn^{2+} alters the distribution of Cu^{2+} coordination modes. *J Am Chem Soc* 129 (50):15440–41.

Watt, N. T., and N. M. Hooper. 2003. The prion protein and neuronal zinc homeostasis. *Trends Biochem Sci* 28 (8):406–10.

Whittal, R. M., H. L. Ball, F. E. Cohen, A. L. Burlingame, S. B. Prusiner, and M. A. Baldwin. 2000. Copper binding to octarepeat peptides of the prion protein monitored by mass spectrometry. *Protein Sci* 9:332–43.

Wopfner, F., G. Weidenhofer, R. Schneider, A. von Brunn, S. Gilch, T. F. Schwarz, T. Werner, and H. M. Schatzl. 1999. Analysis of 27 mammalian and 9 avian Prps reveals high conservation of flexible regions of the prion protein. *J Mol Biol* 289:1163–78.

Section IV

Metalloprotein Design, Simulation, and Models

12 Metallopeptides as Tools to Understand Metalloprotein Folding and Stability

Brian R. Gibney

CONTENTS

Introduction 227
Protein Design 228
Metalloprotein Design 230
Equilibrium Thermodynamics 233
Heme Protein Folding 234
Zinc Protein Folding 238
The Cost of Protein Folding in Natural Zinc Fingers 241
Future Perspective 243
References 244

INTRODUCTION

Structural biology was born over 50 years ago with the revelation of the structures of both deoxyribonucleic acid (DNA) and myoglobin (Mb) in full atomic detail. The structures of DNA and Mb provided evidence for a stark contrast between the structures of nucleic acids and proteins. On the one hand, the simple and elegant structure of DNA clearly illustrated a key structural determinant of nucleic acid structure, the hydrogen-bonding base-pairing network that provides both stability and specificity to the DNA structure. On the other hand, the structure of Mb exposed the inherent complexity of protein structure that is based on a combination of hydrogen bonding, hydrophobic, electrostatic, and metal ligand interactions. The structure of Mb foreshadowed what has now crystallized into one of the largest unsolved problems in science: the protein-folding problem, or the question of how the fundamental chemical structure of a polypeptide chain determines its protein structure, and therefore its biological activity (Dill et al. 2008).

Within a decade of the structural elucidation of Mb, Anfinsen and coworkers provided a key insight into the thermodynamic and kinetic nature of protein folding (Sela, White, and Anfinsen 1957). Working with the enzyme ribonuclease A (RNAse A),

Anfinsen demonstrated that the unfolded state of RNAse A spontaneously and rapidly folds into the lowest free energy conformation, the biologically active native state. Furthermore, Anfinsen showed that the folding of RNAse A does not require assistance from any other biological machinery. Thus Anfinsen hypothesized that all of the kinetic and thermodynamic information required to achieve the native state structure was encoded within the primary structure of the polypeptide chain. This hypothesis has only been recently amended to include the existence of chaperone proteins that prevent and overcome folding errors without actively directing protein folding or altering the free energy of the final folded state.

Following Anfinsen's work, Levinthal provided a critical kinetic insight into the protein-folding problem (Levinthal 1968). He reasoned that given both the vastness of conformational space and the rapid rate at which proteins fold, there existed a kinetic paradox. A polypeptide chain simply does not have sufficient time to randomly search out the native state. Thus Levinthal argued that a kinetic pathway must exist from the unfolded state to the native state, and furthermore that the native state in some cases may not be the lowest energy state as argued by Anfinsen.

The fundamental principles of Anfinsen and Levinthal provide the thermodynamic and kinetic tenets that guide the myriad of modern approaches to solving the protein-folding problem. On the thermodynamic side, researchers are delineating the thermodynamic contributions of individual interatomic interactions to the free-energy landscape of protein folding as shown in Figure 12.1. Kauzmann began the early investigations into the thermodynamic role of amino acid side chain hydrophobicity in determining the thermodynamics of protein folding (Kauzmann 1959). The free-energy change that occurs when analogue compounds are partitioned between water and organic solvents was measured. These thermodynamic values were used to model the free energy provided by the hydrophobic burial of amino acid side chains upon protein folding. These early studies showed that the burial of hydrophobic side chains could contribute between 2 and 5 kcal/mol/side chain, making it a major factor in protein-folding thermodynamics. On the kinetic side, scientists are identifying the intermediates in the pathways of protein folding and the rates of their formation. Most impressively, Englander and coworkers have delineated the kinetic folding pathway of the heme protein ferrocytochrome c that shows several distinct intermediates (Bai et al. 1995). The observation of individual folding units, or foldons, in the folding pathway of ferrocytochrome c clearly illustrates that the formation of local structural units precedes assembly into the final folded structure. These thermodynamic and kinetic data are being used to refine molecular mechanics and molecular dynamics algorithms of protein structure, folding, and dynamics. The ultimate solution of the protein-folding problem not only will solve the question of how a native structure is thermodynamically derived from an amino acid sequence and kinetically obtained from the ensemble of unfolded states, but also promises to allow for the computational prediction and rational design of a native structure from an amino acid sequence.

PROTEIN DESIGN

One approach that has shown considerable success in elucidating the thermodynamic and kinetic aspects of the protein-folding problem is protein design. As an inherently

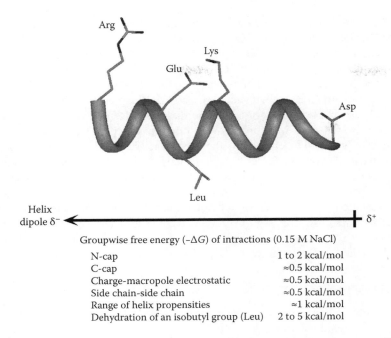

FIGURE 12.1 Protein–protein interaction free energies. (Adapted from Bryson et al.).

constructive experimental approach, the design of protein sequences capable of folding into stable protein structures from first principles provides for the rigorous testing of the ability to predict the native structure from an amino acid sequence. As a hierarchal approach, the early design of isolated secondary-structure elements has led to the design of tertiary structures and quaternary structures from first principles (Bryson et al. 1995). At each stage of this logical progression, the designs have become more sophisticated and more faithful representatives of natural proteins.

Protein design has played a critical role in developing thermodynamic and kinetic concepts for addressing the protein-folding problem (Baldwin 2007). The results of these studies have elucidated the thermodynamic contributions of individual interatomic interactions to the free energy of the folded state as well as provided simplified model systems with which to probe the kinetics of protein folding.

With regard to the thermodynamics of protein folding, protein design has provided critical data on the relative magnitude of various protein-protein interactions. As illustrated in Figure 12.1, the burial of hydrophobic side chains is a major source of protein-folding free energy, but a multitude of other interactions modulate the global stability and aid in providing specificity. One notable thermodynamic example involves the use of a designed α-helix to probe the intrinsic propensities of each amino acid to form α-helical structures (O'Neil and DeGrado 1990). The data revealed the thermodynamic favorability/penalty of each amino acid to form an α-helix. These data, and related data on β-sheet propensities (Kim and Berg 1993), have aided in the design of minimal peptide sequences with stable protein folds (Marqusee, Robbins, and Baldwin 1989; Cochran, Skelton, and Starovasnik

2001). These thermodynamic parameters have also been used to improve molecular mechanics/molecular dynamics energy scoring functions, which has led to the computational design of amino acid sequences that fold into stable three-dimensional structures (Dahiyat and Mayo 1996). The advent of more efficient algorithms a decade ago has propelled computational protein design to yield successful designs with novel protein folds and with catalytic activity (Jiang et al. 2008).

In addition to detailed thermodynamic data on protein folding, protein design has also provided simplified protein structures in which to study the kinetics of protein folding. One key example is the use of the GCN4-p1 coiled-coil protein to study helix formation in a dimeric protein scaffold (Sosnick et al. 1996). Mutations based on helical propensities were used to rationally alter the stability of GCN4-p1. Folding experiments demonstrated that the observed protein stability changes were due to alterations in the unfolding rate rather than the folding rate, and furthermore, that helix formation occurs after hydrophobic collapse. The success of these protein design efforts illustrates how the approach contributes to improving our understanding of both the thermodynamics and kinetics of the protein-folding problem.

METALLOPROTEIN DESIGN

The acceleration in the sophistication of protein design since the introduction of computational algorithms has not been fully realized in the area of cofactor-containing protein design where the requirements of cofactor binding have to be balanced within the context of the protein-folding problem. This observation is due in large part to the current limited understanding of cofactor-protein interactions that are critical to the stability, folding, and function of the resulting proteins. Just as protein design has been applied to the protein-folding problem with great success, metalloprotein design is providing key insights into the thermodynamic and kinetic aspects of the more complex metalloprotein-folding problem.

The introduction of cofactors into protein design presents additional complexities at the design phase. An amino acid sequence must be designed that not only folds into a suitable three-dimensional structure but also accurately positions the constellation of amino acids that affect cofactor binding. The extremes of metalloprotein design are represented by the zinc finger transcription factors and by carbonic anydrase; the former are unfolded and the latter is fully folded in the absence of the metal ion. Figure 12.2 shows the thermodynamics scheme for a typical zinc finger protein (Berg 1995). The apoprotein is unfolded, which implies that the apo-Folded state is higher in energy and there is a free-energy cost to protein folding, a positive value of $\Delta G_{apo}^{Folding}$. Furthermore, the measured thermodynamic contribution of metal ion binding, ΔG^{ML-Obs}, is lower than the actual free energy by the cost to protein folding; i.e., $\Delta G^{ML-Obs} = \Delta G^{ML} - \Delta G_{apo}^{Folding}$. Figure 12.3 presents the protein-folding scheme for carbonic anhydrase, which is fully folded in the absence of the metal ion (Christianson and Fierke 1996). In cases like this, the measured metal ion binding free energy, ΔG^{ML-Obs}, represents the actual free energy of metal ligand binding, ΔG^{ML}, because the protein scaffold is already folded. While these cases illustrate the protein-folding extremes, it is clear that cofactor binding introduces additional complexities into the protein-folding problem.

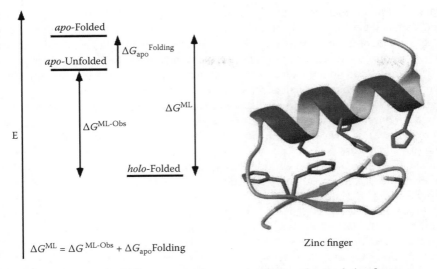

FIGURE 12.2 Protein-folding energy scheme and structure of natural zinc finger.

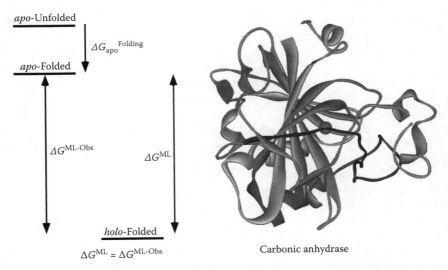

FIGURE 12.3 Protein-folding energy scheme and structure of carbonic anhydrase.

In the case of inorganic cofactors, the earliest of designs relied on protein scaffolds that were otherwise stable in the absence of the cofactor, i.e., situations closer to that observed in carbonic anhydrase (Choma et al. 1994; Robertson et al. 1994). In these systems, the incorporation of the metal ion leads to significant increases in global protein stability with minimal alteration of the global protein structure. These constructs relied on modifying pre-existing solutions to the protein-folding problem either by rational redesign or by combinatorial screening of sequences with randomly placed metal ion binding amino acid ligands (Hecht et al. 1997). The success of these designs has led to more sophisticated computational approaches for metalloprotein design. Many of these approaches rely on molecular mechanics/molecular dynamics design algorithms based on the protein–protein interaction thermodynamics given in Figure 12.1 and do not address the metalloprotein-folding problem directly. One approach is the computational search of a host protein structure for a region that can be mutated to accommodate a fixed-geometry cofactor-binding site (Hellinga 1998). A second approach is to start with a predesigned cofactor-binding site and to build a protein scaffold to accommodate it (Wade, Stayrook, and DeGrado 2006). These first two approaches have had notable success because they treat the metal ion-binding site as a static structure and rely on advances in computational protein design to solve the protein-folding problem around the metal ion–binding site. A third approach is the design of unfolded peptides, which requires the binding of the metal ion to template the folding, and in some cases the assembly, of the protein structure (Farrer and Pecoraro 2003). This approach allows the requirements of the metal and the encoded protein structure to balance each other in the final structure, and thus addresses the metalloprotein-folding problem most directly. Regardless of the approach, the limited understanding of cofactor-protein thermodynamics remains a major stumbling block to the advancement of rational metalloprotein design.

Our laboratory has been incorporating biological cofactors into designed peptide and protein scaffolds in an effort to generate molecular maquettes (Robertson et al. 1994), synthetic protein structures that are simplified relative to their natural counterparts. Figure 12.4 shows a molecular model of the prototype heme protein maquette, $[H10A24]_2$, which binds two hemes via bis-His coordination in the hydrophobic core of a four-helix bundle motif. Compared with natural protein scaffolds, the maquette scaffolds are designed to limit the complexity of the metalloprotein in an effort to isolate a single structure-function relationship for study. This provides the maquettes with the advantage of being able to investigate a key biological structure-function relationship with limited interference from the multitude of others found in natural protein scaffolds. Compared with inorganic model complexes where simplicity and the detailed study of a single structure-function relationship are intrinsic to the approach, the maquettes are designed to coordinate the metal ions using biological ligands under pseudophysiological conditions. This facilitates the direct comparison of maquette with natural metalloproteins and provides physiologically relevant synthetic analogues with which to test biological hypotheses.

Aware of the limitations imposed by our current understanding of metal-protein interaction free energies, our approach to the metalloprotein-folding problem is to evaluate the equilibrium thermodynamics of metal-cofactor affinity in these simplified designed peptides and proteins. By performing detailed measurements of the thermodynamic affinity of designed protein scaffolds for metal ion cofactors, we are elucidating the fundamental free energies of metal-protein interactions. These studies, along with those of other researchers, are beginning to reveal the fundamental thermodynamics of metal–protein interactions and to provide considerable insight into metal-induced protein folding. The results of these studies are critical to the improvement of energy functions for molecular mechanics/molecular dynamics computations of metalloproteins required to solve the metalloprotein-folding problem.

Metallopeptides as Tools to Understand Metalloprotein Folding and Stability 233

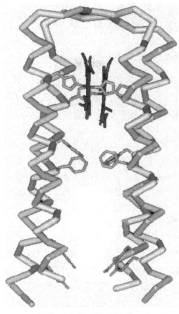

[heme b - H10A24]$_2$

FIGURE 12.4 Molecular model of the prototype heme protein maquette, [H10A24]$_2$.

EQUILIBRIUM THERMODYNAMICS

Our approach to studying the equilibrium thermodynamics of metal-peptide affinity is based in inorganic coordination chemistry (Magyar and Arnold-Godwin 2003). We began with the measurement of conditional stability (or dissociation) constants, K_a (or K_d) values, over as wide a pH range as possible. Each K_d value yields the free energy of metal-peptide complex formation at a single pH since $\Delta G = -RT \ln K_d$. The pH dependence of the K_d values illustrates the proton sensitivity of metal-peptide complex formation and the resulting modulation of the reaction free energies. In addition, we determine the acid dissociation constants for the peptide ligands, the ligand pK_a values, as well as the effective acid dissociation constants for the metal-peptide complexes, the pK_a^{eff} values, using potentiometric titrations. These data reveal the difference in thermodynamic stability between the metal bound to the protonated ligands, [Zn(HS•Cys)$_4$]$^{2+}$, and the metal bound to the deprotonated ligands, [Zn(S•Cys)$_4$]$^{2-}$. The combination of the pK_a, pK_a^{eff}, and K_d values allows for the mathematical solution of the multiple chemical equilibria required to determine the speciation in solution as well as the pH-independent formation constants, or K_f values (Petros et al. 2006). In cases where multiple metal ion oxidation states are accessible, the inclusion of equilibrium reduction potentials, E_m values, provides for a rigorous description of the influence of electrons on the thermodynamic stabilities of these metallopeptides as well (Reddi, Reedy et al. 2007). We will present the results of this approach applied to heme proteins and zinc proteins to illustrate how it addresses fundamental aspects of the metalloprotein-folding problem.

HEME PROTEIN FOLDING

Hemes are one of the most visible and versatile classes of metal ion cofactors utilized in biology. Nature has evolved heme-binding sites within a variety of protein scaffolds to carry out such diverse tasks as electron transfer, substrate oxidation, metal ion storage, ligand sensing, and ligand transport. As such, heme proteins are attractive targets for metalloprotein design (Reedy and Gibney 2004). Synthetic heme proteins have been constructed with a myriad of protein folds ranging from simple histidine-containing unfolded peptides to folded four-helix bundle scaffolds. The earliest heme protein designs were based on the incorporation of bis-His heme-binding sites into folded four-helix bundles (Choma et al. 1994; Robertson et al. 1994). These designs demonstrated that ferric heme dissociation constants in the range of 20 μM to 50 nM could be readily achieved. These K^d values indicate that ferric heme binding can contribute from −6.4 to −10.0 kcal/mol, $\Delta G^{\text{ML-Obs}}$, toward heme protein stability. These free energies appear much larger than those given in Figure 12.1 for typical protein–protein interactions (Bryson et al. 1995) and, at the same time, marginal when compared with the best natural ferric heme-binding protein, myoglobin, whose K^d value is 28 fM, $\Delta G^{\text{ML-Obs}} = -18.5$ kcal/mol (Hargrove and Olson 1996). Furthermore, the ability of a design devoid of the histidine ligands to associate heme strongly enough to be retained after gel permeation chromatography purification illustrates that peptide scaffolds can associate the heme through hydrophobic interactions without the necessity for endogenous axial ligands (Choma et al. 1994). The early success of heme protein design has led to a variety of successful designs that have served to refine our understanding of the thermodynamic contribution of ferric heme binding to protein folding and stability. These results will be discussed in the context of our own studies on ferric and ferrous heme affinity.

Our studies on the thermodynamic contribution of heme binding have concentrated on the differential contribution of the two oxidation states and how that is reflected in the resulting electrochemistry and global protein stabilities. The binding of heme in either oxidation state to the apoprotein contributes to the holoprotein global stability as expressed next in Equations 12.1 and 12.2:

$$\Delta G_{\text{unf}}^{\text{Fe(III) holoprotein}} = \Delta G_{\text{unf}}^{\text{apoprotein}} + \Delta G^{\text{Fe(III) heme binding}} \quad (12.1)$$

$$\Delta G_{\text{unf}}^{\text{Fe(II) holoprotein}} = \Delta G_{\text{unf}}^{\text{apoprotein}} + \Delta G^{\text{Fe(II) heme binding}} \quad (12.2)$$

$$\Delta\Delta G_{\text{unf}}^{\text{Fe(III)} - \text{Fe(II)}} = \Delta G^{\text{Fe(III) heme binding}} - \Delta G^{\text{Fe(II) heme binding}} \quad (12.3)$$

Equation 12.3 shows that free-energy difference between ferric and ferrous heme binding to the apoprotein, $\Delta G_{\text{unf}}^{\text{Fe(III) heme binding}} - \Delta G_{\text{unf}}^{\text{Fe(II) heme binding}}$, is equal to the free-energy difference between the oxidized and reduced heme protein global folding free energies, $\Delta\Delta G_{\text{unf}}^{\text{Fe(III)} - \text{Fe(II)}}$. Thus heme binding is critical to the thermodynamics and kinetics of heme protein folding.

Additionally, since $\Delta G = -nFE_m$ (Equation 12.4), the free-energy difference between ferric and ferrous heme-binding free energies, $\Delta G^{\text{Fe(III) heme binding}} - \Delta G^{\text{Fe(II) heme binding}}$, establishes the midpoint reduction potential of the heme in the protein scaffold, E_m. Using Equations 12.6 and 12.7, $\Delta G = -RT \ln K_{eq}$ and $\Delta E_m = E_m^{\text{bound}} - E_m^{\text{free}}$, the relationship between the ferric and ferrous heme affinities and the bound midpoint reduction potential, E_m^{bound}, can be derived as shown by Equations 12.8 and 12.9. Heme protein reduction potentials, E_m values, are critical to their biological function as they control both the thermodynamic driving force and kinetic rate of electron transfer as well as protein folding. Unfortunately, heme affinity studies are rarely performed using ferrous heme, so the thermodynamic contributions of ferric and ferrous heme binding to protein folding and heme electrochemistry are not well delineated.

$$\Delta G = -nFE_m \tag{12.4}$$

$$\Delta G^{\text{Fe(III) heme binding}} - \Delta G^{\text{Fe(II) heme binding}} = -nF\Delta E_m \tag{12.5}$$

$$\Delta G = -RT \ln K_{eq} \tag{12.6}$$

$$\Delta E_m = E_m^{\text{bound}} - E_m^{\text{free}} \tag{12.7}$$

$$RT \ln[K_a^{\text{Fe(III)}}/K_a^{\text{Fe(II)}}] = -nF\Delta E_m \tag{12.8}$$

$$RT \ln[K_d^{\text{Fe(II)}}/K_d^{\text{Fe(III)}}] = -nF\Delta E_m \tag{12.9}$$

In order to provide thermodynamic data on the stabilization of protein scaffolds by ferric and ferrous heme, we designed a stably folded four-helix bundle related to [H10A24]$_2$ containing two bis-His binding sites, [Δ7-H$_{10}$I$_{14}$I$_{21}$]$_2$ or [Δ7-His]$_2$, shown schematically in Figure 12.5 (Reedy, Kennedy, and Gibney 2003). As observed with many previous designed heme proteins, the spectroscopic properties of heme bound to [Δ7-His]$_2$ are reminiscent of its natural analogue, the bis-His ligated heme in cytochrome b_5 (Silchenko et al. 2000). Using equilibrium titrations, the affinities of this protein for both the first and second ferric and ferrous hemes were determined. The data evinced that the first ferric and ferrous heme affinities of [Δ7-His]$_2$ are K_a of 7 × 10^9 M^{-1} (K_{d1} of 140 pM) and K_{a1} of 2.3 × 10^7 M^{-1} (K_{d1} of 42 nM), respectively. The binding of the second heme is significantly weaker than the first in the same oxidation state due to both steric and electrostatic repulsion. In addition, the free-energy difference between ferric and ferrous heme affinities, 3.4 kcal/mol, indicates that the binding of heme to [Δ7-His]$_2$ lowers its reduction potential by 3.4 kcal/mol relative to unbound heme. Thus the observed E_m^{bound} value of –222 mV versus SHE suggests that E_m^{free} has a value of –76 mV versus SHE. Since the value of E_m^{free} cannot be measured directly, these data illustrate the utility of metalloprotein design in providing novel insight into metalloprotein-folding thermodynamics.

The thermodynamic cycle of [Δ7-His]$_2$ shown in Figure 12.5 represents a significant step in the evolution of our understanding of designed and synthetic heme proteins. This cycle has allowed us to study the effects of coordination sphere alterations

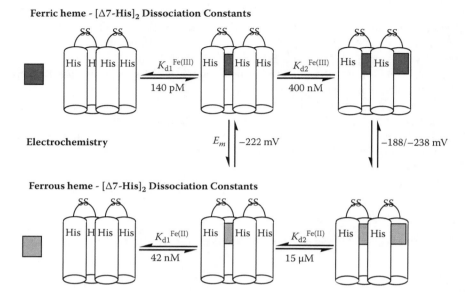

FIGURE 12.5 Thermodynamic cycle of heme binding and electrochemistry for [Δ7-His]$_2$.

on heme affinity and electrochemistry. The effects of primary coordination sphere changes have been analyzed in the [Δ7-His]$_2$ scaffold. The extension of the thermodynamic cycle to another scaffold, the prototype heme protein maquette [H10A24]$_2$, has provided for the investigation of secondary coordination sphere effects. While not yet complete, these studies are yielding new insight into the metalloprotein-folding problem.

In the case of ferric heme binding to [Δ7-His]$_2$, the observed −13.5 kcal/mol value of $\Delta G^{\text{ML-Obs}}$ can be compared against only a few literature values. The ferric heme affinity of [Δ7-His]$_2$ is several kcal/mol tighter than many designed heme proteins. Well-folded four-helix bundle scaffolds show that ferric heme-binding free energies between −6.4 and −10 kcal/mol and peptides that assemble and fold upon heme binding possess smaller free energies. [Δ7-His]$_2$ binds ferric heme only slightly weaker than the natural hemophore protein HasA, K^d of 10 pM ($\Delta G^{\text{ML-Obs}} = -14.9$ kcal/mol) (Izadi et al. 1997), but is significantly weaker than the best natural ferric heme-binding protein, myoglobin, whose K_d value is 28 fM, $\Delta G^{\text{ML-Obs}} = -18.5$ kcal/mol (Hargrove and Olson 1996). Thus the ferric heme affinity of [Δ7-His]$_2$ is slightly tighter than many designed proteins, but substantially weaker than the best natural heme protein.

In the case of ferrous heme binding to [Δ7-His]$_2$, there is a dearth of comparable literature values. The best-characterized ferrous heme affinity in a natural heme protein is that of myoglobin. However, ferrous myoglobin has been found to be kinetically trapped, which precludes measurement of its K_d value. The lack of comparable values serves only to further underscore the importance of using designed proteins to collect thermodynamic data in order to improve our understanding of the metalloprotein-folding problem.

In terms of the primary coordination sphere, the heme protein design literature demonstrates that the type of ligand, its placement, and the type of porphyrin are all critical to heme affinity. The vast majority of all designed heme proteins utilize bis-His coordination. The placement of histidine ligands is critical to heme affinity in designed four-helix bundles with heptad *a* positions, directed at the center of the hydrophobic core being optimal (Gibney and Dutton 1999). The removal of one His ligand is highly detrimental to ferric heme affinity, while still allowing for the pentacoordinate binding of Zn(II)(protoporphyrin IX). (Sharp, Diers et al. 1998) The use of 3-methyl-L-histidine as the axial ligand in the [Δ7-His]$_2$ scaffold results in minimal changes to the heme affinity and electrochemistry; however, the use of 1-methyl-L-histidine results in a loss of ferric heme binding and the generation of a pentacoordinate ferrous heme whose binding affinity is only 125-fold (2.8 kcal/mol) weaker than [Δ7-His]$_2$ (Zhuang et al. 2004). The use of 4-β-(pyridyl)-L-alanine], a pyridine-based amino acid ligand, in place of the histidine residues in [Δ7-His]2 resulted in a 60,000-fold, 6.5 kcal/mol, weakening of the ferric heme affinity, a slight improvement of the ferrous heme affinity, and a +287 mV shift in the E_m^{bound} value (Privett et al. 2002). As expected, the type of porphyrin is also critical to the thermodynamic affinity. Iron(diacetyldeuteroporphyrin IX) binds to a bis-His site 4.6 kcal/mol weaker than heme in the ferric state, but the ferrous affinities are comparable (Zhuang et al. 2006a). The incorporation of heme *a* or heme *o* enhances the binding of both oxidation states by >4.0 kcal/mol due to the hydrophobic interactions between their farnesyl groups and the protein hydrophobic core (Zhuang et al. 2006b).

In terms of secondary coordination sphere effects of heme protein stabilities, the literature is not as robust, but early indications are that it will provide considerable insight into metalloprotein folding as it develops. Using the thermodynamic cycle of [Δ7-His]$_2$, we investigated the influence of solution pH and protein scaffold size on heme affinity (Reddi, Reedy et al. 2007). By measuring the thermodynamic cycle of [Δ7-His]$_2$ at various pH values, we were able to distinguish the thermodynamic contribution of porphyrin association with the protein, ΔG^{assoc}, from metal ligand coordination in [Δ7-His]$_2$, ΔG^{ML}. The results, detailed in Figure 12.6, show that ferric heme binding is largely driven by the −11.2 kcal/mol free energy of metal ligand binding with porphyrin association contributing only −2.3 kcal/mol. The reverse scenario is observed in ferrous heme binding to [Δ7-His]$_2$ where porphyrin association provides −7.4 kcal/mol and metal ligand coordination −2.7 kcal/mol. Figure 12.7 shows the results of a similar study in [H10A24]$_2$ whose ferric and ferrous heme K_{d1} values are 32 pM (ΔG^{ML-Obs} = −14.3 kcal/mol) and 1.9 nM (ΔG^{ML-Obs} = −11.9 kcal/mol) (Reddi, Reedy et al. 2007). While these values are similar in magnitude to those in [Δ7-His]$_2$, the origins of their free energies are quite different. The binding of heme in both oxidation states to [H10A24]$_2$ is due to a combination of porphyrin association, metal ligand coordination, and electrostatic interactions. The coordination of the iron contributes only −3.6 kcal/mol to ferric heme affinity and only −1.9 kcal/mol of ferrous heme affinity. These values are significantly smaller than both the free energy of association of the heme, ΔG^{assoc}, and the free energy due to the electrostatic interactions, $\Delta G^{electro}$, as shown in Figure 12.7. These data illustrate that heme binding is comprised of multiple interactions that each contribute to the overall free energy of binding. A general solution to the hemeprotein-folding problem will

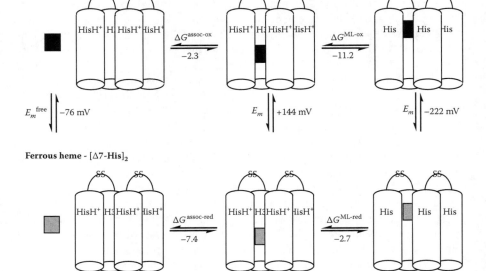

FIGURE 12.6 Thermodynamics of heme association and metal ligand binding in [Δ7-His]$_2$.

require an understanding of each of these and others that may be revealed in future studies. The studies to date illustrate that structural changes in heme proteins can alter the affinity of either or both oxidation states independently. While significant progress is evident, further work with natural and designed heme proteins will be required to delineate these structure-function relationships in heme proteins related to the metalloprotein-folding problem.

ZINC PROTEIN FOLDING

Zinc, the most abundant trace metal in cells, is critical to life as evidenced by its pervasive use in structural and catalytic proteins. Zinc is involved in biological processes ranging from gene transcription and cell proliferation to immune function and defense against free radicals. Not surprisingly, zinc proteins are attractive protein design targets. Indeed, one of the earliest metalloprotein designs involved incorporation of a tetrahedral Cys$_2$His$_2$ site inside a four-helix bundle protein, α$_4$ (Regan and Clarke 1990). The affinity of this design for Zn(II), K_d of 25 nM, demonstrated that Zn(II) biding could contribute −10.3 kcal/mol to protein stability, $\Delta G^{\text{ML-Obs}}$. This affinity, as well as those of other designed and natural proteins, is critical to solving the metalloprotein-folding problem, as the zinc finger transcription factors, the largest class of zinc proteins, require the binding of zinc to fold. The zinc finger proteins are composed of several modular zinc finger domains (there is one shown in Figure 12.2) that contain a single zinc-binding site and that recognize a specific base pair segment of DNA. Zinc fingers are typically unfolded in the absence of zinc and

Metallopeptides as Tools to Understand Metalloprotein Folding and Stability

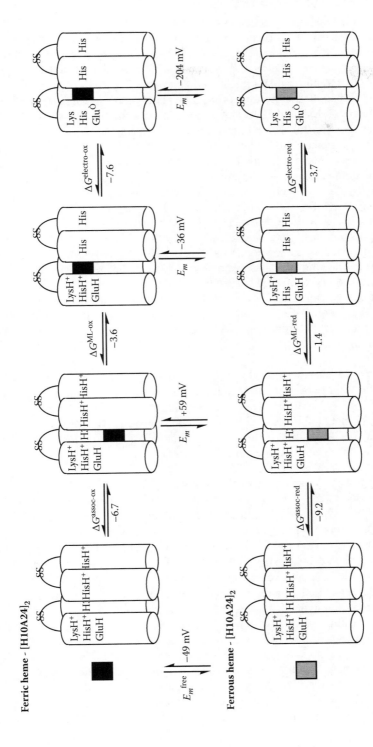

FIGURE 12.7 Thermodynamics of heme association, metal ligand binding, and electrostatic interactions in [H10A24]$_2$.

fold into their biologically active three-dimensional structure upon metal ion binding. This metal-induced protein-folding event has served to obscure the thermodynamic contribution of zinc binding to the free energy of protein folding as discussed earlier with respect to Figure 12.2. Recently, protein design has begun to provide novel insight into this critical protein-folding event.

Our studies on the contribution of metal ion binding to zinc finger protein folding sprang out of our design of a series of ferredoxin maquettes that bind a [4Fe-4S] cluster (Gibney et al. 1996). These ferredoxin maquettes are 16-amino acid peptides that contain four cysteine residues that bind a single [4Fe-4S]$^{2+/+}$ cluster whose spectroscopic and electrochemical properties mimic those of a natural [4Fe-4S]$^{2+/+}$ ferredoxin. Using this synthetic ferredoxin, we showed that the nonligand amino acids were just as critical to [4Fe-4S]$^{2+/+}$ cluster stability as the Cys ligands themselves. Furthermore, we evaluated the thermodynamic selectivity of the tetrahedral tetrathiolate site in the ferredoxin maquette **IGA** for Fe(II), Co(II), Zn(II), and a [4Fe-4S]$^{2+}$ cluster. By analyzing the coupled metal ligand and proton ligand equilibria, we elucidated the pH-independent dissociation constants of Fe(II)-**IGA**, Co(II)-**IGA**, and Zn(II)-**IGA**, which are 2.0 nanomolar, 2.0 picomolar, and 125 attomolar, respectively. These K_d values translate into formation constants of 5.0×10^8 M^{-1}, 4.2×10^{11} M^{-1}, and 8.0×10^{15} M^{-1}, or a potential stabilization of the metalloprotein structure by -11.9, -15.8, and -21.6 kcal/mol for Fe(II), Co(II), and Zn(II) respectively (Petros et al. 2006).

The affinity of the Cys4 site in **IGA** for Zn(II) at pH 9.0, K_d value of 125 aM, is slightly tighter than the 1.5 fM K_d value observed for the zinc sensor/regulator protein ZntR, whose affinity has been described as "'extreme" (Hitomi, Outen, and O'Halloran 2001). However, the Zn(II) affinity of **IGA** weakens below pH 8.3, whereas the affinity of ZntR is pH independent above pH 6.5. Thus at physiological pH values, the 4.0 pM K_d value of Zn(II)-**IGA** more closely resembles the values observed for metallothionein (0.1 pM), carbonic anhydrase II (4 pM), and the natural zinc finger BRCA1 (32 pM). Thus while the binding of Zn(II) to the tetrathiolate site in **IGA** can contribute -21.6 kcal/mol to protein folding and stability, it contributes only -15.5 kcal/mol at pH 7.0.

Our approach to parsing apart the contribution of metal ion binding to protein folding in zinc fingers has been to compare their observed metal ligand free-energy values, $\Delta G^{\text{ML-Obs}}$ values to those measured in a minimal, unstructured peptide scaffold with the same metal-binding coordination motif. As shown in Figure 12.8, our goal was to design a system in which the free-energy difference between the apo-unfolded states and the apo-folded state was negligible; i.e., the cost of protein folding is zero. In such a system, the observed metal ligand binding free energy, $\Delta G^{\text{ML-Obs}}$, would be equal to the actual metal ligand-binding free energy, ΔG^{ML}, just as discussed for the case of carbonic anhydrase in Figure 12.3.

In order to minimize the cost of protein folding in this design, the **GGG** peptide, the potential bias imposed by the iron sulfur sequence of **IGA** was removed by replacing several amino acids with glycines (Reddi and Gibney, 2007). The **GGG** peptide was elaborated with each of the natural zinc finger coordination motifs, Cys$_4$, Cys$_3$His and Cys$_2$His$_2$, to allow for comparison to all natural zinc fingers. Additionally, the affinity of each for zinc was measured over the pH range of 5.0–9.0 to facilitate comparison under a wide range of solution conditions. Figure 12.9 shows that the thermodynamic

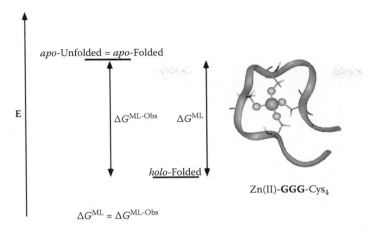

FIGURE 12.8 Protein-folding energy scheme and molecular model of Zn(II)-**GGG**-Cys$_4$.

results demonstrate that Zn(II)-**GGG**-Cys$_4$, Zn(II)-**GGG**-Cys$_3$His, and Zn(II)-**GGG**-Cys$_2$His$_2$ possess pH-independent formation constants of 5.6×10^{16}, 1.5×10^{15}, and 2.5×10^{13} M^{-1}, respectively (Reddi, Guzman et al. 2007). The data show that the substitution of a Cys residue by a His residue lowers the formation constant by a factor of about 50-fold per substitution; i.e., the limiting value at high pH is lower. However, the conditional dissociation constants of Zn(II)-**GGG**-Cys$_4$, Zn(II)-**GGG**-Cys$_3$His, and Zn(II)-**GGG**-Cys$_2$His$_2$ are virtually identical at physiological pH due to proton-based enthalpy-entropy compensation (H$^+$-EEC) observed via calorimetry. Thus while cysteinate is a better ligand than histidine imidazole, cysteine and histidine bind Zn(II) equivalently at physiological pH. Furthermore, H$^+$-EEC appears to be operative in natural zinc finger proteins because Cys to His alterations in their coordination spheres typically result in small K_d value changes. These results demonstrate the insight into the metalloprotein-folding problem that can be gained from the metalloprotein design approach.

THE COST OF PROTEIN FOLDING IN NATURAL ZINC FINGERS

Natural zinc fingers are expected to possess K_d values that are weaker than those of the designed peptide with the same coordination motif by an amount equal to the free-energy cost of protein folding, i.e., $\Delta G^{ML}(\textbf{GGG}) = \Delta G^{ML\text{-}Obs}(\text{zinc finger}) - \Delta G_{apo}^{folding}(\text{zinc finger})$. In addition, since $\Delta G_{apo}^{folding}$ approaches zero in the designed peptide, it should possess the tightest Zn(II) affinity possible for its coordination motif. Literature $\Delta G^{ML\text{-}Obs}$ values for natural zinc finger proteins were compared to the values measured for Zn(II)-**GGG** with the same coordination motif to estimate the cost of protein folding in natural zinc finger proteins. The results of the analysis showed the free-energy cost of protein folding in natural zinc fingers to be relatively small, <4.0 kcal/mol, compared to the thermodynamic contribution of zinc binding, >15.0 kcal/mol at physiological pH values (Reddi, Guzman et al. 2007). This analysis of the free-energy cost of protein folding shows that the thermodynamic

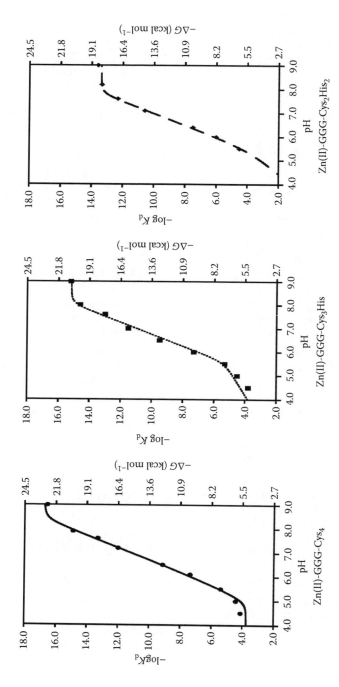

FIGURE 12.9 pH dependencies of the K_d values for Zn(II)-**GGG**-Cys$_4$, Zn(II)-**GGG**-Cys$_3$His, and Zn(II)-**GGG**-Cys$_2$His$_2$.

barrier to folding a zinc finger protein is not as costly as the +16 kcal/mol predicted, but much smaller, ~+4.0 kcal/mol, and perhaps even negligible. These results illustrate that a metalloprotein design can be used to reveal the free-energy cost of protein folding in natural zinc metalloproteins, heretofore unknown values in systems where protein folding is coupled to metal binding.

The idea that there is a minimal free-energy cost to zinc finger protein folding has circumstantial support from several vantage points in the literature. The first line of evidence comes from experimental studies of zinc fingers that have failed to show dramatic increases in metal ion affinities upon apo-state stabilization of the ββα-fold (Struthers, Cheng, and Imperiali 1996). If folding into the ββα-fold required a substantial free-energy contribution from zinc binding, prefolding the apoprotein should increase zinc-binding affinity substantially. A second line of evidence comes from the fact that the first successful computational protein design was the redesign of a zinc finger protein ββα-fold without the metal or its ligands (Dahiyat and Mayo 1997). The ability of relatively few protein-protein interactions to provide sufficient free energy to overcome the cost of protein folding and to fold the protein into a stable ββα-fold argues that the original metal ligand-binding free energy contributed little to the protein-folding reaction. The last line of evidence arises from molecular dynamics simulations of zinc finger proteins that indicate that the unfolded ensemble of a ββα protein fold corresponds to the native folded state in an average sense (Zagrovic et al. 2002). This suggests that thermodynamic contribution of zinc to the folding reaction is limited, but leaves open the question of the kinetic role of zinc binding in zinc finger protein folding.

While there is circumstantial support for the idea that the cost of protein folding in zinc fingers is minimal, there is a recent report that challenges the corollary that the Zn(II)-**GGG** series possesses the tightest possible zinc ion affinities for their coordination spheres (Sénèque et al. 2009). A cyclic peptide that constrains two of the ligands in its Cys_4 zinc-binding site possesses a formation constant of $4.0 \times 10_{20}$ M_{-1}, or K_d value of 2.5 zM. This represents a 7900-fold, or 5.3 kcal/mol, improvement in zinc affinity over Zn(II)-**GGG**- Cys_4. While further study is needed to identify the source of the improvement in the Zn(II) affinity, it is presumably due to the more favorable entropy provided by geometrically constraining two of the Cys ligands.

FUTURE PERSPECTIVE

Just as protein design is an important approach to the protein-folding problem, metalloprotein design is beginning to yield thermodynamic and kinetic insight into the metalloprotein-folding problem. The studies discussed herein illustrate the complexity of the metalloprotein-folding problem and the value of using designed proteins to measure the underlying chemical equilibria. The ability to parse apart the free-energy contributions to metalloprotein stability into its constituent thermodynamic components foreshadows a full understanding of the thermodynamic basis of metal ion binding site affinities, speciation, selectivities, and electrochemistry. While there are only a few rigorous studies of designed metalloprotein thermodynamics, this limited set offers keen insight into the fundamental metal-protein interactions that control metalloprotein folding. These thermodynamic studies are a necessary step toward improving the energy functions for computational metalloprotein design, whose advances promise to accelerate the metalloprotein design approach to solving the metalloprotein-folding problem.

REFERENCES

Bai, Y. W., T. R. Sosnick, L. Mayne, and S. W. Englander. 1995. Protein folding intermediates: native-state hydrogen exchange. *Science* 269:192.

Baldwin, R. L. 2007. Energetics of protein folding. *J Mol Bio* 371:283–301.

Berg, J. M. 1995. Zinc finger domains: from predictions to design. *Acc Chem Res* 28:14–19.

Bryson, J. W., S. F. Betz, H. S. Lu, D. J. Suich, H. X. X. Zhou, K. T. O'Neil, and W. F. DeGrado. 1995. Protein design: a hierarchic approach. *Science* 270:935–41.

Choma, C. T., J. D. Lear, M. J. Nelson, P. L. Dutton, D. E. Robertson, and W. F. DeGrado. 1994. Design of a heme-binding four helix bundle. *J Am Chem Soc* 116:856–65.

Christianson, D. W., and C. A. Fierke. 1996. Carbonic anhydrase: evolution of the zinc binding site by nature and by design. *Acc Chem Res* 29: 331–39.

Cochran, A. G., N. J. Skelton, and M. A. Starovasnik. 2001. Tryptophan zippers: stable, monomeric β-hairpins. *Proc Natl Acad Sci U S A* 98:5578–83.

Dahiyat, B. I., and S. L. Mayo. 1997. De novo protein design: fully automated sequence selection. *Science* 278 82–87.

Dill, K. A., S. B. Ozkan, M. S. Shell, and T. R. Weikl. 2008. The protein folding problem. *Ann Rev Biophys* 37:289–316.

Farrer, B. T., and V. L. Pecoraro. 2003. Hg(II) binding to a weakly associated coiled coil nucleates an encoded metalloprotein fold: kinetic analysis. *Proc Natl Acad Sci U S A* 103:3760–65.

Gibney, B. R., and P. L. Dutton. 1999. Histidine placement in de novo designed heme proteins. *Protein Sci* 8:1888–98.

Gibney, B. R., S. E. Mulholland, F. Rabanal, and P. L. Dutton. 1996. Ferredoxin and ferredoxin-heme maquettes. *Proc Nat Acad Sci U S A* 93:15041–46.

Hargrove, M.S., and J. S. Olson. 1996. The stability of holomyoglobin is determined by heme affinity. *Biochemistry* 35:11310–18.

Hecht, M. H., K. M. Vogel, T. G. Spiro, N. R. L. Rojas, S. Kamtekar, C. T. Simons, J. E. Mclean, and R. S. Farid. 1997. De novo heme proteins from designed combinatorial libraries. *Protein Sci* 6:2512–24.

Hellinga, H. W. 1998. Synthetic type 1 copper design efforts. *J Am Chem Soc* 120:10055–66.

Hitomi, Y., C. E. Outten, and T. V. O'Halloran 2001. Extreme zinc-binding thermodynamics of the metal sensor/regulator protein, ZntR. *J Am Chem Soc* 123:8614–15.

Izadi, N., Y. Henry, J. Haladjian, M. E. Goldberg, C. Wandersman, M. Delepierre, and A. Lecroisey. 1997. Purification and characterization of an extracellular heme-binding protein, HasA, involved in heme iron acquisition. *Biochemistry* 36:7050–57.

Jiang, L., E. A. Altoff, F. R. Clemente, L. Doyle, D. Röthlisberger, A. Zanghellini, J. L. Gallaher, J. L. Betker, F. Tanaka, C. F. Barbas, D. Hilvert, K. N. Houk, B. L. Stoddard, and D. Baker. 2008. De novo computational design of retro-aldol enzymes. *Science* 319:1387–91.

Kauzmann, W. 1959. Some factors in the interpretation of protein denaturation. *Adv Protein Chem* 14:1–63.

Kim, C. K., and J. M. Berg. 1993. Thermodynamic β-sheet propensities measured using a zinc-finger host peptide. *Nature* 362:267–70.

Levinthal, C. 1968. Are there pathways to protein folding? *J Chim Phys* 65:44–45.

Magyar, J. S., and H. Arnold-Godwin. 2003. Spectrophotometric analysis of metal binding to structural zinc-binding sites: accounting quantitatively for pH and metal ion buffering effects. *Anal Biochem* 320:39–54.

Marqusee, S., V. H. Robbins, and R. L. Baldwin. 1989. Unusual stable helix formation in short alanine-based peptides. *Proc Natl Acad Sci U S A* 86 (14):5286–90.

O'Neil, K. T., and W. F. DeGrado. 1990. A thermodynamic scale for the helix-forming tendencies of the commonly occurring amino acids. *Science* 250:646–51.

Petros, A. K., A. R. Reddi, M. L. Kennedy, A. G. Hyslop, and B. R. Gibney. 2006. Femtomolar Zn(II) affinity in a peptide ligand designed to model thiolate-rich metalloprotein active sites. *Inorg Chem* 45:9941–58.

Privett, H.K., C. J. Reedy, M. L. Kennedy, B. R. Gibney. 2002. Nonnatural amino acid ligands in heme protein design. *J Am Chem Soc* 124:6828–29.

Reddi, A. R., and B. R. Gibney. 2007. The role of protons in the thermodynamic contribution of a Zn(II)-Cys4 site toward protein stability. *Biochemistry* 46:3745–58.

Reddi, A. R., T. Guzman, R. M. Breece, D. L. Tierney, and B. R. Gibney. 2007. Deducing the energetic cost of protein folding in zinc finger proteins using designed metallopeptides. *J Am Chem Soc* 129:12815–27.

Reddi, A.R., C. J. Reedy, S. Mui, and B. R. Gibney. 2007. Thermodynamic investigation into the mechanisms of proton-coupled electron transfer in heme protein maquettes. *Biochemistry* 46:291–305.

Reedy, C. J., and B. R. Gibney. 2004. Heme-protein assemblies. *Chem Rev* 101:617–49.

Reedy, C. J., M. L. Kennedy, and B. R. Gibney. 2003. Thermodynamic characterization of ferric and ferrous haem binding to a designed four-α-helix protein. *Chem Commun* 570–71.

Regan, L., and N. D. Clarke. 1990. A tetrahedral zinc (II)-binding site introduced into a designed protein. *Biochemistry* 29:10878–83.

Robertson, D. E., R. S. Farid, C. C. Moser, J. L. Urbauer, S. E. Mulholland, R. Pidikiti, J. D. Lear, A. J. Wand, W. F. DeGrado, and P. L. Dutton. 1994. Design and synthesis of multi-haem proteins. *Nature* 368:425–32.

Sela, M., F. H. White Jr., and C. B. Anfinsen. 1957. Reductive cleavage of disulfide bridges in ribonuclease. *Science* 125:691–92.

Sénèque, O., E. Bonnet, F. L. Joumas, and J.-M. Latour. 2009. Cooperative metal binding and helical folding in model peptides of treble-clef zinc fingers. *Chem Eur J* 19:4798–4810.

Sharp, R. E., J. R. Diers, D. F. Bocian, and P. L. Dutton. 1998. Differential binding of iron (III) and zinc (II) protoporphyrin IX to synthetic four-helix bundles. *J Am Chem Soc* 120:7103–4.

Silchenko, S., M. L. Sippel, O. Kuchment, D. R. Benson, A. G. Mauk, A. Altuve, and M. Rivera. 2000. Hemin is kinetically trapped in cytochrome b5 from rat outer mitochondrial membrane. *Biochem Biophys Res Commun* 5:467–72.

Sosnick, T. R., S. Jackson, R. M. Wilk, S. W. Englander, and W. F. DeGrado. 1996. The role of helix formation in the folding of a fully alpha-helical protein. *Proteins* 24:427–32.

Struthers, M. D., R. P. Cheng, and B. Imperiali. 1996. Economy in protein design: evolution of a metal-independent $\beta\beta\alpha$ motif based on the zinc finger domains. *J Am Chem Soc* 118:3073–81.

Wade, H., S. E. Stayrook, and W. F. DeGrado. 2006. The structure of a designed diiron(III) protein: implications for cofactor stabilization and catalysis. *Angew Chem Int Ed Engl* 45 (30):4951–54.

Zagrovic, B., C. D. Snow, S. Khaliq, M. R. Shirts, and V. S. Pande. 2002. Native-like mean structure in the unfolded ensemble of small proteins. *J Mol Bio* 323:153–64.

Zhuang, J., J. H. Amoroso, R. Kinloch, J. H. Dawson, M. J. Baldwin, and B. R. Gibney. 2004. Design of a five-coordinate heme protein maquette: a spectroscopic model of deoxymyoglobin. *Inorg Chem* 43:8218–20.

———. 2006a. Evaluating the roles of the heme *a* sidechains in cytochrome *c* oxidase using designed heme proteins. *Biochemistry* 45:12530–38.

———. 2006b. Evaluation of electron-withdrawing group effects on heme binding in designed proteins: Implications for heme a in cytochrome c oxidase. *Inorg Chem* 45:4685–94.

13 The Folding Landscapes of Metalloproteins

Patrick Weinkam and Peter G. Wolynes

CONTENTS

Introduction 247
The Energy Landscape of Cytochrome c 252
Chemical Frustration in the Folding Energy Landscape 259
Folding of Azurin 266
Future Perspective 270
References 270

INTRODUCTION

During their evolution, life forms based on carbon have come to exploit the chemical complexity made available by the rest of the periodic table. Proteins containing metals play important roles in energy transduction and information transfer in the cell. How do these metalloproteins assemble themselves into working structures? Are the principles of metalloprotein folding different from those governing the folding of their completely organic relatives? In this chapter we will highlight some lessons that have been learned by combining theoretical and experimental studies to elucidate the mechanisms of metalloprotein folding.

In the main we have found that the mechanisms of metalloprotein folding conform to the same energy landscape principles as does the folding of ordinary, metal-less proteins. This is fortunate because metalloproteins have some special advantages when it comes to experimental investigation of folding and therefore have provided some of the best data on the folding problem. The foremost advantage is that the rich electronic structure of the metal center offers a convenient probe for spectroscopy. Folding experiments on metalloproteins such as the cytochromes can exploit changes in their relatively intense color. The tunable redox characteristics of the metal centers also allow the binding free energies to be modulated and therefore allow large changes in the overall folded protein stability. Finally, the rich coordination chemistry of the metal center, which again can be modulated by changing solvent conditions, allows the investigation of the roles of very specific, often nonnative contacts in the ultimate protein structure formation. These chemically rich features that are especially important in the energy landscapes of metalloproteins do cause some differences in the details of the preferred folding pathways of metalloproteins when compared with folding routes for proteins without metals. In this chapter we will highlight these differences.

Most experiments provide a macroscopic, ensemble averaged view of the folding process. Despite the complexity of the protein energy landscape, experimental data commonly give the appearance that folding is rather simple. Experimental studies of folding equilibria often indicate that proteins exhibit a simple two-state transition when titrated with a chemical denaturant such as guanidinium chloride (Figure 13.1b). This cooperative folding behavior is signaled by a two-state transition that is conventionally represented schematically with a single transition state separating two free-energy basins (Figure 13.1a). Very often it is indeed correct that a single bottleneck occurs along the flow toward the folded state, but sometimes folding is a more complicated process kinetically even when two-state thermodynamics prevails. Theory can address this kinetic complexity because proteins can be modeled computationally with a rather more complete level of detail that cannot be accessed experimentally. Statistical mechanical calculations that are based on the configurations of individual proteins can be used to predict and therefore understand experimental data that are demonstrative of an ensemble of proteins.

Statistical mechanical techniques also allow access to the rare states that act as bottlenecks in the folding process. Computational studies of protein folding often have distinct goals: (1) predicting the native folded protein structure from a protein's sequence and (2) predicting how a protein folds to its native structure. The more difficult problem for today's approaches remains the problem of, given only a protein's sequence, predicting its structure. Even here energy landscape theory has proved helpful (Goldstein, Luthey-Schulten, and Wolynes 1992; Papoian et al. 2004). Structure prediction is further helped by the surprising fact that the number of unique folds is small compared to the immense number of possible sequences composed

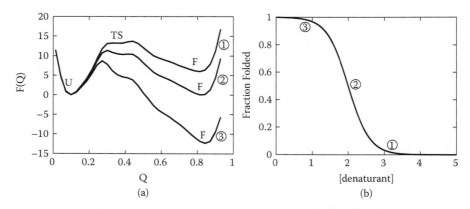

FIGURE 13.1 (a) Free-energy profiles plotted as a function of the reaction coordinate Q. The curves are calculated from simulations of cytochrome c at different temperatures. The folded (F) and unfolded (U) basins are labeled along with the transition state (TS). (b) A hypothetical equilibrium experimental denaturation curve. Increasing the concentration of denaturant disrupts stabilizing interactions analogously to increasing the temperature in a simulation. Numbers that label the free-energy profiles also are placed on the denaturation curve in the position expected given the free-energy differences between folded and unfolded basins.

The Folding Landscapes of Metalloproteins

of the 20 common amino acids. Predicting the structure from the sequence alone can therefore often be based on information about known folds where the detailed interactions between different amino acids are described by taking advantage of the "knowledge" in the database (Haney et al. 1997; Simons et al. 1997). This chapter will however focus on trying to understand how proteins fold to a unique structure. We will usually take the native structure as given, an important clue to the protein's overall energy landscape. By developing different types of theoretical models, we can figure out which interactions are most important in describing the motions of proteins. By taking the derivative at any point of the solvent averaged free-energy function, we can obtain the forces that will describe the protein motions. These protein motions, when complemented by the random thermal influences of the solvent, lead to Brownian diffusive motions of the chain. In the end, the question we address in this chapter will be: What effect do metal ions and cofactors have on the energy landscape of proteins?

The energy landscape describes the pattern of free energies averaged over the solvent for all of the possible configurations for a protein with a given sequence (Bryngelson et al. 1995; Oliveberg and Wolynes 2005). Of all the possible protein sequences, only a small fraction can actually fold robustly to a unique structure. Most random sequences have ground-state structures that are very sensitive to environmental conditions and ground states that are hard to get to kinetically because of trapping in wrong structures. The energy landscapes of the foldable sequences found naturally are characterized by a large energy gap (ΔE) between the native and unfolded structures and small energetic barriers between any traps that are encountered during folding (δE) (Figure 13.2b). An energy landscape of this type may be described as being funneled, since such a landscape naturally allows thermal motions to be guided to the native structure. Funneling arises from what has been called the "principle of minimal frustration" (Bryngelson et al. 1995), which

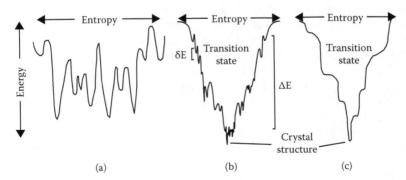

FIGURE 13.2 (a) A representation of an energy landscape for a typical random, frustrated protein sequence that does not fold robustly to a single structure. (b) A representation of an energy landscape for a protein that will fold to a distinct structure. The landscape is characterized by a large energy gap between the native state and the unfolded state (ΔE) and only minimal amounts of frustration, or traps, due to nonnative contacts (δE). (c) An idealized energy landscape for a protein with a perfectly funneled energy landscape and no energetic frustration due to nonnative contacts.

suggests that the reason there are few energetic barriers to impede folding is because there are very few alternative low-energy structures that would come from conflicting local interactions. In this way the protein can reach the energetically optimized native state without getting confused or "frustrated." When multiple structures do satisfy the local energetic rules equally well, we would say the protein is frustrated. Frustration arises from favorable but nonnative interactions that must be broken and then remade in an alternative way so that the protein can fold. The energy landscape of most random sequences that cannot fold is characterized by large amounts of frustration (Figure 13.2a). Unfoldable proteins at low temperatures would populate structures that are energetically similar but are nevertheless conformationally diverse. In the opposite extreme, with minimal frustration, nonnative interactions form rarely during folding and hardly influence the folding routes at all. For an idealized protein with no frustration, the energy landscape would be perfectly funneled (Figure 13.2c). Certainly proteins can exhibit behavior across this spectrum of energy landscapes from perfectly funneled to those with some significant frustration. Localized frustration is sometimes an evolutionary necessity to accommodate binding sites (Ferreiro et al. 2007) or hinges of allosteric motion (Whitford, Onuchic, and Wolynes 2008). There is much to be gained by examining the physicochemical sources that give rise to frustration and the role that frustration plays during folding.

Energy landscape theory provides a general statistical mechanical framework that can be used to describe many different folding mechanisms and relate them to the native structure and interactions of the protein molecule. Landscapes are often illustrated using cartoons such as in Figure 13.2b, which shows how the large number of unfolded conformations gradually decreases until a narrow set of conformationally related structures is obtained. These cartoons, especially if impressionistically drawn, do not always make clear the features that give rise to free-energy barriers that are important to describe the simple patterns of kinetics observed experimentally. Energy and entropy largely compensate each other during folding. For this reason, the free-energy barriers are not apparent on the entropy-energy diagrams used to draw the funnel due to differences in energy scale (Figure 13.3). The energy and entropy represented in the energy landscape are each on the order of hundreds of thermal units (kT) while the free energy, which is related to the difference in these numbers, is on the scale of tens of kT. In fact the fluctuations in the level of compensation of the energy and entropy along the folding funnel is what gives rise to bottlenecks in the funnel. The imperfect tradeoff of entropy for stabilization energy gives rise to the transition state for folding when there is two-state thermodynamics. The free energy of the transition state ensemble can be used to approximate the folding rate through a formula that resembles transition state theory but with an important dynamical prefactor.

$$k = De^{-\Delta F^{\ddagger}/kT} \tag{13.1}$$

In addition to the free-energy barrier describing the folding bottleneck, the folding time actually depends on the rate at which the chain can diffuse and reorganize in order to form the key structural regions needed for folding to proceed, eventually downhill in free energy. Reorganizational chain diffusion rates are affected

The Folding Landscapes of Metalloproteins

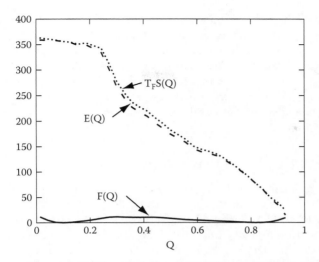

FIGURE 13.3 The energy (E(Q)), scaled entropy ($T_F S(Q)$), and free energy (F(Q)) in units of kT calculated from simulations of cytochrome c. The plot shows how small imbalances between energy and entropy, in which one compensates for the other, gives rise to the free-energy barrier. The folding bottleneck occurs when there is the largest imbalance in the trade-off between energy and entropy ($Q = 0.3$).

by the rate of chain diffusion as well as the rate that water can be expelled from the protein's interior. Energetic frustration due to nonnative contacts can also slow chain diffusion by placing traps on the free-energy landscape. Diffusion rates over the barrier are accounted for using the prefactor. A direct route to the prefactor relies on simulating how the protein fluctuates along the reaction coordinate (Socci, Onuchic, and Wolynes 1996), however there are other methods to calculate the prefactor from polymer physics (Portman, Takada, and Wolynes 1998, 2001a, 2001b). Typically the bottleneck or transition state ensemble can be described as a set of structures in which a significant but incomplete amount of nativelike structure has formed. We may describe the regions of nearly completed structure as a nucleation cluster, in the language of crystallization. The nucleation cluster or critical nucleus is quite large usually encompassing about one-third of the protein. The structural ensemble at the transition state depends on the relative stability of different regions of the protein. The transition state ensemble therefore usually involves the association of the more stable regions of the protein. For metalloproteins the metal center can be a crucial contributor to stability. Basically the metal can act as a "seed" for nucleation. Transition state structural ensembles are at least as varied as the types of folds that exist.

In this chapter we will discuss several systems where the metal cofactors play a role in the structure and formation of the transition state and consequently influence the folding mechanism. While the basic principles of folding physics are clear, the difficulty of developing models that capture the chemical details of protein folding can be underestimated. While in some cases obtaining results from simulations or calculations is as easy as pressing a few buttons on the keyboard and letting the computer do the work, there are still many subtle details involved with developing and utilizing computational models. For metalloproteins, the process of understanding how to capture the key chemical details can be often more important than the results themselves. We will describe in this chapter a few computational techniques that are used to study folding transitions for proteins without cofactors. We will then show how to modify these methods to deal with cofactors and apply them to the heme protein cytochrome c and the copper protein azurin.

THE ENERGY LANDSCAPE OF CYTOCHROME C

As described in Chapter 3, cytochrome c is an electron transport protein that contains a covalently bound heme cofactor. It has been studied experimentally for decades as a model system for protein folding and dynamics (Babul and Stellwag 1972; Schejter and Eaton 1984; Roder, Elove, and Englander 1988; Jones et al. 1993; Pascher et al. 1996; Tezcan, Winkler, and Gray 1998). Primarily because of the complications involving the heme cofactor, the folding of cytochrome c has been seriously modeled computationally only in more recent years (Erman 2001; Cardenas and Elber 2003; Weinkam, Zong, and Wolynes 2005; Weinkam, Romesberg, and Wolynes 2009; Weinkam et al. 2009). A key question is: What is the effect of the heme moiety on the folding mechanism? Kinetic studies showed that different residues can compete for the heme ligation site thereby decreasing the folding rate (Sosnick, Mayne, and Englander 1996). This strong, chemically mediated source of nonnative traps was therefore one of the first identified biochemical manifestations of the concept of frustration. It was easy to find because of the large energy scale associated with the coordination chemistry in relation to the scale of interresidue interactions. This chemical frustration is not the only way the heme enters the scene, however: The hydrophobic surface of the heme acts to guide the folding when there is no misligation. To explore the effect of the heme on the energy landscape, we will therefore first consider the most simple situation: protein folding without heme misligation. We will then go on to consider how the folding behavior changes when there are frustrated interactions and thus nonnative traps on the energy landscape.

Many features of the folding of proteins without cofactors can be predicted using models that do not include specific possible sources of energetic stabilization from nonnative structural elements. These are called "structure-based models" since the native structure of the final folded state is the key input. In these models, the energy function is primarily made up of favorable interactions only between those residues that are in contact in the native crystal structure along with some generic constraints characterizing the backbone connectivity. Such models were originally introduced using a lattice representation for the protein (Ueda, Taketomi, and Go 1978) in two and three dimensions since the entire lattice conformational space can be easily enumerated. The energy landscapes of structure-based models are "perfectly funneled" since the energetic barriers from nonnative contacts that give rise to frustration and therefore traps with structures different from the native are absent. Folding routes for a variety of different proteins are successfully predicted using free-energy functionals based on funneled energy landscapes (Shoemaker,

Wang, and Wolynes 1997, 1999; Shoemaker and Wolynes 1999). Free-energy functionals give precise numerical answers but are approximate and can be difficult to formulate rigorously. In some cases, the simplest free-energy functionals can miss alternative folding pathways that are accessible. These can usually be uncovered using more complicated variational methods that better account for chain entropy (Portman, Takada, and Wolynes 1998). Computer simulations are simpler to implement than the free-energy functional methods but are always subject to numerical, statistical inaccuracies. Nonetheless, simulations based on models with perfectly funneled landscapes are commonly used to predict folding routes and mechanisms (Clementi, Jennings, and Onuchic 2000; Levy et al. 2005; Cho, Weinkam, and Wolynes 2008). The success of these structure-based models in explaining many details of folding kinetics strongly suggests that the energy landscape for many proteins is indeed highly funneled, justifying using such simple models as a starting point. The actual degree of funneling of protein landscapes in nature appears to be larger, in fact, than is captured in most current all-atom models, which exhibit landscape complexity that seems kinetically invisible, if present at all.

Most of the structure-based models that accurately predict kinetic folding routes are coarse grained. Coarse-grained models use a reduced description of the protein in which each amino acid is represented by only one or a few pseudoatoms. While there are many ways to develop a coarse-grained model, using even a single atom per residue is sufficient to describe the topology of a protein's fold. All-atom representations, in principle, model protein structure in full detail. Simulations using all-atom models are helpful to examine still more detailed conformational changes, such as those involving side chain motions and minor backbone movements. Calculating all of the interatom interactions however is computationally expensive. By averaging over the side-chain degrees of freedom, coarse-grained models can be used to study longer timescale motions, such as for protein folding, while still capturing the general topology of protein structures. The computational efficiency of coarse-grained models can therefore be used to explore the entire conformational space accessible to a protein. Adequate sampling of the conformational space is a prerequisite for any discussion or calculation of a protein's energy landscape. For instance, we cannot say that one folding route is preferred over another if we do not sample enough trajectories of each route to allow for statistically meaningful calculations.

The use of coarse-grained models is not justified on grounds of expediency alone. Protein structure remains robust but not invariant to the many mutations of individual residues involved in the natural history of a protein family. This robustness goes along with the funneled nature of the energy landscape. This funnel-like aspect is therefore captured at the scale of resolution appropriate to the differences between evolutionary homologues. Clearly this funneled aspect of the landscapes may be therefore compromised at the all-atom level even though it persists in coarse-grained residue-level models.

Models for cytochrome c that are funneled at the coarse-grained level successfully predict the features of a variety of experimental results (Weinkam, Zong, and Wolynes 2005; Weinkam et al. 2008, 2009; Weinkam, Romesberg, and Wolynes 2009). The actual computer code used for the cytochrome simulations carried out by our group was based on a protein structure prediction program using the associative memory Hamiltonian (AMH) (Friedrichs and Wolynes 1989). The associative memory Hamiltonian generally uses many known local structures as "memories." Memory sequences that provide a good alignment to the protein's sequence are used to parameterize the interactions during structure prediction. The interresidue contacts from all memories describe the pairwise interactions within the protein and guide fragments of the sequence to the likely range of local structures. While local structures are used to parameterize interactions within short fragments, sequence alignment of memories to the protein sequence cannot provide statistically meaningful information to describe interactions between residues far apart in sequence. Generic long-range potentials must therefore be used. One memory corresponding to the native crystal structure can be used as input to describe interactions between residues both close and far apart in the sequence. In this case the resulting structure-based model corresponds with a perfectly funneled energy landscape. In the single-memory AMH, each residue is represented by three atoms (C^α, C^β, and O) and the heme is represented by five pseudoatoms arranged in a square planar geometry. The energy function, or Hamiltonian, has the form:

$$E = (V_{Rama} + V_{chiral} + V_{E.V.} + V_{SHAKE}) + V_{contact} \tag{13.2}$$

A detailed description of all the components of the AMH energy function would go beyond the scope of this chapter but has been published previously (Eastwood and Wolynes 2001). The terms indicated in the parentheses do not depend on the sequence but act to maintain a realistic geometry of the protein backbone. V_{Rama} biases the phi and psi angles to be in the sterically allowed region. They also act to ensure a realistic chain stiffness. V_{chiral} maintains the chirality so that only right-handed helices form. Global chirality is also biased in this way although somewhat weakly. $V_{E.V.}$ is an excluded volume potential that ensures that the atoms do not overlap so the chain does not pass through itself. V_{SHAKE} maintains the backbone bond lengths and angles. The interesting part of the energy function that guides the protein to fold is the contact potential $V_{contact}$. The contact potential is a sum of Gaussian well potentials between C^α and C^β atoms with distances between each atom taken from the native crystal structure. In this most simple model, all contacts are given the same energy and only residues less than 8 Å apart in the crystal structure are included. Although in many structure-based models this contact potential is pairwise additive, in our calculations the contact potential uses nonadditive and many body forces. The forces that guide folding are nonadditive because they have a big component coming from averaging over the solvent and from averaging over the side chain reorientation degrees of freedom. In both cases, when two residues come into contact there is a significant entropic cost from rearranging the neglected degrees of freedom. Neither side chain in the pair can move around freely, and solvent surrounding a pair usually changes its available configurations. If a third side chain comes into contact with both residues, the entropic cost for freezing the first two residues has already been realized and the solvent has already been partially reorganized. Owing to this there will be an additional energetic contribution to account for the decrease in entropic cost for adding the third residue. This contribution means the energy is not a sum over pairs; i.e., it is nonadditive. Nonadditivity significantly increases the folding free-energy barrier (Kolinski, Galazka, and Skolnick 1996). Nonadditivity also tends to make intermediates crisper in their structured portions and more distinct in the free-energy profile. With sufficient amount of nonadditivity, predicted folding rates and mechanisms have been shown to rather accurately correspond with experiment (Kolinski, Galazka, and Skolnick 1996; Plotkin, Wang, and Wolynes 1997).

To elucidate folding mechanisms, both straightforward direct simulations and more subtle statistical mechanical sampling procedures must be used. Rare species can be fleeting in direct simulations, but we can probe them in great detail by partitioning the ensemble of structures into discrete bins: $X_1, X_2, X_3, \ldots X_n$, which might characterize intermediates corresponding to predetermined quantities. This partitioning can be carried out by adding constraint potentials. The equilibrium probability, $P(X_i)$, that X attains a certain value X_i can always be calculated using the Boltzmann distribution.

$$P(X_i) = \frac{e^{-E_i/kT}}{\sum_i e^{-E_i/kT}} \tag{13.3}$$

The summation in the denominator, the partition function Z, is not directly calculated, but the ratios of Z for various values of the X_i constraint can be found to describe the statistical distribution of states at equilibrium in the absence of constraints. These in turn can be turned into free energies of constrained ensembles by taking the negative log of $Z(X)$.

$$F(X) = -kT\log(Z(X)) \tag{13.4}$$

For kinetics we partition the ensemble of states by a coordinate X that we expect to at least approximate the reaction coordinate. There are many choices for approximate reaction coordinates, each with its own strength and weaknesses, such as P_{fold}, root mean square deviation (RMSD) from the native structure, and the fraction of native contacts (Q). In some respects, the ideal coordinate for a two-state folder is P_{fold}, which is calculated by finding the probability that a specific structure will fold. This reaction coordinate magnifies the dynamics in the transition region where up to 80% of the coordinate focused on structures just 1 kT of free energy from the transition state. Much of the interesting folding dynamics occurs well past the transition state and would be missed using P_{fold}. $P_{fold} = 1/2$ would define the transition state for a two-state folding protein. For a three-state folding protein however, the usefulness of P_{fold} breaks down since the intermediate is likely to not be resolved from other structures. Another problem with P_{fold} is that it is computationally expensive since one must run many trajectories arising from any given structure. RMSD and Q are both simple structural calculations quickly computed from each structure. RMSD while easy to calculate is often misleading for partially folded structures. One can imagine for instance a protein structure in which both halves are in fact already well structured but have not associated so that the interface between the regions has not been formed. An RMSD calculation would give a bad result because only one-half can be aligned at a time even though the majority of contacts in the protein have been made, thereby making it appear no structure had formed at all! We have found that the fraction of native contacts Q, which is intuitively related to the guiding energy, works well. It has been shown to be generally well correlated with P_{fold} for two-state folders but also works when intermediates exist, where P_{fold} actually fails even qualitatively (Cho, Levy, and Wolynes 2006). Another advantage to using Q is that it can be easily applied to local regions of the protein and can be used to monitor specific types of contacts. Such a strategy can provide a great deal of useful detail for interpreting folding mechanisms structurally.

Figure 13.1a shows free-energy profiles for cytochrome c with heme, each curve corresponding to a different temperature. In order to generate the free-energy profiles, there needs to be sufficient sampling along the reaction coordinate here chosen as Q. One strategy to achieving sufficient number of structures along the reaction coordinate is to run many constant temperature simulations near the folding temperature. Quickly this strategy fails as many folding barriers are too high in free energy to be adequately sampled. The weighted histogram analysis method (WHAM) is a technique that utilizes artificial biasing potentials that restrict the simulation to specific regions of the coordinate space thereby allowing sufficient sampling. By knowing the form of this biasing potential, its energetic contribution can be subtracted out, allowing the free energy to be calculated as if there were no biasing potential at all. In Figure 13.1a, curve number three is the free energy calculated from simulations at room temperature showing two-state folding for cytochrome c. The free-energy difference between the folded and unfolded basins is greater than 10 kT, showing that cytochrome c is a very stable protein. A straightforward simulation without constraints would sample either folded or unfolded configurations, making such a large free-energy difference hard to determine.

Without the heme the protein is completely destabilized with a free-energy profile resembling the curve labeled "1" in Figure 13.1a. Taking out the heme has analogous effects to denaturing the protein with guanidinium chloride. Chemical denaturants such as urea are thought to bind to hydrophobic residues, thereby decreasing the strength of stabilizing hydrophobic contacts. It is also postulated that denaturants like guanidinium chloride can compete with water for binding sites on the protein's surface, which causes unfolding (Mande and Sobhia 2000). Binding of denaturant to the protein shifts the equilibrium toward the unfolded basin. By taking out the heme, for steric reasons there are fewer hydrophobic contacts that can be made. This paucity of contacts in turn destabilizes the protein by 30 kT. Under normal thermodynamic conditions, the apo-form (without heme) of cytochrome c cannot fold and therefore no complete solution structures of the protein exists (Stellwag, Rysavy, and Babul 1972). In fact, apocytochrome c has typically been studied only when it is associated with a hydrophobic membrane and/or proteins with which it may bind.

In contrast, the folding of apomyoglobin has been studied extensively in solution and several crystal and nuclear magnetic resonance (NMR) structures exist. Myoglobin is larger than cytochrome c (having 150 residues compared to about 100), and its heme is located on one side of the structure (Figure 13.4). Apomyoglobin

The Folding Landscapes of Metalloproteins

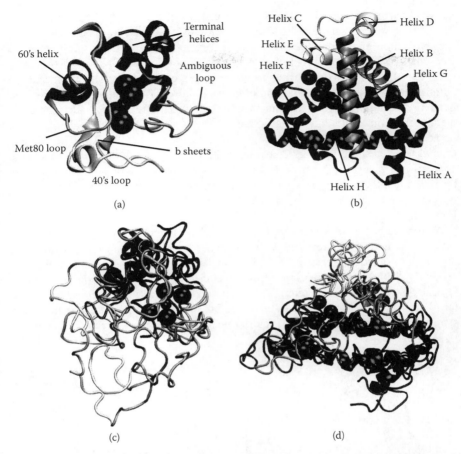

FIGURE 13.4 The crystal structures of (a) cytochrome c and (b) myoglobin with folding units shaded from black to white corresponding to early versus late structure formation during the kinetic folding pathway. The heme is represented with five pseudoatoms. Also shown are three aligned structures representing members of the transition state ensembles for both (c) cytochrome c and (d) myoglobin. The folding nucleus in these partially unfolded structures consists of the more stable regions of the protein, which are shaded black.

can fold since there is still significant stabilization energy in those regions of the protein that do not have heme contacts. Since the protein is stable without heme, the cofactor tends to fall out, making experimental studies of the holo protein difficult. Without the heme to provide stabilizing contacts with the protein, there is more structural heterogeneity in the residues near the heme pocket for apomyoglobin. These structural fluctuations are evident in temperature factors obtain by crystallography (Wagner et al. 1995) and are noticeable in computer simulations. No independently stable region exists for apocytochrome c since the heme is in the center of the structure and serves as the main stabilizing hydrophobic core for the protein.

With the heme's key role in stabilization for cytochrome c recognized, it is perhaps not surprising that the heme's presence dominates in the folding mechanism. To see this, we can partition the simulated ensembles using coordinates describing the interaction of the protein chain with the heme and generate multidimensional free-energy surfaces (Figure 13.5a). A second dimension is provided by the order parameter Q_H, which is analogous to Q but whose contribution is restricted to the heme contacts. Q_H allows us to monitor how structure forms in the proximity of the heme as folding progresses with increasing total Q. The free-energy surface generated by the funnel simulation shows that in order for the protein to fold, almost all heme contacts must be formed at the folding transition state (Q = 0.3). The heme is crucial for nucleating folding. Coincident with the increase in the number of heme contacts is a general collapse of the protein chain. The radius of gyration decreases by 80% when the protein passes through the transition state ensemble. The collapse is accompanied by the sudden drop in energy and entropy at Q = 0.3 as seen in Figure 13.3.

While the heme helps bring regions of the chain together via hydrophobic interactions, the folding barrier is high due to the entropic cost of forming a sufficient number of contacts to create the folding nucleus (Figure 13.4c). We can obtain a residue specific view of folding by monitoring a set of local folding coordinates Q_i, one for each residue i (Figure 13.5b). These curves show that the residues in the terminal helices not only dock to the heme but become structured at the transition state thereby forming the critical folding nucleus. Once these crucial contacts are

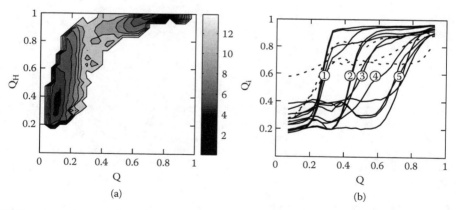

FIGURE 13.5 (a) Free-energy surface showing the role of heme during the folding process. The surface is plotted as a function of the folding reaction coordinate Q and the heme-folding reaction coordinate Q_H. The depth of the surface corresponds to free energy in units of kT. (b) The ensemble average of Q_i, a residue-specific coordinate, demonstrates how regions of the protein fold as cooperative units. Q_i is plotted as a function of Q for several residues grouped according to their folding unit: (1) terminal helices (residues 95, 96, 98, and 99), (2) 60's helix (residue 68), (3) β sheets (residues 60 and 64), (4) Met80 loop (residues 74 and 75), and (5) 40's loop (residues 43, 46, and 52). Shown with dashed lines are Q_i's from several residues in a loop (residues 16 to 33) that cannot be unambiguously assigned to a folding unit because their Q_i's do not have a single sigmoidal transition.

made, folding can continue downhill in free-energy terms throughout the protein as substructures build onto the interfaces formed by the nucleus. The folding route from this perfectly funneled energy landscape is thus seen to involve the sequential stabilization of individual folding units. The order of these sequential events agrees with the order of events inferred from hydrogen exchange experiments (Maity, Maity, and Englander 2004). Why are these folding units sequentially stabilized in this particular order? The native structure-based funnel landscape gives us insight. The key is that there is an asymmetry to the density of contacts in the native structure of cytochrome c. The terminal helices have the highest contact density of any folding unit in the protein. Thus in the absence of strongly destabilizing residues, one would expect these helices to fold first. They then form a large interface for the 60's helix to build from. The 60's helix itself also has a high contact density. Once formed, the 60's helix can provide stable interfaces for the Met80 loop and β sheets to build onto. The last subregion to fold, the 40's loop, has the lowest contact density. By itself it cannot form a fixed structure but requires other substructures to form a stable surface to which it may dock and thus fold. The success of the prediction of the folding route from a perfectly funneled energy landscape suggests that the energy landscape of cytochrome c is highly funneled despite many opportunities for frustration through nonspecific interactions of the hydrophobic core.

We see that the heme molecule provides a hydrophobic core, which helps stabilize the protein, as well as an iron that ligates a histidine and a methionine in the native structure, again lowering the entropy of the search. In addition we will see that the folding process can be complicated by the heme cofactor because other residues, and some small molecules, can compete for the heme ligation site thereby affecting the folding mechanism and protein stability through "chemical frustration." The folding of cytochrome c can therefore be modulated by choosing solvent conditions that favor one set of heme ligands over another.

CHEMICAL FRUSTRATION IN THE FOLDING ENERGY LANDSCAPE

Structure-based models successfully predict experimental results for folding, suggesting that the energy landscapes for these proteins are actually highly funneled in nature. Nonnative contacts that give rise to frustration on the energy landscape nevertheless can play a significant role in modeling the folding mechanism. Some of the first evidence that there is some frustration on the energy landscape was provided by kinetic experiments that showed there was a trap coming from heme misligation in cytochrome c (Jones et al. 1993). As expected from a trap arising from frustration, the folding rate was shown to increase by a couple orders of magnitude when misligation is inhibited (Sosnick, Mayne, and Englander 1996). The amount of frustration in cytochrome c, being chemically specific in origin, can be controlled by the solvent conditions! Histidine misligation can be inhibited by decreasing the pH below the pK_a of histidine or by adding imidazole, which preferentially binds to the heme. Furthermore, changing the charge state of titratable residues can destabilize favorable interactions, which causes more heterogeneity of structure in the proximity of those residues that now have nonphysiologically appropriate charge. This kind of frustration, which can be monitored by a change of solvent conditions and which modifies the detailed chemistry of specific interactions, is called "chemical frustration." Cytochrome c under physiological conditions has a very funneled landscape so most of the native interactions remain dominant. We can therefore model chemical frustration by using structure-based models, which are perturbed in only a few places. The crystal structure of the protein at pH 7 involves contacts that are optimized for the particular electrostatic environment of titratable residues with their charge at that pH. Upon an increase to pH 11, lysine residues deprotonate, thereby destabilizing the previously favorable electrostatic interactions. Native contacts involving lysine must be decreased in strength to capture the effects of deprotonation thereby promoting local structural heterogeneity. Similar perturbations of the effective charge of a residue have been applied to account for the effect on protein structure caused by phosphorylation where a phosphoserine more strongly resembles a superglutamate than the original serine (Latzer, Shen, and Wolynes 2008). As we will see in the following section, the effects of chemical frustration are specific to the exact solvent conditions.

One of the most well-studied examples of denaturation due to chemical frustration involves the alkaline-induced unfolding of cytochrome c. Already in 1941, Theorell and Akesson showed that the change in the ultraviolet (UV)/visible heme absorption spectra (the soret band at 400 nm) due to increasing pH was caused by unfolding (Theorell andAkesson 1941). The soret band is sensitive to structure around the heme but otherwise provides very limited structural information. Improved experimental techniques revealed more conformational details of the alkaline-induced states. By systematically mutating different lysine residues and observing the corresponding NMR signals, several studies demonstrate that two lysine residues (Lys73 and Lys79) misligate around pH 11, another example of chemical frustration (Hong and Dixon 1989; Rosell, Ferrer, and Mauk 1998). Hydrogen exchange measurements (Hoang et al. 2003) suggest that only the two least stable folding units (out of five total folding units) are unstructured in these lysine misligated intermediates. This is confirmed by the NMR "structure" for this partially folded state (Assfalg et al. 2003). A study of alkaline-induced unfolding over a more broad pH range (7 to 13.5) suggests that cytochrome c populates six structurally distinct states (Weinkam et al. 2008). Finding these additional states was made possible through site-specific probes, carbon-deuterium chromophores, inserted at five specific residues through semisynthesis. While the experiments provide useful structural information in the proximity of the observed probes, the procedure is time consuming so significant structural detail potentially available from this method remains unavailable. The identity of the heme ligands is not known in all six observed states. Simulations of native structure-based models with specific frustrated interactions can extend our understanding of the experiments.

The specific nature of the frustration makes modeling the alkaline-induced unfolding of cytochrome c much more complex than the purely native structure-based approximation methods previously discussed in this chapter. In addition to simulating protein-folding transitions, we must also model the acid/base chemistry of protonation and the coordination chemistry of heme ligation. Simulations that

The Folding Landscapes of Metalloproteins

take into account the quantum chemical detail for the acid/base or coordination chemistries have been applied to relatively small and static biomolecules (Yang and Honig 1993) but applying such detailed calculations to protein-folding simulations is currently impossible. A grand canonical formalism provides a unified way to model a system with multiple conformational transitions involving numerous chemically distinct species. Especially when dealing with large biomolecules, this strategy saves tremendous amounts of computing time over explicitly treating each chemical transition or misligation as a molecular event.

In the grand canonical formalism, separate simulations are performed for each chemical species that correspond with a different protonation or ligation state (Figure 13.6). Discrete free-energy profiles calculated from each of these simulations are combined using the free-energy profiles of the specific chemical species. Absolute free energies can be related to each other using the completely unfolded ensembles ($Q = 0$), which are near-random coils in these models. Thus for $Q = 0$, free-energy differences reflect the chemical equilibrium constants involving free amino acids binding to the ligands or their pK_a's. The free energies of these unfolded ensembles, before accounting for the chemical equilibria, are expected to be the roughly identical. The grand canonical partition function relates the free energies of the open system to the chemical equilibria between the ligation and protonation states.

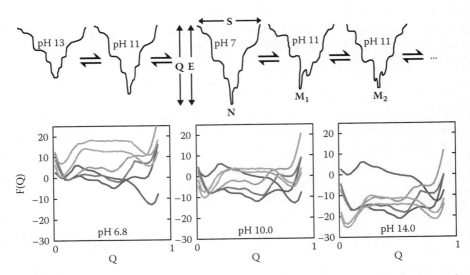

FIGURE 13.6 A schematic representation demonstrating the principles of the grand canonical ensemble method. Distinct energy landscapes exist for each chemical species, according to the pH and the ligation state M_1, M_2, etc. Separate simulations are carried out for each chemical species resulting in a unique free-energy curve. The free-energy curves are reweighted by the chemical equilibria for the acid/base and coordination chemistries, as described in the grand canonical partition function $\Xi(Q, pH)$. The resulting plots are shown below. The free-energy plots show how cytochrome c populates folded, partially unfolded, and fully unfolded conformations between pH 7 and 14.

$$\Xi(Q, pH) = \sum_s [e^{-F_s(Q)/kT}][e^{-\mu_s/kT} K_{lig,s}] \qquad (13.5)$$

The sum in the grand canonical partition function runs over all relevant chemical species s, while the term in the first set of brackets describes the energy landscape and the terms in the second set of brackets refer to the chemical equilibria. $K_{lig,s}$ is the heme ligand dissociation constant. Individual native structure-based simulations were performed for all possible ligands being bound to the heme: methionine, lysine, tyrosine, and hydroxide/water. The chemical potential for the acid/base chemistry is given by:

$$\mu_s = \sum_j kT(pH - pK_{a,j}) \qquad (13.6)$$

where the sum is over all deprotonated residues and the pK_a is chosen to be that of the free amino acid. The parameters corresponding to the chemical equilibria are more or less constrained by experiment within this method. Not as clear are the details of the energy landscape needed to reproduce the experimental observations.

In addition to ligation effects, the native structure-based energy function was modified to account for deprotonation effects and side chain hydrogen bonding. Residues whose protonation state changes are no longer as stabilizing as they were under physiological conditions are essentially represented as mutated residues.

$$E = V_{back} + V_{contact} + V_{local\ elec} + V_{H-bond} \qquad (13.7)$$

For contacts involving a deprotonated residue some of the contact energy ($\varepsilon_{ij}(r_{ij})$) was added back to account for the destabilization due to the change in local electrostatic interactions.

$$V_{local\ elec} = -\sum_{i\ or\ j\ deprot} 0.1\varepsilon_{ij}(r_{ij}) \qquad (13.8)$$

A term to account for side chain hydrogen bonding was needed to account for the large energetic change of disrupting a hydrogen bond due to deprotonation.

$$V_{H-bond} = \sum_{H-bonded,\ not\ deprot} 3\varepsilon_{ij}(r_{ij}) \qquad (13.9)$$

In the hydrogen-bonding term, native hydrogen bonds involving a side chain are given additional strength if the side chain remains protonated. If deprotonation occurs, the energetic boost is taken away. For simulations of lysine misligates, a harmonic potential was added between the ligating residue and the center heme atom.

The Folding Landscapes of Metalloproteins

Combining the site specifically perturbed native structure-based model with the grand canonical ensemble method reproduces the changing structural details of the observed partially folded ensembles. Figure 13.7a shows that under a given solvent condition only a few of the possible species are predicted actually to be populated, as experiments have also demonstrated. It may be surprising that only a handful of states are populated considering that there are many more possible species given that there are 19 lysine residues. In fact the nonpopulated misligates are only slightly less stable (less than 10 kT) and could become populated upon mutation of one or more residues, such as Lys73 and/or Lys79. The observed presence of only 2 of the 19 lysine misligates is a direct consequence of the funneled nature of the energy landscape, which selects those residues for misligation that preserve as much of the favorable nativelike interactions as possible.

The structures of the populated species are reasonably well predicted by the weakly perturbed native structure-based models. The simulations can be compared to hydrogen exchange experiments that give estimates of the average number of contacts made by each residue. Calculations based on the number of contacts per residue have been used to determine protection factors (Hilser and Freire 1996) and are taken to indicate degree of solvent exposure. In agreement with experiment, the simulations predict only the two least stable folding units solvent exposed at intermediate alkaline pH (Hoang et al. 2003) (Figure 13.7b). The other three folding units become fully exposed only at very high pH as the protein unfolds completely. A detailed comparison of the Lys73 misligated structure obtained by NMR (Assfalg et al. 2003) can be made to the predicted structures from simulation. Comparing several of the most stable Lys73 misligated structures (in terms of free energy) from simulation to the NMR structure of this partially unfolded state resulted in an

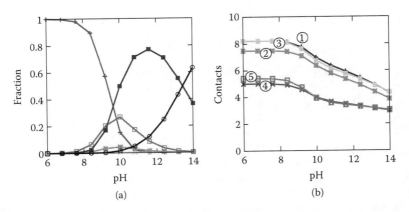

FIGURE 13.7 (a) The populated species predicted using the grand canonical ensemble method. These species include the native state at pH 7 (+), a partially unfolded hydroxide bound state (closed boxes), a Lys73 misligated state (*), a Lys79 misligated state (open squares), and a fully unfolded state (open circles). (b) The number of contacts per residue is an indication of the amount of local solvent exposure in the vicinity of specific residues. The number of contacts per residue is averaged over each folding unit as a function of pH. The folding units are labeled with numbers in order from (1) the most stable to (5) the least stable.

RMSD of 3.5 ± 0.4 Å. Furthermore, calculations of the number of contacts per residue agree with the observed Fourier transform infrared spectroscopy (FTIR) signals for carbon-deuterium probes placed at specific side chains (Weinkam et al. 2008).

The effects of chemical frustration can differ depending on the pH conditions. Cytochrome c forms a molten globule at pH 2 and high salt. The acid-induced molten globule appears to be structurally similar to the native state in a gross sense but is more heterogeneous as indicated by the radius of gyration (R_g), which is somewhat greater than the native state (Akiyama et al. 2002). Measurements of time-resolved fluorescence energy transfer (TRFET) reveals detailed structural features of such heterogeneous protein ensembles. By fitting the fluorescence decays curves to the expressions from the Förster model (Figure 13.8c), TRFET

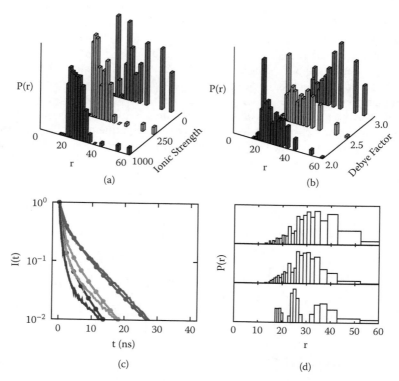

FIGURE 13.8 (a) Residue 39 to heme distance distributions extracted from TRFET experiments at different salt concentrations at pH 2. (b) Residue 39 to heme distance distributions calculated from simulations at pH 2 with different Debye screening factors to account for salt effects. (c) The fluorescence decay curves obtained from TRFET experiments are shown as lines, and fluorescence decay curves calculated from simulation are shown as lines with dots. The shading of the lines correspond to the distance distributions in plots a and b. (d) Residue 39 to heme distance distributions for the GuHCl denatured ensemble. The distributions shown are: calculated from a simulation without funneled, structure-based contacts (upper, $n_{contact} = 0$), calculated from a simulation with funneled contacts (middle, $n_{contact} = 1$), and extracted from TRFET experiments at 3.0 M GuHCl are also shown (lower).

The Folding Landscapes of Metalloproteins

experiments provide distance distributions (P(r)) from dansyl fluorophores to the heme in the molten globule (Pletneva, Gray, and Winkler 2005). The most probable distance in the molten globule is generally close to the separation found in the native state, but the distribution becomes more broad under acidic conditions (Figure 13.8a). A collapsed acid-induced molten globule is found only at high salt concentration. According to the TRFET experiments, decreasing the salt concentration reduces electrostatic screening. The resulting like-charge repulsion unfolds the protein. The acid-induced unfolded ensemble contains both collapsed and extended structures with R_g between 15 and 40 Å. The acid-induced ensembles at pH 2 do not contain the chemically distinct species found in alkaline-induced unfolding. Most forms of misligation are inhibited as all side chains are fully protonated. In contrast to alkaline-induced unfolding of cytochrome c, which involves several distinct chemical species, the acid-induced ensembles, while conformationally heterogeneous, are chemically homogeneous. In yet another case of denatured ensembles, the GuHCl unfolded state at pH 7 involves histidine misligation. Distance distributions extracted from TRFET experiments at high GuHCl concentration with and without imidazole clearly demonstrate that misligation occurs (Pletneva, Gray, and Winkler 2005). Since GuHCl destabilizes most contacts, the GuHCl denatured ensemble is more extended than the acid-induced unfolded ensemble.

Experiments clearly show that the conformations in partially and fully denatured ensembles depends strongly on solvent conditions. While developing separate models to reproduce the experimental results under each solvent condition would be straightforward, there would be many degrees of freedom during parameterization, ensuring good agreement with experiments. In fact it is possible to construct a single model, with a minimal number of parameters, that provides reasonably good agreement with *all* experiments. The construction of such a general model clarifies the physical origin of the most important features of the energy landscape. In other words, its not as useful if the computer can understand the energy landscape than if *we* can understand it.

A general energy function that works well contains nonnative electrostatics and hydrophobic collapse in addition to native structure-based terms (Weinkam et al. 2009).

$$H = (V_{back} + V_{heme}) + n_{contact} \times V_{contact} + n_{Rg} \times V_{Rg} + n_{elec} \times (V_{local\,elec} + V_{long\,elec}) \quad (13.10)$$

The relative contribution of funneled contacts, general hydrophobic collapse, and electrostatics can be varied by changing the parameters: $n_{contact} = 0,1$, $n_{Rg} = 0,1$, and $n_{elec} = 0,1,4$. General collapse is added by constraining the radius of gyration:

$$V_{Rg} = 0.01(Rg - Rg^N)^2 \quad (13.11)$$

in which Rg^N is the native radius of gyration. In addition to local electrostatic effects as described in Equation 13.8, long-range electrostatic contributions were included via a screened Columb potential:

$$V_{long\,elec} = \sum_{|i-j|>2} \gamma V_{Rg} \times \frac{q_i q_j}{4\pi\varepsilon_0 \varepsilon r_{ij}} \times e^{(-r_{ij}/\lambda_D)} \tag{13.12}$$

where λ_D follows from the electrolyte concentration. In the simulations of the GuHCl denatured ensemble, λ_D was set to 1.7 and the temperature was raised above the folding temperature to approximate 3.0 M denaturant.

This general, largely funneled but electrostatically perturbed model reproduces the experimental TRFET results quite well for both the acid-induced and GuHCl-induced unfolded ensembles. While the calculated P(r) distributions resemble the experimental distributions (Figure 13.8b), we can also directly compare the unprocessed fluorescence decay data to the calculated decay curves (Figure 13.8c) without uncertainty due to the fits of the experimental data. The best agreement is found when both funneled and nonnative interactions are included: $n_{contact} = 1$, $n_{Rg} = 1$, and $n_{elec} = 4$. In fact, this simple unified model was able to predict the alkaline-induced lysine misligated NMR structure better than does the earlier model that accounts for hydrogen bonding shown in Equation 13.7 (RMSD 3.2 ± 0.4 Å).

While varying the contributions of funneled versus nonnative interactions, interesting patterns emerge that explain what role each specific interaction plays in the configurations of unfolded ensembles. There is an interplay between electrostatics and hydrophobic collapse in cytochrome c because of its net positive charge. Significant charge repulsion causes the C-terminal end to fray with increasing strength of the electrostatic terms. Residue 39, on the other hand, comes closer to the heme upon the addition of collapse since it is in a more hydrophobic region of the sequence (Figure 13.8). Funneled interactions, resulting from the inclusion of structure-based contacts, cause the chain to collapse in a more specific fashion than the collapse potential alone would do. Funneled contacts considerably improve agreement of the predicted distance distributions with experimental data for the GuHCl ensemble, primarily for distances less than 30 Å (Figure 13.8d). The results suggest that while there are few contacts overall in the unfolded ensembles, native contacts do form transiently to affect the structures. In the GuHCl unfolded ensembles, contacts are sufficiently destabilized so that transiently formed contacts merely shift the distribution. In the acid-induced unfolded ensemble, at high salt native-like interactions form readily, resulting in rather compact structures that form when the funneled energetic terms overcome destabilizing like-charge repulsion.

FOLDING OF AZURIN

While much work has been done on cytochrome c, another metalloprotein of interest is azurin, a β sandwich copper protein with 128 residues. In laboratory folding studies, while copper is its natural cofactor, Cu^{2+} is generally substituted with Zn^{2+} in order to avoid redox chemistry of the metal-coordinating cysteines, which complicates the kinetics. Azurin folds both with and without the metal. The kinetic studies show that there are apparently large differences in the folding mechanisms of the apo- and holo-forms. The kinetic folding mechanism of the metallated protein shows a dependence on denaturant concentration while the apo-form does not. Energy landscape models can capture this complex folding behavior of azurin and reveal further structural details about the mechanism. Structure-based models allow us to elucidate the role the metal plays in shifting the transition state and how the ligating residues become structured during folding. Simulations provide a direct way to sample the energy landscape and ultimately can address these questions, but there are practical issues regarding statistics. Comparisons of rates and folding mechanisms between theory and experiments require the calculation of free-energy curves. Calculation of these curves using simulations corresponding to many temperatures and denaturants concentrations is rather cumbersome. Since simulations require large amounts of sampling to reduce statistical errors, resolving delicate effects on the free-energy profile, for instance from point mutations, is difficult. These often involve getting relative free energies accurate to a few kT. We have used a more efficient calculation scheme to study azurin. This scheme employs the variational free-energy functional.

Portman, Takada, and Wolynes (PTW) introduce a method for computing free-energy profiles based on a variational free-energy functional for using a residue-based order parameter for minimally frustrated models of proteins (Portman, Takada, and Wolynes 1998). Similar to the structure-based simulation models for cytochrome c discussed in this chapter, the PTW method is based on an energy function that uses only native contacts to define the interactions between residues. In the PTW method, however, the free energy is calculated analytically based on performing Gaussian averages on the coordinates of protein residues in near-native locations. Instead of simulating the motions of the protein using Newton's laws of motion, a pseudotrajectory can be calculated analytically by determining local minima and local maxima in the free-energy landscape based on the degree of ordering of each residue. Once several minima and maxima are calculated, the average folding route can be found by connecting each local minima and maxima along the folding reaction coordinate. The transition state is determined by the highest maxima, or saddle point, along the folding route. Unlike simulation methods no initial choice of reaction coordinate is needed; rather the best reaction coordinate comes out of the variational calculation. Calculations based on the PTW variational free-energy functional have been shown to efficiently and accurately reproduce results from simulations on the same underlying energy function. The method, however, allows calculations to be made without the statistical difficulties that arise from using simulations and therefore allows the study of small changes in the free-energy profile and allows precise structural-order parameters to be found at the single-residue level.

The Hamiltonian used in the PTW method models the protein as a collapsed, stiff chain of monomers with residue specific interactions. The Hamiltonian can be separated into backbone and contact interaction terms: $H = H_{\text{chain}} + H_{\text{int}}$. The backbone is modeled so that monomers are held together by harmonic-like potentials.

$$H_{chain} = \frac{3}{2a^2}\sum_{ij} r_i \Gamma_{ij} r_j + \frac{3}{2a^2} B \sum_i r_i^2 \quad (13.13)$$

Here a is the mean square distance between adjacent monomers in the chain, B is the conjugate to the radius of gyration of the chain, r_i is the position of monomer i in the chain, and Γ_{ij}^{-1} represents the correlations between monomer positions expected for naturally occurring denatured protein chains. The specific Γ used reflects that of a "wormlike" chain of a proper persistence length. The interactions between nonadjacent monomers is given by a pairwise potential $u(r_{ij})$ with interaction strength ε_{ij} : $H_{int} = \Sigma_{ij}\varepsilon_{ij}u(r_{ij})$. The interaction strengths are based on the Miyazawa-Jernigan contact energies. The contact potential for each pairwise interaction is composed of three Gaussian wells.

$$u(r) = \sum_{k=(s,m,l)} \gamma_k \exp[-\frac{3}{2a^2}\alpha_k r^2] \quad (13.14)$$

The long-distance interactions are attractive, and the medium- and short-distance interactions are repulsive so the monomers are represented by hard spheres (i.e. $\gamma_s, \gamma_m > 0$ and $\gamma_l < 0$). In order to construct the free energy, a variational calculation is made in which a reference Hamiltonian-containing residue-specific parameters must be created: $H_0 = H_{chain} + (3/2\alpha^2) \Sigma_i C_i(r_i - r_i^N)^2$ in which r_i^N is the position of a monomer in the native structure. The constraint parameters C_i describe the fluctuations of each monomer around its native position: It therefore reflects the amount of local structure formation. The variational free energy is then calculated using the Peierls bound:

$$F[C] = -kT\log(Z_0) + <H - H_0>_0 \quad (13.15)$$

where Z_0 is the partition function of the reference Hamiltonian having the parameter C and $<H - H_0>_0$ is the average with respect to H_0.

Because of its explicit calculational nature, the variational free-energy functional can be used to obtain free-energy curves at multiple temperatures without the statistical errors implicit in simulation approaches. As discussed earlier in the chapter, and shown in Figure 13.1, changing the temperature has the same effect as the presence of denaturant. Both raising the temperature and increasing GuHCl destabilize contacts fairly homogeneously. Using Equation 13.1, folding rates can be calculated from the free-energy profiles (Figure 13.9) at different temperatures and can be assembled to create chevron plots. The calculated chevron plots bear a strong resemblance to the experimentally observed chevron curves for apo-azurin (Zong et al. 2006). The free-energy curves can also be used to calculate phi (ϕ) values. In fact the calculated and experimental ϕ values agree across a broad range of denaturant concentrations with a correlation coefficient of 0.77 (Zong et al. 2006). Again these calculations are based on a structure-based energy function, which strongly argues that the energy landscape of azurin is indeed highly funneled. The model shown to work well for the simple apo-azurin case can be extended to the metallated protein by adding the energetics accompanying coordination chemistry.

The variational method must be slightly modified to include the effects of the metal that will be explicitly represented in the free-energy functional (Zong et al. 2007). To account for coordination chemistry, the five coordinating residues were

The Folding Landscapes of Metalloproteins

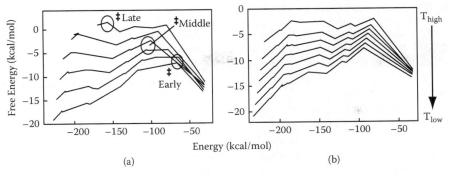

FIGURE 13.9 Free-energy profiles for (a) metallated azurin and (b) apo-azurin calculated using the variational method (Equation 13.15). The transition state shifts when the temperature is varied in calculations for the metallated protein but no such shift is present when examining the apo-form. There are three different transition state ensembles for the folding of metallated azurin (labeled early, middle, and late), which occur at different temperature regimes.

given contacts to the metal, in the same format as for protein-protein interactions, but with contact strengths calibrated to experiment. Two of the five coordinating residues are known to remain ligated to the metal in the unfolded state; therefore these were given large contact energies of interaction with the metal. The contact strengths of the other three ligands were tuned to fit the known binding constants for these ligands in peptide systems, which allows formation and breakage of the bonds during the folding process. In agreement with experiment, the variational calculations predict that the transition state is not invariant but shifts as temperature is varied (Figure 13.9a). Three different specific transition states can be identified in the metallated protein (labeled $\ddagger_{early}$, $\ddagger_{middle}$, and $\ddagger_{late}$) but only one transition state exists for the apo-form. The shifts between these transition states are a direct consequence of the metal. The structures of the three transition states differ slightly throughout the protein but are most profoundly different in the proximity of the metal. While only two residues are ligated in the early transition state ensemble, there are three ligands coordinating in the middle bottleneck and all five ligands coordinate in the last one. The shifts between these transition state ensembles results in a nonlinear relationship between the free-energy barrier height and stability (i.e., a curved chevron plot). The curved chevron plot predicted for the metallated protein is also observed experimentally. The calculated and experimental ϕ values generally agree with each other and show that the shifting transition state involves a change in the folding nucleus (Wilson and Wittung-Stafshede 2005; Zong et al. 2007). At low GuHCl concentrations, the folding nucleus is dispersed. At high GuHCl concentrations however, the folding nucleus resembles that of the apo-protein, with significant contacts between the β sandwiches formed in the transition state. Curved chevron plots are oftentimes observed for three-state folding proteins that have a populated intermediate. Both the experiments and the theoretical calculation on azurin have shown that a metal coordination site may be all that is necessary to induce a shifting transition state.

FUTURE PERSPECTIVE

Metalloproteins have already provided wonderful test beds for the energy landscape theory of protein folding. There is still more to do. The electronic spectroscopy of the colorful metal ions makes them exquisitely responsive in a temporal sense. They are powerful probes of the faster processes in folding. While direct NMR probes of denatured states are limited in their timescale, coupling to paramagnetic or optical centers opens up new ways of probing reconfiguration rates.

The tunability of the dynamics of the coordination processes at the metal center remains to be exploited. The models we have discussed for cofactor seeding of folding assume that coordination processes are fast and enter into the free-energy profile in a quasi-equilibrium fashion. Modulating the coordination kinetics can give a handle on the configurational diffusion motion through the elusive transition state. As we have seen, coordination chemistry is a main source of frustration in metalloprotein chemistry. Tuning the chemistry with different metals will allow us to more fully explore the role of residual frustration in a quantitative fashion.

The folding of proteins generally is more subtle than RNA folding precisely because of the lower degree of specificity of interactions between amino acids compared to the nucleotide bases, which pair by the Watson-Crick rules. Coordination chemistry provides interactions of intermediate strength—making nucleation easier but also making frustration more problematic kinetically for metalloproteins. Which of these two trends was more important in natural history? Did ordinary proteins come first, later acquiring useful inorganic capabilities, or did some kind of metalloprotein emerge first in the primordial soup as in the popular "mineral origin of life" scenarios (Cairns-Smith 1987)? More study of artificial metalloproteins may help resolve this issue (Salgado et al. 2008). In the future we can be sure metalloproteins will remain a major meeting ground for theorists and experimentalists who want to understand how life is assembled.

REFERENCES

Akiyama, S., S. Takahashi, T. Kimura, K. Ishimori, I. Morishima, Y. Nishikawa, and T. Fujisawa. 2002. Conformational landscape of cytochrome c folding studied by microsecond-resolved small-angle X-ray scattering. *Proc Natl Acad Sci U S A* 99 (3):1329–34.

Assfalg, M., I. Bertini, A. Dolfi, P. Turano, A. G. Mauk, F. I. Rosell, and H. B. Gray. 2003. Structural model for an alkaline form of ferricytochrome c. *J Am Chem Soc* 125:2913–22.

Babul, J., and E. Stellwag. 1972. Participation of protein ligands in folding of cytochrome c. *Biochemistry* 11 (7):1195–1200.

Bryngelson, J. D., J. N. Onuchic, N. D. Socci, and P. G. Wolynes. 1995. Funnels, pathways, and the energy landscape of protein-folding—a synthesis. *Proteins* 21 (3):167–95.

Cairns-Smith, A G. 1987. *Genetic takeover: and the mineral origins of life*. Cambridge: Cambridge University Press.

Cardenas, A. E., and R. Elber. 2003. Kinetics of cytochrome c folding: atomically detailed simulations. *Proteins* 51 (2):245–57.

Cho, S. S., Y. Levy, and P. G. Wolynes. 2006. P versus Q: structural reaction coordinates capture protein folding on smooth landscapes. *Proc Natl Acad Sci U S A* 103 (3):586–91.

Cho, S. S., P. Weinkam, and P. G. Wolynes. 2008. Origins of barriers and barrierless folding in BBL. *Proc Natl Acad Sci U S A* 105:118–23.

Clementi, C., P. A. Jennings, and J. N. Onuchic. 2000. How native-state topology affects the folding of dihydrofolate reductase and interleukin-1. *Proc Natl Acad Sci U S A* 97:5871–76.

Eastwood, M. P., and P. G. Wolynes. 2001. Role of explicitly cooperative interactions in protein folding funnels: a simulation study. *J Chem Phys* 114 (10):4702–16.

Ejtehadi, M. R., S. P. Avall, and S. S. Plotkin. 2004. Three-body interactions improve the prediction of rate and mechanism in protein folding models. *Proc Natl Acad Sci U S A* 101 (42):15088–93.

Erman, B. 2001. Analysis of multiple folding routes of proteins by a coarse-grained dynamics model. *Biophys J* 81 (6):3534–44.

Ferreiro, D. U., J. A. Hegler, E. A. Komives, and P. G. Wolynes. 2007. Localizing frustration in native proteins and protein assemblies. *Proc Natl Acad Sci U S A* 104 (50):19819–24.

Friedrichs, M. S., and P. G. Wolynes. 1989. Toward protein tertiary structure recognition by means of associative memory Hamiltonians. *Science* 246 (4928): 371–73.

Galzitskaya, O. V., and A. V. Finkelstein. 1999. A theoretical search for folding/unfolding nuclei in three-dimensional protein structures. *Proc Natl Acad Sci U S A* 96 (20):11299–304.

Goldstein, R. A., Z. A. Luthey-Schulten, and P. G. Wolynes. 1992. Optimal protein-folding codes from spin-glass theory. *Proc Natl Acad Sci U S A* 89 (11):4918–22.

Haney, P., J. Konisky, K. K. Koretke, Z. LutheySchulten, and P. G. Wolynes. 1997. Structural basis for thermostability and identification of potential active site residues for adenylate kinases from the archaeal genus *Methanococcus*. *Proteins* 28 (1): 117–30.

Hilser, V. J., and E. Freire. 1996. Structure-based calculation of the equilibrium folding pathway of proteins. Correlation with hydrogen exchange protection factors. *J Mol Biol* 262 (5):756–72.

Hoang, L., H. Maity, M. M. G. Krishna, Y. Lin, and S. W. Englander. 2003. Folding units govern the cytochrome c alkaline transition. *J Mol Biol* 331 (1):37–43.

Hong, X. L., and D. W. Dixon. 1989. NMR-study of the alkaline isomerization of ferricytochrome-c. *FEBS Lett* 246 (1–2):105–8.

Jones, C. M., E. R. Henry, Y. Hu, C. K. Chan, S. D. Luck, A. Bhuyan, H. Roder, J. Hofrichter, and W. A. Eaton. 1993. Fast events in protein-folding initiated by nanosecond laser photolysis. *Proc Natl Acad Sci U S A* 90 (24):11860–64.

Kolinski, A., W. Galazka, and J. Skolnick. 1996. On the origin of the cooperativity of protein folding: implications from model simulations. *Proteins* 26 (3):271–87.

Latzer, J., T. Shen, and P. G. Wolynes. 2008. Conformational switching upon phosphorylation: a predictive framework based on energy landscape principles. *Biochemistry* 47 (7):2110–22.

Levy, Y., S. S. Cho, T. Shen, J. N. Onuchic, and P. G. Wolynes. 2005. Symmetry and frustration in protein energy landscapes: a near degeneracy resolves the Rop dimer-folding mystery. *Proc Natl Acad Sci U S A* 102 (7):2373–78.

Maity, H., M. Maity, and S. W. Englander. 2004. How cytochrome c folds, and why: submolecular foldon units and their stepwise sequential stabilization. *J Mol Biol* 343 (1):223–33.

Mande, S. C., and M. E. Sobhia. 2000. Structural characterization of protein-denaturant interactions: crystal structures of hen egg-white lysozyme in complex with DMSO and guanidinium chloride. *Protein Eng* 13 (2):133–41.

Oliveberg, M., and P. G. Wolynes. 2005. The experimental survey of protein-folding energy landscapes. *Q Rev Biophys* 38:245–88.

Papoian, G. A., J. Ulander, M. P. Eastwood, Z. Luthey-Schulten, and P. G. Wolynes. 2004. Water in protein structure prediction. *Proc Natl Acad Sci U S A* 101 (10):3352–57.

Pascher, T., J. P. Chesick, J. R. Winkler, and H. B. Gray. 1996. Protein folding triggered by electron transfer. *Science* 271 (5255):1558–60.

Pletneva, E. V., H. B. Gray, and J. R. Winkler. 2005. Many faces of the unfolded state: conformational heterogeneity in denatured yeast cytochrome c. *J Mol Biol* 345 (4): 855–67.

Plotkin, S. S., J. Wang, and P. G. Wolynes. 1997. Statistical mechanics of a correlated energy landscape model for protein folding funnels. *J Chem Phys* 106 (7):2932–2948.

Portman, J. J., S. Takada, and P. G. Wolynes. 1998. Variational theory for site resolved protein folding free energy surfaces. *Phys Rev Lett* 81 (23):5237–40.

———. 2001a. Microscopic theory of protein folding rates. I. Fine structure of the free energy profile and folding routes from a variational approach. *J Chem Phys* 114 (11):5069–81.

———. 2001b. Microscopic theory of protein folding rates. II. Local reaction coordinates and chain dynamics. *J Chem Phys* 114 (11):5082–96.

Roder, H., G. A. Elove, and S. W. Englander. 1988. Structural characterization of folding intermediates in cytochrome-c by H-exchange labeling and proton NMR. *Nature* 335 (6192):700–704.

Rosell, F. I., J. C. Ferrer, and A. G. Mauk. 1998. Proton-linked protein conformational switching: definition of the alkaline conformational transition of yeast iso-1-ferricytochrome c. *J Am Chem Soc* 120 (44):11234–45.

Salgado, E. N., R. A. Lewis, J. Faraone-Mennella, and F. A. Tezcan. 2008. Metal-mediated self-assembly of protein superstructures: Influence of secondary interactions on protein oligomerization and aggregation. *J Am Chem Soc* 130 (19):6082–85.

Schejter, A., and W. A. Eaton. 1984. Charge-transfer optical-spectra, electron-paramagnetic resonance, and redox potentials of cytochromes. *Biochemistry* 23 (6): 1081–84.

Shoemaker, B. A., J. Wang, and P. G. Wolynes. 1997. Structural correlations in protein folding funnels. *Proc Natl Acad Sci U S A* 94 (3): 777–82.

———. 1999. Exploring structures in protein folding funnels with free energy functionals: the transition state ensemble. *J Mol Biol* 287 (3):675–94.

Shoemaker, B. A. and P. G. Wolynes. 1999. Exploring structures in protein folding funnels with free energy functionals: the denatured ensemble. *J Mol Biol* 287 (3):657–74.

Simons, K. T., C. Kooperberg, E. Huang, and D. Baker. 1997. Assembly of protein tertiary structures from fragments with similar local sequences using simulated annealing and Bayesian scoring functions. *J Mol Biol* 268 (1):209–25.

Socci, N. D., J. N. Onuchic, and P. G. Wolynes. 1996. Diffusive dynamics of the reaction coordinate for protein folding funnels. *J Chem Phys* 104 (15):5860–68.

Sosnick, T. R., L. Mayne, and S. W. Englander. 1996. Molecular collapse: the rate-limiting step in two-state cytochrome c folding. *Proteins* 24 (4): 413–26.

Stellwag, E., R. Rysavy, and G. Babul. 1972. Conformation of horse heart apocytochrome-c. *J Biol Chem* 247 (24):8074–77.

Tezcan, F. A., J. R. Winkler, and H. B. Gray. 1998. Effects of ligation and folding on reduction potentials of heme proteins. *J Am Chem Soc* 120 (51): 13383–88.

Theorell, H., and A. J. Akesson. 1941. Studies on cytochrome c. II. The optical properties of pure cytochrome c and some of its derivatives. *J Am Chem Soc* 63:1812–18.

Ueda, Y., H. Taketomi, and N. Go. 1978. Studies on protein folding, unfolding, and fluctuations by computer-simulation .2. 3-dimensional lattice model of lysozyme. *Biopolymers* 17 (6):1531–48.

Wagner, U G, N. Muller, W. Schmitzberger, H. Falk, and C. Kratky. 1995. Structure determination of the biliverdin apomyoglobin complex - crystal-structure analysis of 2 crystal forms at 1.4 and 1.5 angstrom resolution. *J Mol Biol* 247 (2):326–37.

Weinkam, P., E. V. Pletneva, H. B. Gray, J. R. Winkler, and P. G. Wolynes. 2009. Electrostatic effects on funneled landscapes and structural diversity in denatured protein ensembles. *Proc Natl Acad Sci U S A* 106 (6):1796–1801.

Weinkam, P., F. E. Romesberg, and P. G. Wolynes. 2009. Chemical frustration in the protein folding landscape: grand canonical ensemble simulations of cytochrome i. *Biochemistry* 48 (11):2394–2402.

Weinkam, P., J. Zimmermann, L. B. Sagle, S. Matsuda, P. E. Dawson, P. G. Wolynes, and F. E. Romesberg. 2008. Characterization of alkaline transitions in ferricytochrome c using carbon-deuterium IR probes. *Biochemistry* 47 (51):13470–80.

Weinkam, P., C. H. Zong, and P. G. Wolynes. 2005. A funneled energy landscape for cytochrome c directly predicts the sequential folding route inferred from hydrogen exchange experiments. *Proc Natl Acad Sci U S A* 102 (35):12401–6.

Whitford, P. C., J. N. Onuchic, and P. G. Wolynes. 2008. Energy landscape along an enzymatic reaction trajectory: hinges or cracks? *HSFP J* 2 (2):61–64

Wilson, C. J., and P. Wittung-Stafshede. 2005. Snapshots of a dynamic folding nucleus in zinc-substituted *Pseudomonas aeruginosa* azurin. *Biochemistry* 44 (30):10054–62.

Yang, A., and B. Honig. 1993. On the pH dependence of protein stability. *J Mol Biol* 231:459–74.

Zong, C. H., C. J. Wilson, T. Y. Shen, P. Wittung-Stafshede, S. L. Mayo, and P. G. Wolynes. 2007. Establishing the entatic state in folding metallated *Pseudomonas aeruginosa* azurin. *Proc Natl Acad Sci U S A* 104 (9):3159–64.

Zong, C. H., C. J. Wilson, T. Y. Shen, P. G. Wolynes, and P. Wittung-Stafshede. 2006. Phi-value analysis of apo-azurin folding: comparison between experiment and theory. *Biochemistry* 45 (20): 6458–66.

Appendix

DERIVATIONS AND EXPRESSIONS FOR CHAPTER 6

Derivation A: three-state model of apo-Mb unfolding:

$$N \xrightleftharpoons{K_{NI}} I \xrightleftharpoons{K_{IU}} U$$

$$K_{NI} = \frac{[I]}{[N]} = K_{NI}{}^0 \exp\left(\frac{M_{NI}[X]}{RT}\right)$$

$$K_{IU} = \frac{[U]}{[I]} = K_{IU}{}^0 \exp\left(\frac{M_{IU}[X]}{RT}\right)$$

$$[I] = [N]K_{NI}$$

$$[U] = [N]K_{NI}K_{IU}$$

$$P_0 = [N] + [I] + [U] = [N](1 + K_{NI} + K_{NI}K_{IU})$$

$$Y_N = \frac{1}{(1 + K_{NI} + K_{NI}K_{IU})}$$

$$Y_I = \frac{K_{NI}}{(1 + K_{NI} + K_{NI}K_{IU})}$$

$$Y_U = \frac{K_{NI}K_{IU}}{(1 + K_{NI} + K_{NI}K_{IU})}$$

$$S = S_N Y_N + S_I Y_I + S_U Y_U = \frac{S_N + S_I K_{NI}{}^0 \exp\left(\frac{M_{NI}[X]}{RT}\right) S_U K_{NI}{}^0 K_{IU}{}^0 \exp\left(\frac{(M_{NI}+M_{IU})[X]}{RT}\right)}{1 + K_{NI}{}^0 \exp\left(\frac{M_{NI}[X]}{RT}\right) + K_{NI}{}^0 K_{IU}{}^0 \exp\left(\frac{(M_{NI}+M_{IU})[X]}{RT}\right)}$$

Derivation B: unfolding of monomeric holo-Hb with the dissociated heme remaining monodisperse:

$$mHb \underset{}{\overset{K_{NH}}{\rightleftharpoons}} N_{apo} + H$$

$$N_{apo} \underset{}{\overset{K_{NU}}{\rightleftharpoons}} U_{apo}$$

$$K_{NH} = \frac{[N][H]}{[mHb]} = K_{NH}^{0} \exp\left(\frac{M_{NH}[X]}{RT}\right)$$

$$K_{NU} = \frac{[U]}{[N]} = K_{NU}^{0} \exp\left(\frac{M_{NU}[X]}{RT}\right)$$

$$[N] = \frac{[mHb]K_{NH}}{[H]}$$

$$[U] = \frac{[mHb]K_{NH}K_{NU}}{[H]}$$

$$P_0 = [mHb] + [N] + [U] = [mHb]\left(1 + \frac{K_{NH}}{[H]} + \frac{K_{NH}K_{NU}}{[H]}\right)$$

$$[mHb] = P_0 Y_{mHb} = P_0 \left(\frac{1}{1 + \frac{K_{NH}}{[H]} + \frac{K_{NH}K_{NU}}{[H]}}\right)$$

$$[H] = [N] + [U] = [mHb]\left(\frac{K_{NH}}{[H]} + \frac{K_{NH}K_{NU}}{[H]}\right)$$

$$= P_0 \left(\frac{1}{1 + \frac{K_{NH}}{[H]} + \frac{K_{NH}K_{NU}}{[H]}}\right)\left(\frac{K_{NH}}{[H]} + \frac{K_{NH}K_{NU}}{[H]}\right)$$

$$[H]^2 = [H](K_{NH} + K_{NH}K_{NU}) - (P_0)(K_{NH} + K_{NH}K_{NU}) = 0$$

Appendix

$$[H] = \frac{-(K_{NH} + K_{NH}K_{NU}) + \sqrt{(K_{NH} + K_{NH}K_{NU})^2 + 4P_0(K_{NH} + K_{NH}K_{NU})}}{2}$$

$$Y_{mHb} = \frac{1}{1 + \frac{K_{NH}}{[H]} + \frac{K_{NH}K_{NU}}{[H]}}$$

$$Y_N = \frac{\frac{K_{NH}}{[H]}}{1 + \frac{K_{NH}}{[H]} + \frac{K_{NH}K_{NU}}{[H]}}$$

$$Y_U = \frac{\frac{K_{NH}K_{NU}}{[H]}}{1 + \frac{K_{NH}}{[H]} + \frac{K_{NH}K_{NU}}{[H]}}$$

$$S = S_{mHb}Y_{mHb} + S_N Y_N + S_U Y_U = \frac{S_{mHb}[H] + S_N K_{NH} + S_U K_{NH}K_{NU}}{[H] + K_{NH} + K_{NH}K_{NU}}$$

Derivation C: unfolding of monomeric holo-Hb with heme binding to U:

$$\begin{array}{ccc} mHb & \xrightleftharpoons{K_{NH}} & N_{apo} + H \\ \updownarrow & & \updownarrow K_{NU} \\ UH & \xrightleftharpoons{K_{UH}} & U_{apo} + H \end{array}$$

$$K_{NH} = \frac{[N][H]}{[mHb]} = K_{NH}{}^0 \exp\left(\frac{M_{NH}[X]}{RT}\right)$$

$$K_{UH} = \frac{[U][H]}{[UH]} = K_{UH}{}^0 \exp\left(\frac{M_{UH}[X]}{RT}\right)$$

$$K_{NU} = \frac{[U]}{[N]} = K_{NU}{}^0 \exp\left(\frac{M_{NU}[X]}{RT}\right)$$

$$P_0 = [mHb] + [N] + [U] + [UH]$$

$$[N] = \frac{[mHb]K_{NH}}{[H]}$$

$$[U] = \frac{[mHb]K_{NH}K_{NU}}{[H]}$$

$$[UH] = \frac{[U][H]}{K_{UH}} = \frac{[mHb]K_{NH}K_{NU}}{K_{UH}}$$

$$Y_{mHb} = \frac{[mHb]}{[mHb]+[N]+[U]+[UH]} = \frac{1}{1+\frac{K_{NH}}{[H]}+\frac{K_{NH}K_{NU}}{[H]}+\frac{K_{NH}K_{NU}}{K_{UH}}}$$

$$[H] = [N]+[U] = [mHb]\left(\frac{K_{NH}}{[H]}+\frac{K_{NH}K_{NU}}{[H]}\right)$$

$$= P_0\left(\frac{1}{1+\frac{K_{NH}}{[H]}+\frac{K_{NH}K_{NU}}{[H]}+\frac{K_{NH}K_{NU}}{K_{UH}}}\right)\left(\frac{K_{NH}}{[H]}+\frac{K_{NH}K_{NU}}{[H]}\right)$$

$$[H]^2\left(1+\frac{K_{NH}K_{NU}}{K_{UH}}\right)+[H](K_{NH}+K_{NH}K_{NU})-P_0(K_{NH}+K_{NH}K_{NU})=0$$

$$[H] = \frac{-(K_{NH}+K_{NH}K_{NU})+\sqrt{(K_{NH}+K_{NH}K_{NU})^2+4P_0\left(1+\frac{K_{NH}K_{NU}}{K_{UH}}\right)(K_{NH}+K_{NH}K_{NU})}}{2\left(1+\frac{K_{NH}K_{NU}}{K_{UH}}\right)}$$

$$Y_{mHb} = \frac{1}{1+\frac{K_{NH}}{[H]}+\frac{K_{NH}K_{NU}}{[H]}+\frac{K_{NH}K_{NU}}{K_{UH}}}$$

$$Y_N = \frac{\frac{K_{NH}}{[H]}}{1+\frac{K_{NH}}{[H]}+\frac{K_{NH}K_{NU}}{[H]}+\frac{K_{NH}K_{NU}}{K_{UH}}}$$

Appendix

$$Y_U = \frac{\dfrac{K_{NH}K_{NU}}{[H]}}{1 + \dfrac{K_{NH}}{[H]} + \dfrac{K_{NH}K_{NU}}{[H]} + \dfrac{K_{NH}K_{NU}}{K_{UH}}}$$

$$Y_{UH} = \frac{\dfrac{K_{NH}K_{NU}}{K_{UH}}}{1 + \dfrac{K_{NH}}{[H]} + \dfrac{K_{NH}K_{NU}}{[H]} + \dfrac{K_{NH}K_{NU}}{K_{UH}}}$$

$$S = S_{mHb}Y_{mHb} + S_N Y_N + S_U Y_U$$

$$+ S_{UH}Y_{UH} = \frac{S_{mHb}[H] + S_N K_{NH} + S_U K_{NH}K_{NU} + S_{UH}\dfrac{[H]K_{NH}K_{NU}}{K_{UH}}}{[H] + \dfrac{K_{NH}}{[H]} + \dfrac{K_{NH}K_{NU}}{[H]} + \dfrac{[H]K_{NH}K_{NU}}{K_{UH}}}$$

Derivation D: unfolding of monomeric holo-Hb with self-assembly of dissociated heme:

$$mHb \xrightleftharpoons{K_{NH}} N_{apo} + H$$

$$\Big\updownarrow K_{NU}$$

$$U_{apo}$$

$$2H \xrightleftharpoons{K_{H_2}} H_2$$

$$K_{NH} = \frac{[N][H]}{[mHb]} = K_{NH}{}^0 \exp\left(\frac{M_{NH}[X]}{RT}\right)$$

$$K_{NU} = \frac{[U]}{[N]} = K_{NU}{}^0 \exp\left(\frac{M_{NU}[X]}{RT}\right)$$

$$K_{H_2} = \frac{[H_2]}{[H]^2} = K_{H_2}{}^0 \exp\left(\frac{m_{H_2}[X]}{RT}\right)$$

$$P_0 = [mHb] + [N] + [U] = [mHb]\left(1 + \frac{K_{NH}}{[H]} + \frac{K_{NH}K_{NU}}{[H]}\right)$$

$$[N] + [U] = \frac{[mHb]}{[H]}(K_{NH} + K_{NH}K_{NU})$$

$$[H] + 2[H_2] = [H] + 2K_{H_2}[H]^2$$

$$[N] + [U] = [H] + 2[D]$$

$$\frac{[mHb]}{[H]}(K_{NH} + K_{NH}K_{NU}) = [H] + 2K_{H_2}[H]^2$$

$$[mHb] = Y_{mHb}P_0 = \left(\frac{1}{1 + \frac{K_{NH}}{[H]} + \frac{K_{NH}K_{NU}}{[H]}}\right)P_0$$

$$\left(\frac{1}{1 + \frac{K_{NH}}{[H]} + \frac{K_{NH}K_{NU}}{[H]}}\right)P_0\left(\frac{1}{[H]}\right)(K_{NH} + K_{NH}K_{NU}) = [H] + 2K_{H_2}[H]^2$$

$$\left(\frac{P_0(K_{NH} + K_{NH}K_{NU})}{[H] + K_{NH} + K_{NH}K_{NU}}\right) = [H] + 2K_{H_2}[H]^2\left([H] + 2K_{H_2}[H]^2\right)$$

$$([H] + K_{NH} + K_{NH}K_{NU}) - P_0(K_{NH} + K_{NH}K_{NU}) = 0$$

$$2K_{H_2}[H]^2 + \left(1 + 2K_{H_2}(K_{NH} + K_{NH}K_{NU})\right)[H]^2$$
$$+ (K_{NH} + K_{NH}K_{NU})[H] - P_0(K_{NH} + K_{NH}K_{NU}) = 0$$

$$a = 2K_{H_2}$$

$$b = 1 + 2K_{H_2}(K_{NH} + K_{NH^*}K_{NU})$$

$$c = K_{NH} + K_{NH}K_{NU}$$

$$d = -P_0(K_{NH} + K_{NH}K_{NU})$$

Appendix

$$[H] = -\frac{b}{3a} - \frac{1}{3a}\left(\sqrt[3]{\frac{2b^2 - 9abc + 27a^2d + \sqrt{(2b^2 - 9abc + 27a^2d)^2 - 4(b^2 - 3ac)^2}}{2}}\right)$$

$$-\frac{1}{3a}\left(\sqrt[3]{\frac{2b^2 - 9abc + 27a^2d - \sqrt{(2b^2 - 9abc + 27a^2d)^2 - 4(b^2 - 3ac)^2}}{2}}\right)$$

$$Y_{mHb} = \frac{1}{1 + \frac{K_{NH}}{[H]} + \frac{K_{NH}K_{NU}}{[H]}}$$

$$Y_N = \frac{\frac{K_{NH}}{[H]}}{1 + \frac{K_{NH}}{[H]} + \frac{K_{NH}K_{NU}}{[H]}}$$

$$Y_U = \frac{\frac{K_{NH}K_{NU}}{[H]}}{1 + \frac{K_{NH}}{[H]} + \frac{K_{NH}K_{NU}}{[H]}}$$

$$S = S_{mHb}Y_{mHb} + S_N Y_N + S_U Y_U$$

Derivation E: equilibrium unfolding of holo-Mb based on the three states of apo-Mb and heme binding to all three of them:

$$\begin{array}{ccc}
Mb & \xrightleftharpoons{K_{NH}} & N + H \\
\updownarrow & & \updownarrow \\
IH & \xrightleftharpoons{K_{IH}} & I + H \\
\updownarrow & & \updownarrow K_{IU} \\
UH & \xrightleftharpoons{K_{UH}} & U + H
\end{array}$$

$$K_{NH} = \frac{[N][H]}{[Mb]} = K_{NH}{}^0 \exp\left(\frac{M_{NH}[X]}{RT}\right)$$

$$K_{IH} = \frac{[I][H]}{[IH]} = K_{IH}^{0} \exp\left(\frac{M_{IH[X]}}{RT}\right)$$

$$K_{UH} = \frac{[U][H]}{[UH]} = K_{UH}^{0} \exp\left(\frac{M_{UH[X]}}{RT}\right)$$

$$K_{NI} = \frac{[I]}{[N]} = K_{NI}^{0} \exp\left(\frac{M_{NI[X]}}{RT}\right)$$

$$K_{IU} = \frac{[U]}{[I]} = K_{IU}^{0} \exp\left(\frac{M_{IU[X]}}{RT}\right)$$

$$P_0 = [Mb] + [N] + [I] + [U] + [IH] + [UH]$$

$$[N] = \frac{[Mb]K_{NH}}{[H]}$$

$$[I] = \frac{[Mb]K_{NH}K_{NI}}{[H]}$$

$$[U] = \frac{[Mb]K_{NH}K_{NI}K_{IU}}{[H]}$$

$$[IH] = \frac{[I][H]}{K_{IH}} = \frac{[Mb]K_{NH}K_{NI}}{K_{IH}}$$

$$[UH] = \frac{[U][H]}{K_{UH}} = \frac{[Mb]K_{NH}K_{NI}K_{IU}}{K_{UH}}$$

$$Y_{Mb} = \frac{[Mb]}{[Mb] + [N] + [I] + [IH] + [UH]}$$

$$Y_{Mb} = \frac{1}{1 + \dfrac{K_{NH}}{[H]} + \dfrac{K_{NH}K_{NI}}{[H]} + \dfrac{K_{NH}K_{NI}K_{IU}}{[H]} + \dfrac{K_{NH}K_{NI}}{K_{IH}} + \dfrac{K_{NH}K_{NI}K_{IU}}{K_{UH}}}$$

Appendix

$$[H] = [N] + [I] + [U] = [Mb]\left(\frac{K_{NH}}{[H]} + \frac{K_{NH}K_{NI}}{[H]} + \frac{K_{NH}K_{NI}K_{IU}}{[H]}\right)$$

$$= Y_{Mb}P_0\left(\frac{K_{NH}}{[H]} + \frac{K_{NH}K_{NI}}{[H]} + \frac{K_{NH}K_{NI}K_{IU}}{[H]}\right)$$

$$[H] = \frac{P_0\left(\frac{K_{NH}}{[H]} + \frac{K_{NH}K_{NI}}{[H]} + \frac{K_{NH}K_{NI}K_{IU}}{[H]}\right)}{1 + \frac{K_{NH}}{[H]} + \frac{K_{NH}K_{NI}}{[H]} + \frac{K_{NH}K_{NI}K_{IU}}{[H]} + \frac{K_{NH}K_{NI}}{K_{IH}} + \frac{K_{NH}K_{NI}K_{IU}}{K_{UH}}}$$

$$[H]^2\left(1 + \frac{K_{NH}K_{NI}}{K_{IH}} + \frac{K_{NH}K_{NI}K_{IU}}{K_{UH}}\right) + [H](K_{NH} + K_{NH}K_{NI} + K_{NH}K_{NI}K_{IU})$$
$$- P_0(K_{NH} + K_{NH}K_{NI} + K_{NH}K_{NI}K_{IU}) = 0$$

$$[H] = \frac{-(K_{NH} + K_{NH}K_{NI} + K_{NH}K_{NI}K_{IU})}{2\left(1 + \frac{K_{NH}K_{NI}}{K_{IH}} + \frac{K_{NH}K_{NI}K_{IU}}{K_{UH}}\right)} +$$
$$\frac{\sqrt{(K_{NH} + K_{NH}K_{NI} + K_{NH}K_{NI}K_{IU})^2 + 4P_0\left(1 + \frac{K_{NH}K_{NI}}{K_{IH}} + \frac{K_{NH}K_{NI}K_{IU}}{K_{UH}}\right)(K_{NH} + K_{NH}K_{NI} + K_{NH}K_{NI}K_{IU})}}{2\left(1 + \frac{K_{NH}K_{NI}}{K_{IH}} + \frac{K_{NH}K_{NI}K_{IU}}{K_{UH}}\right)}$$

$$Y_{Mb} = \frac{1}{1 + \frac{K_{NH}}{[H]} + \frac{K_{NH}K_{NI}}{[H]} + \frac{K_{NH}K_{NI}K_{IU}}{[H]} + \frac{K_{NH}K_{NI}}{K_{IH}} + \frac{K_{NH}K_{NI}K_{IU}}{K_{UH}}}$$

$$Y_N = \frac{\frac{K_{NH}}{[H]}}{1 + \frac{K_{NH}}{[H]} + \frac{K_{NH}K_{NI}}{[H]} + \frac{K_{NH}K_{NI}K_{IU}}{[H]} + \frac{K_{NH}K_{NI}}{K_{IH}} + \frac{K_{NH}K_{NI}K_{IU}}{K_{UH}}}$$

$$Y_I = \frac{\frac{K_{NH}K_{NI}}{[H]}}{1 + \frac{K_{NH}}{[H]} + \frac{K_{NH}K_{NI}}{[H]} + \frac{K_{NH}K_{NI}K_{IU}}{[H]} + \frac{K_{NH}K_{NI}}{K_{IH}} + \frac{K_{NH}K_{NI}K_{IU}}{K_{UH}}}$$

$$Y_U = \frac{\frac{K_{NH}K_{NI}K_{IU}}{[H]}}{1 + \frac{K_{NH}}{[H]} + \frac{K_{NH}K_{NI}}{[H]} + \frac{K_{NH}K_{NI}K_{IU}}{[H]} + \frac{K_{NH}K_{NI}}{K_{IH}} + \frac{K_{NH}K_{NI}K_{IU}}{K_{UH}}}$$

$$Y_{IH} = \frac{\dfrac{K_{NH}K_{NI}}{K_{IH}}}{1 + \dfrac{K_{NH}}{[H]} + \dfrac{K_{NH}K_{NI}}{[H]} + \dfrac{K_{NH}K_{NI}K_{IU}}{[H]} + \dfrac{K_{NH}K_{NI}}{K_{IH}} + \dfrac{K_{NH}K_{NI}K_{IU}}{K_{UH}}}$$

$$Y_{UH} = \frac{\dfrac{K_{NH}K_{NI}K_{IU}}{K_{UH}}}{1 + \dfrac{K_{NH}}{[H]} + \dfrac{K_{NH}K_{NI}}{[H]} + \dfrac{K_{NH}K_{NI}K_{IU}}{[H]} + \dfrac{K_{NH}K_{NI}}{K_{IH}} + \dfrac{K_{NH}K_{NI}K_{IU}}{K_{UH}}}$$

$$S = S_{Mb}Y_{Mb} + S_N Y_N + S_I Y_I + S_U Y_U + S_{IH} Y_{IH} + S_{UH} Y_{UH}$$

Derivation F: equilibrium unfolding of apo-Hb dimer in a simple two-step process:

$$D \underset{K_{2U,D}}{\rightleftharpoons} 2U$$

$$K_{2U,D} = \frac{[D]}{[U]^2} = K_{2U,D}{}^0 \exp\left(\frac{-M_{2U,D}[X]}{RT}\right)$$

$$[D] = K_{2U,D}[U]^2$$

$$P_0 = 2[D] + [U] = 2K_{2U,D}[U]^2 + [U]$$

$$[U] = \left(\frac{-1 + \sqrt{1 + 8P_0 K_{2U,D}}}{4K_{2U,D}}\right)$$

$$Y_D = \frac{2[D]}{2[D] + [U]} = \frac{2K_{2U,D}\left(\dfrac{-1+\sqrt{1+8P_0 K_{2U,D}}}{4K_{2U,D}}\right)^2}{2K_{2U,D}\left(\dfrac{-1+\sqrt{1+8P_0 K_{2U,D}}}{4K_{2U,D}}\right)^2 + \left(\dfrac{-1+\sqrt{1+8P_0 K_{2U,D}}}{4K_{2U,D}}\right)}$$

Derivation G: equilibrium unfolding of apo-Hb dimer involving a first dissociation step followed by unfolding of folded subunits, via an intermediate:

$$D \underset{K_{2N,D}}{\rightleftharpoons} 2N \underset{K_{I,N}}{\rightleftharpoons} 2I \underset{K_{U,I}}{\rightleftharpoons} 2U$$

Appendix

$$K_{U,I} = \frac{[I]}{[U]} = K_{U,I}^{\ 0} \exp\left(\frac{-M_{U,I}[X]}{RT}\right)$$

$$K_{I,N} = \frac{[N]}{[I]} = K_{I,N}^{\ 0} \exp\left(\frac{-M_{I,N}[X]}{RT}\right)$$

$$K_{2N,D} = \frac{[D]}{[N]^2} = K_{2N,D}^{\ 0} \exp\left(\frac{-M_{2N,D}[X]}{RT}\right)$$

$$[I] = K_{U,I}[U]$$

$$[N] = K_{I,N}[I] = K_{I,N}K_{U,I}[U]$$

$$[D] = K_{2N,D}[N]^2 = K_{2N,D}K_{I,N}^{\ 2}K_{U,I}^{\ 2}[U]^2$$

$$P_0 = 2[D]+[N]+[I]+[U] = 2K_{2N,D}K_{I,N}^{\ 2}K_{U,I}^{\ 2}[U]^2 + K_{I,N}K_{U,I}[U] + K_{U,I}[U] + [U]$$

$$[U] = \left(\frac{-(1+K_{I,N}K_{U,I}+K_{U,I}) + \sqrt{(1+K_{I,N}K_{U,I}+K_{U,I})^2 + 8P_0 K_{2N,D}K_{I,N}^{\ 2}K_{U,I}^{\ 2}}}{4K_{2N,D}K_{I,N}^{\ 2}K_{U,I}^{\ 2}}\right)$$

$$Y_D = \frac{2[D]}{2[D]+[N]+[I]+[U]} = \frac{2K_{2N,D}K_{I,N}^{\ 2}K_{U,I}^{\ 2}[U]^2}{2K_{2N,D}K_{I,N}^{\ 2}K_{U,I}^{\ 2}[U]^2 + K_{I,N}K_{U,I}[U] + K_{U,I}[U] + [U]}$$

Derivation H: equilibrium unfolding of apo-Hb dimer involving a dimer intermediate step:

$$D \underset{K_{I_D,D}}{\rightleftharpoons} I_D \underset{K_{2U,I_D}}{\rightleftharpoons} 2U$$

$$K_{2U,I_D} = \frac{[I_D]}{[U]^2} = K_{2U,I_D}^{\ 0} \exp\left(\frac{-M_{2U,I_D}[X]}{RT}\right)$$

$$K_{I_D,D} = \frac{[D]}{[I_D]} = K_{I_D,D}^{\ 0} \exp\left(\frac{-M_{I_D,D}[X]}{RT}\right)$$

$$[I_D] = K_{I_D,2U}[U]^2$$

$$[D] = K_{I_D,D} K_{2U,I_D} [U]^2$$

$$P_0 = 2[D] + 2[I_D] + [U] = 2K_{I_D,D} K_{2U,I_D} [U]^2 + 2K_{2U,I_D} [U]^2 + [U]$$
$$= 2(K_{I_D,D} K_{2U,I_D} + 2K_{2U,I_D})[U]^2 + [U]$$

$$[U] = \left(\frac{-1 + \sqrt{1 + 8P_0 \left(2K_{I_D,D} K_{2U,I_D} + K_{2U,I_D} \right)}}{4 \left(K_{I_D,D} K_{2U,I_D} + 2K_{2U,I_D} \right)} \right)$$

$$Y_D = \frac{2[D]}{2[D] + 2[I_D] + [U]} = \frac{2K_{I_D,D} K_{2U,I_D} [U]^2}{2K_{I_D,D} K_{2U,I_D} [U]^2 + 2K_{2U,I_D} [U]^2 + [U]}$$

Index

A

Acididanus ambivalens, 85, 86
AD, *see* Alzheimer's disease
AFM, *see* Atomic force microscopy
Aggregation
    calorimetric analysis, 85
    copper-binding proteins, 65
    in vitro, 135
    neurodegenerative diseases, 176
    off-pathway mechanism, 200
    oligomerization-prone protein, 7
    peptide, 40
    propensity, 171, 185
    Pseudomonas aeruginosa, 19
    self-, 103
    α-synuclein, 170, 180, 182
    toxic, 131
    transient events, 25, 67
    zinc, 134, 194, 201
Aluminum
    exposure linked to PD pathology, 174
    mirror, 45
    α-synuclein and, 174–175
Alzheimer's disease (AD), 7
    perturbed neuronal Ca^{2+} homeostasis in, 176
    prion protein and, 210
    relevance of metal homeostasis in, 7
    α-synuclein and, 169
AMH, *see* Associative memory Hamiltonian
Aminothiotyrosine, 40
Amyloid deposition syndromes, 7
1-Anilino naphthalene-8-sulfonic acid (ANS), 90
ANS, *see* 1-Anilino naphthalene-8-sulfonic acid
Apoprotein, attachment of heme group to, 13
Apoptosis
    cytochromes c, 13, 52
    MPP+-induced iron signaling, 180
Apotransferrin, 72
Aquifex aeolicus, 76
Archaea, 86, 97, 132
Archaeoglobus fulgidus, 152
Associative memory Hamiltonian (AMH), 254
Atomic force microscopy (AFM), 176, 181
Azurin
    apo-, 268
    folding of, 266–269
    kinetic behavior, 68

B

Bacillus subtilis, 9, 69
Biogenesis, 5, 132
Boltzmann distribution, 255
Brain
    aluminum exposure, 174
    copper concentrations in, 177, 210
    elderly Down syndrome, 170
    iron homeostasis, 179
    magnesium, 182
    PD, 171
    prion protein, 209
    α-synuclein, 169
    WD patients, 146
    zinc concentration in, 214
Bulk metals, 3
Burst phase collapse, 22

C

Caenorhabditis elegans, 135
Calcium
    Ca^{2+} binding proteins, 8
    frataxin and, 135
    sensor proteins, 8
    α-synuclein and, 171, 175–177
Calmodulin, 8
Cancer, 96, 119, 292, 293, 304, 307, 308, 309, 310, 311
    anticancer agent, 66
    p53 function and, 193, 202, 204
    zinc bioavailability and, 202
Carbonic anydrase, 230
CATH, 13
CB, *see* Circular birefringence
CCD, *see* Charge-coupled device
CD, *see* Circular dichroism
Ceruloplasmin (CP), 72
    apo-form, 73
    thermal stability, 73
Charge-coupled device (CCD), 45
Chevron plot
    apo-azurin, 268
    c-type cytochromes, 18
    folding pathways, 30, 31
    Hydrogenobacter thermophilus, 25
    kinetic traps, 18
    metalloprotein folding, 63, 64

Pseudomonas aeruginosa, 19, 20
  redcyt c unfolding rates, 54
Rhodobacter capsulatus, 27, 28
  rollover effects, 18
Circular birefringence (CB), 43
Circular dichroism (CD)
  far-UV, 63, 85
  frataxin, 139
  magnetic, 45
  nanosecond spectroscopy, 43
  p53 stability, 196
  spectrometer, 22
  α-synuclein, 173
  time-resolved, 38
Coarse-grained models, 253
Cobalt, α-synuclein and, 177
COMMD1, 160
Conformational states, *see* Metal ions, protein folding, and conformational states
Copper, *see also* Human copper transporter
  saturation, 213
  α-synuclein and, 177–179
Copper-binding proteins, 61–80
  anticancer agent, 66
  apotransferrin, 72
  azurin, kinetic behavior, 68
  blue-copper protein as far-reaching model system, 65–68
  chevron plot, 63, 64
  copper uptake and delivery, 62
  distribution of copper in living cells, 61–63
  electron and oxygen management, 62
  enzymes, 62
  eukaryotes, 70
  experimental approaches to study metalloprotein folding *in vitro*, 63–65
  extended X-ray absorption fine structure experiments, 66
  future perspective, 76
  human ceruloplasmin, 72
  molecular dynamics, 65
  multicopper oxidases, 71–74
  mutated residue, 64
  prokaryotes, 70
  protein engineering, 64
  protein flexibility, 71
  proteins facilitating cytoplasmic copper transfer, 68–71
  quantum-mechanics molecular-mechanics methods, 65
  structural determinants, 67
  studies on other copper-binding proteins, 74–76
  tryptophan fluorescence, 63
  visible absorption, 73
COX, *see* Cytochrome c oxidase
CP, *see* Ceruloplasmin

Creutzfeldt-Jacob disease, 210
C-terminal helix, 13
c-type cytochromes, folding mechanism of, 13–36
  base elimination mechanism, 23
  burst phase collapse, 22
  chevron plot, 18
  circular dichroism spectrometer
  collapse in sub-ms time window, 22–23
  consensus folding mechanism, 26–29
  cytochrome c_{552} from *Hydrogenobacter thermophilus*, 25–26
  cytochrome c_{552} from *Thermus thermophilus*, 23–25
  engineering folding pathways, 30–33
  equilibrium studies, 14–17
  EX1 limit, 16
  EX2 limit, 16
  fast unfolding process, 32
  fluorescence quenching, 25
  folding mechanism of cytochrome c_{551} from *Pseudomonas aeruginosa*, 19–22
  foldons, 16
  future perspective, 33
  heme iron, 13
  hydrogen exchange, 15
  infra-red experiments, 15
  kinetic studies, 17–19
  misfolded off-pathway state, 24
  on-pathway obligatory intermediate, 24
  photo-induced experiments, 14
  refolding channels, kinetic partitioning between, 21
  rollover effect, 18, 19
  solvent denaturation studies, 15
  Tanford β-value, 25
  tryptophan fluorescence changes, 24
CysXaaXaaCysHis motif, 13
Cytochrome c folding, mechanism of (early events and kinetic intermediates), 37–59
  aminothiotyrosine, 40
  azobenzene groups, 40
  charge-coupled device, 45
  circular birefringence, 43
  denaturant viscosity, 54
  early events in reduced cytochrome *c* folding, 45–56
    conformational diffusion, 48–50
    earliest secondary-structure formation events, 45–48
    folding intermediates, 52–55
    implications for redcyt c folding mechanism, 55–56
    nonnative conformational states, 50–52
  flash photolysis methods, 37
  fluorescence spectroscopy, 42

Index

future studies, 56
heme-ligation states, 50
Kramers-Kronig transform mates, 44
laser-induced photoexcitation, 39
magnetic CD, 45
NADH photoexcitation, 48
pH change lifetime, 39
photoreduction trigger, 41
protein-folding triggers, 38–42
    endogenous photosensitivity of heme group in cytochrome c, 40–42
    laser-induced pH jump, 39
    laser-induced temperature jump, 38–39
    photoexcitation with exogenous photosensitive groups, 40
    rapid mixing, 39–40
strain plate, 43
time-resolved polarization spectroscopy, 42–56
    magnetic time-resolved CD and ORD, 45
    nanosecond circular dichroism spectroscopy, 43–44
    nanosecond optical rotatory dispersion spectroscopy, 44–45
transition state theory folding regime, 49
triggers, 37
turbulence-free mixing method, 39
van der Waals interactions, 52
Cytochrome c oxidase (COX), 61

D

DBD, *see* DNA binding domain
Dementia with Lewy bodies (DLB), 169
Deoxyribonucleic acid (DNA), 227
Desulfovibrio vulgaris, 28
Dextran, 56
Differential scanning calorimetry (DSC), 73, 92
DLB, *see* Dementia with Lewy bodies
DNA, *see* Deoxyribonucleic acid
DNA binding domain (DBD), 193
Drag and Drop cloning, 150
DSC, *see* Differential scanning calorimetry

E

Electron paramagnetic resonance (EPR), 85, 178, 212
Electrostatic interactions, 88–90
Energy landscape
    chemical frustration in, 259–266
    cytochrome c, 252–259
    folding routes, 259
    metalloproteins, 249
Enterococcus hirae, 63, 160
Entropy–energy diagrams, 250, 251
EPR, *see* Electron paramagnetic resonance

Escherichia coli, 6, 13, 75, 118
EXAFS, *see* Extended x-ray absorption fine structure
EX1 limit, 16
EX2 limit, 16
Extended x-ray absorption fine structure (EXAFS), 66, 134

F

Ferredoxin
    stability, *see* Iron-sulfur clusters, protein folds, and ferredoxin stability
    unfolding analysis, 92
Ficoll, 56
FIDA, *see* Fluorescence intensity distribution analysis
Fluorescence
    Acididanus ambivalens, 85
    ANS emission, 90, 93
    apo-Mb unfolding, 100
    continuous-flow experiments, 22
    copper saturation, 213
    frataxin, 133, 137
    Hydrogenobacter thermophilus, 25
    hyperfluorescence, 101
    instruments, 39
    intensity distribution analysis (FIDA), 181
    intrinsic, 73
    natural quencher, 14
    PD (aluminum), 174
    Pseudomonas aeruginosa, 19, 20
    quenching, 25
    redcyt c folding, 47
    resonance energy transfer, 155
    spectroscopy, 38, 42
    stopped-flow studies, 54
    time-resolved fluorescence energy transfer, 264
    tryptophan fluorescence changes, 24, 63
    zinc, 196
Folding route
    energy landscape, 259
    metalloproteins, 247
    nonnative interactions, 250
    parallel, 31
    predicted, 252
    pseudotrajectory, 267
Foldons, 16
Fourier transform infrared spectroscopy (FTIR), 85
    ferredoxin unfolding analysis, 92
    Fe-S cluster formation, 94
    metalloproteins, 264
    protein structure changes monitored by, 85
Frataxin, 125–144
    bacterial, 126
    cellular localization and maturation, 126

enzymatic cluster formation, 138
ferritin-like model, 135
as ferritin-like protein implicated in oxidative damage, 134–135
fluorescence, 133, 137
as inhibitor, 137–139
as iron chaperone, 136–137
iron scavenger, 132
protein nanocages, 132
same fold, different stabilities, 130–131
same fold, similar function(s), 131–132
sequence and structure conservation, 126–130
as unusual iron-binding protein, 132–134
yeast knockout studies, 131
FRDA, see Friedreich's ataxia
Friedreich's ataxia (FRDA), 125, see also Frataxin
FTIR, see Fourier transform infrared spectroscopy

G

Gaussian well potentials, 254
GCIs, see Glial cytoplasmic inclusions
Gerstmann-Sträussler-Scheinker syndrome, 210
Glial cytoplasmic inclusions (GCIs), 170, 176
Glutamatergic synapse, 7
Glycophosphatidylinositol (GPI), 210
GPI, see Glycophosphatidylinositol

H

Heinz bodies, 105
Heme
  affinity, 107, 237
  apo-, 6
  attachment, 13, 29
  -binding cytochromes, 4
  -binding protein, 64
  biosynthesis, 137
  covalently bound, 14, 22
  dissociation, 103, 109, 279
  globin structure, 98
  group
    attachment of to apoprotein, 13
    endogenous photosensitivity of in cytochrome c, 40–42
  iron, c-type cytochromes, 13
  ligation, 42, 51, 260
  loss, 99
  mammalian hemoglobins, 112
  molten globule, 265
  monomers, 102
  nonnative coordination, 18
  oxidation, 55
  photosensitivity, 40
  propionate groups, 19
  pseudoatoms, 254
  structural asymmetry of moiety, 43
  visible band TROD spectroscopy, 47
Hemoglobins, see Myoglobins and hemoglobins
High-performance liquid chromatography (HPLC), 178
HiPIP, 83
HPLC, see High-performance liquid chromatography
Human copper transporter, 145–165
  advantage of oocytes, 150
  ATP7B as Cu-transporting P-type ATPase, 146–149
  Drag and Drop cloning, 150
  functional expression of ATP7B, 149–151
  intracellular localization and trafficking of ATP7B, 157–158
  structural studies of ATP7B, 151–156
    A-domain, 154
    cytosolic domains, 155–156
    N-domain, 152–154
    P-domain, 154
  unique feature of P_{1B}-ATPases mechanism, 156–157
  Wilson's disease-causing mutations, 158–161
    A-domain, 160
    COMMD1 interactions with ATP7B, 160–161
    His1069Gln affects protein dynamics and placement of ATP in binding pocket, 159–160
    mutations in ATP-binding domain affect residues in ATP vicinity, 158–159
  Wilson's disease protein, 145–146
HX, see Hydrogen exchange
Hydrogen exchange (HX), 15
Hydrogenobacter thermophilus, 25–26

I

IDP, see Intrinsically disordered protein
Intrinsically disordered protein (IDP), 170
Iron
  -binding protein, see Frataxin
  heme, 13
  α-synuclein and, 179–182
Iron-sulfur clusters, protein folds, and ferredoxin stability, 81–96
  apo-ferredoxin state, 93
  biogenesis of iron-sulfur proteins, 82–83
  di-cluster ferredoxins as case study, 86–94
    effects of electrostatic interactions and metal centers on ferredoxin stability, 88–90
    ferredoxin hyperstability and unfolding pathways, 86–88

Index

molten globule state of ferredoxin, 92–94
monitoring ferredoxin unfolding at different levels of metalloprotein organization, 91–92
role of His/Asp zinc center in ferredoxin stability, 90–91
EDTA, 88
folding and stability of small iron-sulfur proteins, 83–86
future perspective, 94
holo ferredoxin, 93
infinite heating rate, 85
interplay between protein folding and iron-sulfur cluster binding, 83
metal cross-linker, 91
molten globules, 92
rubredoxin, 83, 85, 86
thermoacidophilic archaea, 86

K

Kramers-Kronig transform mates, 44
Kuru, 210

L

LBs, *see* Lewy bodies
LBVAD, *see* Lewy body variant of AD
Left elliptically polarized (LEP) light, 43
LEP light, *see* Left elliptically polarized light
Lewy bodies (LBs), 169
Lewy body variant of AD (LBVAD), 169
Lewy neurites (LNs), 169
LNs, *see* Lewy neurites
Lumbricus terrestris, 99

M

Mad cow disease, 209
Magnesium, α-synuclein and, 182–183
Magnetic CD (MCD), 45
MALDI-TOF MS, *see* Matrix-assisted laser desorption/ionization time-of-flight mass spectrometry
Manganese, α-synuclein and, 183
Matrix-assisted laser desorption/ionization time-of-flight mass spectrometry (MALDI-TOF MS), 178
MCD, *see* Magnetic CD
MCOs, *see* Multicopper oxidases
MD, *see* Molecular dynamics
Menke's syndrome, 7
Metal ions, protein folding, and conformational states, 3–11
amino acids, 3
amyloid deposition syndromes, 7
bulk metals, 3
Ca^{2+} binding proteins, 8
calcium sensor proteins, 8
c-type cytochromes, 6
EF-hand motif, 8
incorporation of metals into appropriate proteins, 5–6
metal binding clipping effects, 8
metals as modulators of protein structure and conformation, 79
metals in proteins, 3–5
mitochondrial superoxide dismutase, 4
photosynthesis processes, 4
protein affinity for metals, 5
respiration processes, 4
roles of metals in protein folding and misfolding, 6–7
superacid center in metalloenzymes, 4
trace metals, 3
transferrin, 6
zinc finger motifs, 4
Metallochaperone, 5, 156, 157, 202
Metallopeptides, 227–245
carbonic anydrase, 230
cost of protein folding in natural zinc fingers, 241–243
equilibrium thermodynamics, 233–234
future perspective, 243–244
heme protein folding, 234–238
metalloprotein design, 230–232
protein design, 228–230
protein maquette, prototype heme, 232, 233, 236
zinc protein folding, 238–241
Metalloproteins, folding landscapes of, 247–273, *see also* Metallopeptides
associative memory Hamiltonian, 254
Boltzmann distribution, 255
chemical frustration in folding energy landscape, 259–266
coarse-grained models, 253
database knowledge, 249
energy landscape, 249, 252–259
entropy–energy diagrams, 250, 251
folding of azurin, 266–269
future perspective, 270
phosphoserine, 260
principle of minimal frustration, 249
reorganizational chain diffusion rates, 250
structure-based models, 252
Methanocaldococcus jannaschii, 86
Minor groove, 194, 197
Misfolding
apo-protein, 73
metal release, 5
prion protein, 209

prolyl peptide bonds, 18
PrP^C, 210
roles of metals, 6
α-synuclein, 185
zinc, 200
Molecular dynamics (MD), 65
Molten globule
    apo-protein, 73
    copper-binding proteins, 73
    cytochrome *c* folding, 47
    ferredoxin, 90, 93
    heme in, 265
    metal binding, 8
    myoglobins and hemoglobins, 102
Mössbauer studies, 94, 139
MSA, *see* Multiple system atrophy
Multicopper oxidases (MCOs), 71–74, *see also* Copper-binding proteins
Multiple system atrophy (MSA), 170
Myoglobins and hemoglobins, 97–122
    basic globin structures, 98
    derivations and expressions, 275–286
    factors governing Mb and Hb stabilities, 99
    globins in biology, 97–98
    Heinz bodies, 105
    hemin binding, 105
    hemoglobin assembly scheme, 114
    mammalian hemoglobins, 112–117
        complete mechanism for holo-HbA unfolding, 116–117
        erythropoiesis, 112–113
        mechanism for apo-Hb dimer unfolding, 113–116
        proposed mechanism for Hb assembly *in vivo*, 113
    myoglobin and monomeric hemoglobin unfolding, 99–112
        general equilibrium mechanisms, 102–107
        holo-Mb unfolding studies, 107–112
        previous studies of apo-Mb unfolding, 100–102
        structure of apo-Mb, 99–100
    perspectives, 117–118
    vertebrate hemoglobins, 97

N

NADH, 46, 48
Neurotransmitter(s)
    biosynthesis, 61
    synaptic release of, 169
Nickel, 3
NMR, *see* Nuclear magnetic resonance
NTD, *see* N-terminal domain
N-terminal domain (NTD), 193
N-terminal helix, 13

Nuclear magnetic resonance (NMR), 75, 212
    A-domain, 154
    chemical exchange of labile protons with deuterons, 16
    chemical shift perturbation, 137
    copper-binding proteins, 75
    equilibrium studies, 15
    frataxin, 128, 134
    iron-sulfur proteins, 85
    metalloproteins, 256, 260
    myoglobins and hemoglobins, 100
    prion protein, 212
    structural resolution, 15
    α-synuclein, 173
    Wilson's disease protein, 157, 159
    zinc, 197, 198

O

Octarepeat (OR) domain, 209, *see also* Prion protein, octarepeat domain of
Oocytes, 150
OR domain, *see* Octarepeat domain
Oxidative damage, 134–135

P

Parkinson's disease (PD), 7, 169
    dopaminergic neurodegeneration in, 179
    fluorescence, 174
    prion protein and, 210
    relevance of metal homeostasis in, 7
    α-synuclein and, 169
PD, *see* Parkinson's disease
PDB, *see* Protein structure database
Φ-value analysis, 29, 64
Phosphoserine, 260
Photoexcitation
    cytochrome *c* folding, 39, 40
    NADH, 46, 48
Physeter catodon, 99
p53 misfolding, *see* Zinc and p53 misfolding
Presynaptic vesicles, 169
Principle of minimal frustration, 249
Prion protein, octarepeat domain of, 209–223
    component 3, 212
    electron paramagnetic resonance, 212
    future perspective, 221
    imidazole coordination, 217
    nuclear magnetic resonance, 212
    octarepeat domain, 211–214
    octarepeat expansion disease, 217–220
    PrP^C misfolding, 210
    zinc uptake in octarepeat domain, 214–217

Index

Protein
  design, 228
  maquette, prototype heme, 232, 233, 236
  structure database (PDB), 129
PrP^C misfolding, 210
Pseudomonas aeruginosa, 19–22
Pyrococcus furiosus, 86

Q

QM-MM methods, *see* Quantum-mechanics molecular-mechanics methods
Quantum-mechanics molecular-mechanics (QM-MM) methods, 65

R

Raman
  resonance Raman spectroscopy, 38, 85
  selection rules, 42
  vibrational transitions, 42
Reactive oxygen species (ROS), 178
REP light, *see* Right elliptically polarized light
Resonance Raman (RR), 38, 85, 139
Rhodobacter capsulatus, 27
Ribonuclease A (RNAse A), 227, 228
Right elliptically polarized (REP) light, 43
RMSD, *see* Root mean square deviation
RNAse A, *see* Ribonuclease A
Rollover effect, c-type cytochromes, 18, 19
Root mean square deviation (RMSD), 70
ROS, *see* Reactive oxygen species
RR, *see* Resonance Raman
Rubredoxin, 83, 85, 86

S

S100, 8
Saccharomyces cerevisiae, 74, 149, 150
SAX monitoring, *see* Small angle x-ray scattering monitoring
Scanning for intensely fluorescent targets (SIFT), 181
SCOP, 13
Scrapie, 209
SDS monomers, *see* Sodium dodecyl sulfate monomers
Secretory pathway, 62, 63
SIFT, *see* Scanning for intensely fluorescent targets
Small angle x-ray scattering (SAX) monitoring, 48
SOD, *see* Superoxide dismutase
Sodium dodecyl sulfate (SDS) monomers, 52
SP, *see* Strain plate
Spectroscopy
  fluorescence, 38, 42

Fourier transform infrared spectroscopy, 85
  nanosecond, 43
  resonance Raman, 38
  time-resolved polarization spectroscopy, 42–56
  X-ray absorption spectroscopy, 134
Strain plate (SP), 43
Sulfolobus
  metallicus, 90
  tokodaii, 86
Superoxide dismutase (SOD), 74
  copper, 74, 213
  Cu/Zn, 61
  denatured state, 75
  prion protein, 213, 217
  zinc ions, 4
α-Synuclein and metals, 169–191
  aluminum, 174–175
  calcium, 175–177
  cobalt, 177
  copper, 177–179
  dopaminergic neurodegeneration, 179
  free-radical formation, 171
  future perspective, 185
  intrinsically disordered protein, 170
  iron, 179–182
  magnesium, 182–183
  manganese, 183
  neurodegenerative diseases related to aging, 176
  (non)-specific interactions of α-synuclein with metal ions, 171–174
  presynaptic vesicles, 169
  reactive oxygen species, 178
  terbium, 184
  zinc, 184–185

T

Tanford β-value, 25, 28
Terbium, α-synuclein and, 184
TGN, see *trans*-Golgi network
Thermoacidophilic archaea, 86
Thermotoga maritima, 153
Thermus thermophilus, 23–25, 129
Time-resolved circular dichroism (TRCD), 38
Time-resolved fluorescence energy transfer (TRFET), 264
Trace metals, 3
Transcription factor, zinc finger, 194, 230, 238
Transferrin, 6
trans-Golgi network (TGN), 147
Transition state theory (TST), 49
  conformation equilibrium, 50
  dynamical prefactor, 250
  prerequisite to folding regime, 49

Transmissible spongiform encephalopathies (TSEs), 209
TRCD, see Time-resolved circular dichroism
TRFET, see Time-resolved fluorescence energy transfer
Tryptophan fluorescence, 24, 63
TSEs, see Transmissible spongiform encephalopathies
TST, see Transition state theory

V

van der Waals interactions, 52
Visible absorption
 copper-binding proteins, 73
 iron-sulfur cluster binding, 85, 86
 metal environment, 63

W

WD, see Wilson's disease
Weighted histogram analysis method (WHAM), 256
WHAM, see Weighted histogram analysis method
Wilson's disease (WD), 145
 human copper transporter, 145–146
 metal–protein interactions, 7
 nuclear magnetic resonance, 157, 159

X

XAS, see X-ray absorption spectroscopy
X-ray absorption spectroscopy (XAS), 134

Y

Yfh1, 127
 acidic ridge, 130
 affinity purified, 136
 ferroxidase activity, 135
 frataxin interactions, 133
 NMR structures, 128
 oligomerized, 134
 studies, 127

Z

Zinc
 fluorescence, 196
 protein folding, 238–241
 superacid center in metalloenzymes, 4
 α-synuclein and, 184–185
 uptake in octarepeat domain, 214–217
Zinc, p53 misfolding and, 193–207
 clinical perspective, 203–204
 folding mechanisms of DBD and apoDBD, 198–199
 methodology and experimental approaches, 194–197
 rescue of misfolding by metallochaperones, 200–202
 structure and function of apoDBD, 197–198
 zinc binding, zinc loss, and misfolding, 200
 zinc bioavailability, p53, and cancer, 202–203
Zinc finger
 Cys to His alterations, 241
 domain folding process, 8
 free-energy cost, 243
 metal ion binding, 240
 motifs, 4
 protein conformations, 7
 protein-folding energy scheme, 231
 transcription factor, 194, 230, 238